MW00358598

3e

NJATC

DC THEORY

DC Theory is intended to be an educational resource for the user and contains procedures commonly practiced in industry and the trade. Specific procedures vary with each task and must be performed by a qualified person. For maximum safety, always refer to specific manufacturer recommendations, insurance regulations, specific job site and plant procedures, applicable federal, state, and local regulations, and any authority having jurisdiction. The electrical training ALLIANCE assumes no responsibility or liability in connection with this material or its use by any individual or organization.

3 4 5 6 7 8 9 – 10 – 15 14 13 12

Printed in the United States of America

A C K N O W L E D G M E N T S

PRINCIPAL WRITER

Stan Klein, NJATC Staff

CONTRIBUTING WRITER

Jim Paladino, Omaha, NE

TECHNICAL EDITORS

Ken Haden, NJATC Director
Jeff Keljik, Dunwoody College of Technology

ADDITIONAL ACKNOWLEDGMENTS

This material is continually reviewed and evaluated by industry professionals who are members of NJATC Curriculum workgroups. The invaluable input provided by these individuals allows for the development of instructional material that is of the absolute highest quality. At the time of this printing, the NJATC DC Theory Curriculum Workgroup, chaired by John Biondi, who is also a member of the NJATC Inside Education Committee, was composed of the following members: Gary Beckstrand; John Brockenbrough; Anthony D'Anna, Jr.; Gerald Goddard; Steve Harper; Mike Hartranft; Carl Latona; John Sabaliauskas; Robert Scheller; Valari Spence; Jim Weimer; Bob Withers; and T.J. Woods.

CONTENTS

KEY FEATURES

This text has been strengthened from top to bottom with many new features and enhancements to existing content. All-new chapter features provide structure and guidance for learners. Enhanced and concrete Chapter Objectives are complemented by solid and reinforcing Chapter Summaries, Review Questions, and Practice Problems. Chapter contents are introduced at the beginning of each chapter and then bolstered before moving on to the next chapter. Throughout each chapter, concepts are explained from their theoretical roots to their application principles, with reminders about safety, technology, professionalism, and more.

DC Theory, Third Edition, has been expanded to more fully explore a number of concepts through a major reorganization of the chapters. After working with the new language of electricity to enable the student to get started, the book goes back to explain the nature of electricity and all the atomic theory that is associated with a working knowledge of why electricity works. You will experience the different ways used to explain DC circuits so that you will be able to understand the concepts of your electrical profession, no matter where you practice your skill. Once you have a thorough understanding of the basic DC, we will begin to control the circuits, then design circuits to provide the desired results.

A complete electrical circuit is needed to complete a system to allow electricity to do work for us. There are many different forms of electrical circuits. This book will construct, analyze, and determine how and why they work. As any physical science has rules for application, so do electrical circuits. Each type of electrical system has a set of rules to follow to determine what will happen as we apply the science and theory to practical applications. Each chapter adds more to the understanding of theory and practical uses of electrical circuits.

See the following pages for examples of these new features.

FEATURES

SUMMARY

In this chapter you have been introduced to the standard method of drawing circuit diagrams as schematic diagrams. The symbols used are almost universal, with just a few variations. This allows electrical personnel worldwide to understand each other's work. As the series DC circuit is the most basic, we started with the rules for series circuits. These rules stay constant for series DC circuits and are only slightly modified for AC circuits.

We know that there is only one path for current to flow in a series circuit. The total resistance of the circuit is the sum of the individual resistors. The voltage drop on the individual loads is proportional to the resistance, and all voltage must be accounted for (as in Kirchhoff's law). Power consumed or dissipated by the loads is added to determine the total wattage load. Series circuits can include multiple voltage sources, as in batteries connected in series. The electrical polarity of the DC sources dictates whether the voltages add or subtract from the total.

Resistor styles and types were introduced to allow you to conveniently work with resistors in testing circuits. Resistors have several parameters of which to be aware. Often the values of these parameters are coded by a color stripe system and the standard color code system.

CHAPTER GLOSSARY

Component diagram (pictorial diagram) A component diagram is a drawing that shows the interconnection of system components by using photographs or drawings of the actual components. This is also known as a pictorial diagram.

Composition carbon resistor A resistor that derives its resistance from a combination of carbon graphite and a resin bonding material.

Metal film resistor A resistor that derives its resistance from a thin metal film applied to a ceramic rod.

Metal glaze resistor Similar to a metal film resistor except that the film is much thicker and is made of metal and glass.

Schematic diagram A schematic is a structural or procedural diagram, especially of an electrical or mechanical system, using special symbols to represent the actual physical components. The schematic shows the "scheme" of the current flow in a systematic representation.

Wire-wound resistor A resistor made by winding resistive wire around an insulating form.

REVIEW QUESTIONS

1. Review the circuit symbols shown in Table 4–1. Draw a circuit that shows a battery in series with an ammeter, in series with three resis-

discussion are resistor values, tolerances, and power capabilities.

4. Carefully review Figure RQ4–1. What would be the color code for a ...n resis- ... A stan- ... resistor ... or vari- ...ge does ...r have ...istor?

Glossaries are included in each chapter along with a comprehensive glossary at the end of the book.

This extra energy forces the "hit" electron to leave orbit and jump to a neighboring electron's orbit, and the process is repeated. The electron that impacted the first valence electron now takes the leaving electron's place as the new valence electron. Figures 2–11 and 2–12 detail this "electron flow."

An important fact to note here is that not all of the energy is transferred when the electrons collide. Some of the energy is lost in the form of resistance. This resistance comes from the original valence electron's attraction to its own nucleus. The energy expended in moving electrons is released as heat (Figure 2–11). That is why conductors (wires) that conduct electricity get warm. Too much electron flow (electrical current) in too small of a conductor (wire) or with too few electrons can overheat the conductors (wires) and the electrical insulating covering and may melt or ignite them.

When one electron strikes a valence shell with two valence electrons, the energy of the incoming electron is divided between the two valence electrons.

Energy Transfer

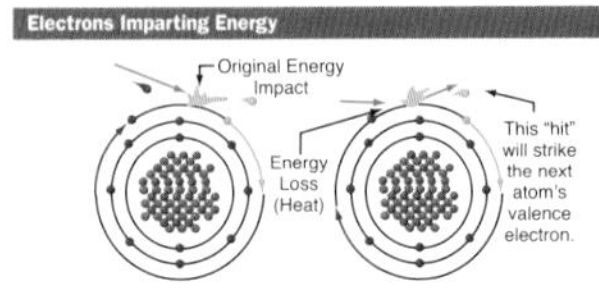

Figure 2–10 As one ball strikes the next, the energy is transferred to the next ball.

Electrons Imparting Energy

Original Energy Impact
Energy Loss (Heat)
This "hit" will strike the next atom's valence electron.

Figure 2–11 Electrons are moved from one atom's shell to another atom's shell, imparting energy.

Divided Energy

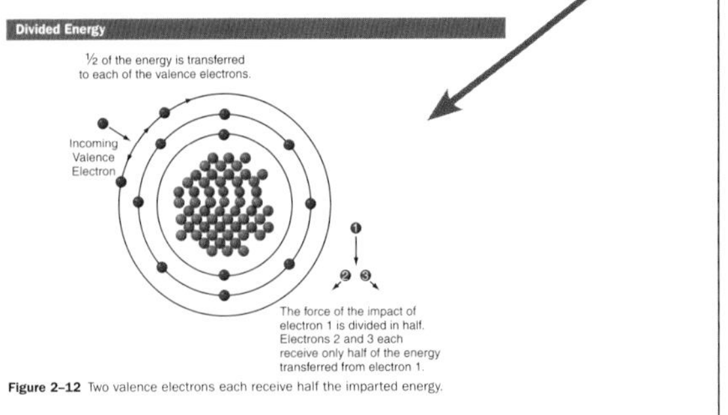

Figure 2–12 Two valence electrons each receive half the imparted energy.

High-contrast images give a clear and colorful view that "pops" off the page.

FieldNote!

English physicist James Prescott Joule (1818–1889) dealt with this problem of nonlinear proportionality. He wrote a paper in 1840 called "On the Production of Heat by Voltaic Electricity." In that paper, he explained his experiments—which we will not go into here. His conclusions are called Joule's law.

Joule's law states that "the total amount of heat produced in a conductor is directly proportional to the resistance times the square of the current" (see TechTip below). Mathematically, the heat (power) produced can be represented as:

$$P = I^2 \times R$$

TechTip!

Some wires, or conductors, rather than being one solid piece of copper or aluminum, are actually made up of many separate small-gauge wires, bundled together to form one "stranded" conductor.

Small signals that pass through fine stranded wire cables of this type may exhibit variations in amplitude. Even though the cable appears to be in good condition, it may be defective. Over time, especially where stretching and bending occur, strands of a wire conductor may break. These broken strands "make," and "break" as environmental conditions change. The amplitude fluctuations, or noise, experienced at the cable output results from the variations of series-circuit resistance within the faulty conductor. Conductor corrosion may produce the same outcome.

High-interest content is given in FieldNote and TechTip sidebars to make real-world connections to lessons learned in the chapters.

New photos and illustrations are located near their text references and clarify explanations.

Each distance has special conditions, with the "prohibited" being the category that assumes you may be in contact with the live electrical conductor. This means that you need specific "voltage-rated" tools (Figure 1–44), including voltage-rated gloves (Figure 1–45) and safety glasses.

Another hazard of working on live parts is the risk of arc flash and arc blast. As you already know, when electricity moves through the air it creates an arc (such as lightning). It creates extreme heat—up to 35,000 degrees Fahrenheit at the arc. This extreme heat melts copper conductors and just about everything else near the arc. The extreme heat and light are known as the arc flash. This flash can burn unprotected skin, and its heat can set clothes on fire. The flash can cause temporary blindness or long-term eye problems. You must be protected from this extreme heat and light.

Use fire-rated (FR) clothing designed for the amount of heat that could be present. This heat is expressed as calories per square centimeter of "incident energy." Different categories of FR clothes create the needed protection. To protect yourself adequately, a "flash hazard analysis" is calculated on the basis of the amount of current present if a short circuit occurs, the time the arc is allowed to burn, and the distance the person is from the arc.

The distance you may see on a rating is called the "flash protection boundary." This distance refers to how far you must be from the flash to sustain a burn that is "just curable" according to OSHA. "Just curable" means that it might be a second-degree burn but not a third-degree burn. Ask your safety representative about clothing and gear that protects you from arc flash injuries.

The other dimension of an arc is the blast or the explosion that occurs as the gases expand around the superheated arc. The blast propels materials and molten metal toward you. It may blow you out of the equipment or pelt you with molten material, creating bodily injury. The hot gases that are expanding may knock the air from your lungs. Your natural reaction is to try to breathe again, possibly inhaling hot gases and molten metal.

Voltage-Rated Tools

Voltage Rating

Voltage Rating

Figure 1–44 Voltage-rated tools must have a voltage rating marked on them.

Personal Protective Equipment

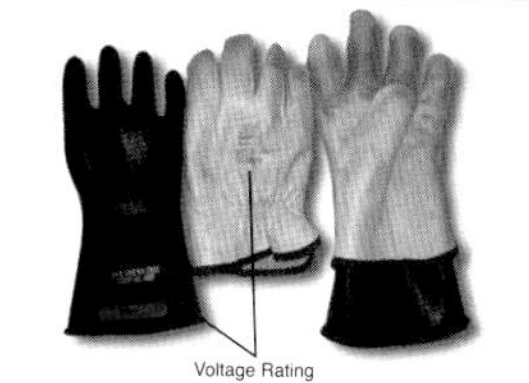

Figure 1–45 Gloves for working on live equipment must have voltage rating marks and must be tested.

attempting to measure resistance. Circu[it] power sources will not only affect th[e] reading on resistance measuring device[s] but potentially pose a safety hazard i[f] left connected.

CAUTION: Never attempt to measure the resistance of a component in a circuit while power is applied to that circuit. To do so not only will damage or destroy the meter used for the test but might injure the individual performing the test.

Other components connected in a circuit might also affect the accuracy of resistance measurements. When measuring resistance, you must verify that the only current path available is the one through the component whose resistance you wish to measure. Generally, when measuring the resistance of components in a circuit, the polarity of the ohmmeter is not important. Most resistive components are not sensitive to the polarity of the voltage applied to those components during a resistance test.

Electronics devices such as diodes and transistors are polarized. Therefore, caution must be applied to ensure that the polarity of the meter's internal source is correct for the test to be accurate. Figure 1–31 shows the proper way to measure the resistance of resistor R_1. The power source and all other current paths have been isolated from the component whose resistance is to be measured.

Figure 1–32 shows a circuit in which the resistance measurement would be in error because the power is still connected to the circuit.

Ohmmeter: Proper Connection

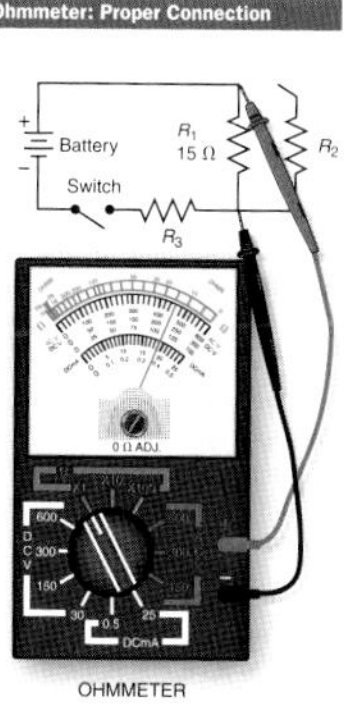

Figure 1–31 Correct ohmmeter connection on a component is illustrated above. When the switch is open, the power source will not affect the measurement.

Ohmmeter: Improper Connection

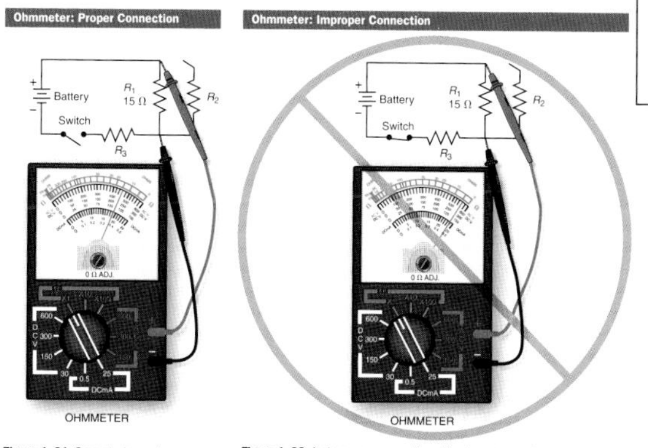

Figure 1–32 An improper connection with ohmmeter and power source still connected is shown. When the switch is closed, the power source will affect the measurement.

Step-by-step sample problems and solutions relate to the end-of-chapter exercises.

Example

If a circuit has a total resistance of 20 ohms (Ω) and a current of 3 amps, what is the circuit voltage?

Solution:

Using Figure 1–38, cover the E with your finger and you have $I \times R$ left over. Thus, the equation is:

$$E = 3\text{ A} \times 20\ \Omega$$

$$E = 60\text{ V}$$

Example

If a circuit has an applied voltage of 24 volts and has a current of 2 amps, what is the resistance of the circuit?

Solution:

Again, looking at Figure 1–38, cover the R and the equation becomes:

$$R = \frac{E}{I}$$

$$R = \frac{24\text{ V}}{2\text{ A}}$$

$$R = 12\ \Omega$$

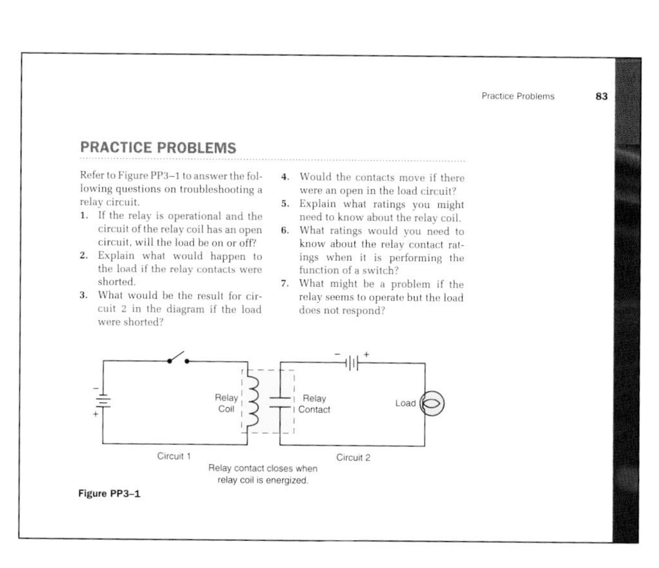

PRACTICE PROBLEMS

Refer to Figure PP3–1 to answer the following questions on troubleshooting a relay circuit.

1. If the relay is operational and the circuit of the relay coil has an open circuit, will the load be on or off?
2. Explain what would happen to the load if the relay contacts were shorted.
3. What would be the result for circuit 2 in the diagram if the load were shorted?
4. Would the contacts move if there were an open in the load circuit?
5. Explain what ratings you might need to know about the relay coil.
6. What ratings would you need to know about the relay contact ratings when it is performing the function of a switch?
7. What might be a problem if the relay seems to operate but the load does not respond?

Figure PP3–1

End-of-chapter problems reinforce critical concepts and relate to the worked-out examples in the chapter.

INTRODUCTION

Welcome to the third edition of *DC Theory,* which has been redesigned and updated to provide knowledge of the fundamentals to electrical technologists in apprenticeship programs, vocational-technical schools and colleges, and community colleges. The text emphasizes a solid foundation of classroom theory supported by on-the-job hands-on practice. Every project, every piece of knowledge, and every new task will be based on the experience and information acquired as each technician progresses through his or her career. This book, along with the others in this series, contains a significant portion of the material that will form the basis for success in an electrical career.

This text was developed by blending up-to-date practice with long-lived theories in an effort to help technicians learn how to better perform on the job. It is written at a level that invites further discussion beyond its pages while clearly and succinctly answering the questions of *how* and *why.*

SUPPLEMENTAL PACKAGE

The Instructor Resource is geared to provide instructors with all the tools they need in one convenient package. Instructors will find that this resource provides them with a far-reaching teaching partner that includes:

- PowerPoint® slides for each book chapter that reinforce key points and feature illustrations and photos from the book
- The Computerized Test Bank in ExamView format, which allows for test customization for evaluating student comprehension of noteworthy concepts
- An electronic Instructor's Manual, with supplemental lesson plans and support

ABOUT THIS BOOK

The efforts for continuous enhancement have produced the product you see before you: this technically precise, academically superior edition of *DC Theory*. The essential terms such as voltage, current, resistance, and power are defined and used. These four concepts are thoroughly explained, and their inter-connection through Ohm's law is tested. The units of measure and the measuring methods are defined as needed in the field. The equations used to predict circuit quantities are explained and tested. The use of mathematics is essential but kept to the basic algebra problems. Safety with electrical circuits is explained early, but the theme of working safely is used throughout the book.

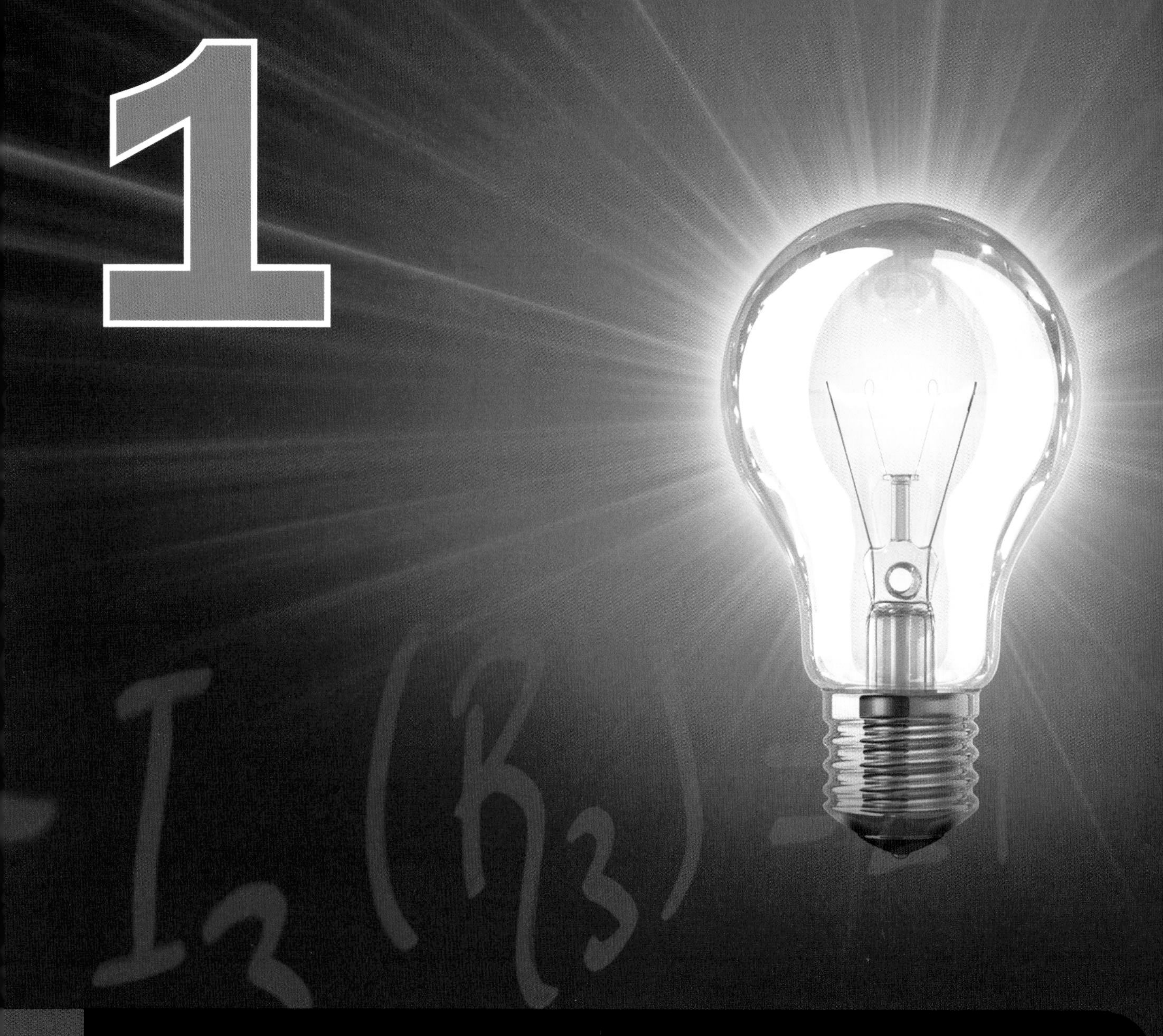

1 Electrical Basics

TOPICS

- INTRODUCTION
- ELECTRICAL TERMS
- CURRENT
- THE AMPERE
- DIRECTION OF CURRENT FLOW AND METERS
- MEASUREMENT OF CURRENT
- THE VOLT
- MEASURING VOLTAGE
- THE OHM
- USING AN OHMMETER
- MEASURING RESISTANCE IN A CIRCUIT
- OHM'S LAW OF PROPORTIONALITY
- PREFIX MULTIPLIERS
- ELECTRICAL POWER
- SAFETY AND ARCS

OVERVIEW

A question asked by many people not educated in electrical theory is: What is electricity? You cannot see electricity but can witness its effects on electrical apparatuses around the world. You cannot hear electricity but can hear electric discharges in the form of thunder, or the production of sound in speakers. You cannot smell or taste electricity, but it is a real force in getting work done. Electricity lights lamps, heats and cools buildings, makes motors turn, and drives electronic devices.

Once you learn the basics of electrical theory, you gain a new understanding of many of the devices you see around you each day. The key to using electricity is to be able to predict what will happen as you connect electrical power to electrical and electronic circuits and apparatuses and use the phenomenon of electricity. However, you must take precautions to use it carefully. It is a very versatile form of energy, but it can be lethal if not used correctly. As an electrician or electrical worker, it is your job to apply the benefits of electrical power while keeping yourself and the general public free of dangerous situations.

Very early in the 20th century, a sleeping student was awakened by his teacher's question, "James, what is electricity?" Sleepy and confused, the student tried vainly to recover. "Er-r-r, well, um-m-m, I *did* know, sir. But I forgot." "What?" cried the smiling professor. "You are the only person in history to know what electricity is, and you forgot!"

Intensive research and investigation in the 18th and 19th centuries had shown much about the behavior of electricity but little about its roots. Consequently, the unfortunate student's statement that "I *did* know" predated anyone knowing the true nature of electricity. The real understanding of electrical fundamentals began with J. J. Thompson's 1897 discovery of the electron. However, a comprehensive knowledge of electricity was slow in developing, requiring many subsequent 20th-century revelations about the fundamental structure of the atom.

In this chapter you will learn the basic measurements used in the study of electricity. We start with direct current (DC) first because it is easier to understand. To get you started, we begin with electrical terms that allow you to understand the basic features of any electrical circuit. In Chapter 2 you will learn the molecular theory of electricity. Each chapter in this book adds to your knowledge of electricity.

OBJECTIVES

After completing this chapter, you will be able to:

- Describe the units of measurement of current, voltage, resistance, and power
- Demonstrate your knowledge of the units ampere, volt, ohm, and watt by giving examples of their usage
- Explain the electron theory of current flow versus conventional current flow
- Solve electrical problems using Ohm's law and Watt's law
- Describe some safety procedures and equipment used for electrical safety requirements

INTRODUCTION

Electricity is an invisible force that can produce heat, motion, light, and many other physical effects. This invisible force provides power for lighting, radios, motors, heating and cooling, and many other applications. The common link among these applications is the electrical charge.

Electricity, *electron*, and *electronics* are English words that come from a word with a Greek background, *elektron*. The literal English translation of this word is "to be like amber." More than 2,500 years ago, the Greeks found that by rubbing amber with other materials it became charged with this invisible force and could attract bird feathers, hair, cloth, and other materials. In the 1600s, William Gilbert found that along with amber, other materials could be charged with this invisible force. He categorized those materials that could be charged as "electriks" and those materials that could not be charged as "non-electriks."

About a hundred years later, in 1733, Charles DuFay discovered that some charged materials would attract other objects and that other charged objects would repel different objects. Benjamin Franklin suggested the convention that two types of charges existed, positive (+) and negative (–), and that "like" charges repel (positive from positive and negative from negative) and "unlike" charges attract (positive to negative). An example of how these charged materials were classified and how they react to each other is shown in Figure 1–1.

To use electricity, you must be able to measure its effects and be able to control how the electricity is manipulated. In this chapter you will learn how to apply Ohm's law in your day-to-day electrical responsibilities. As you work through the chapter, remember that although the examples are for DC circuits, Ohm's law will find application in AC as well. In fact, Ohm's law will probably be the formula you use most often as an electrician.

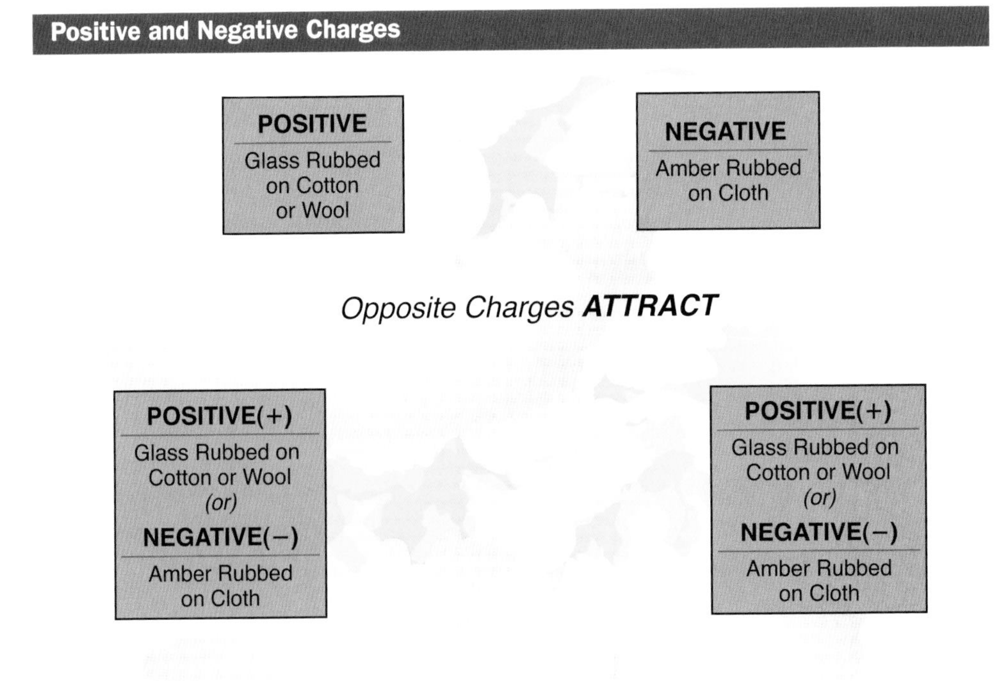

Figure 1–1 Positive and negative charges are caused by various material interactions.

Remember that when Ohm was doing his greatest work there were no computers, no calculators, and no electric lights! He could use only earlier published manuscripts on physics, a slide rule, and his mind. Without the aid of modern technology, he published Ohm's law, which describes in detail the relationship among voltage, current, and resistance in a simplified form. In Ohm's nomenclature, I is the current of the circuit, E is the voltage applied to the circuit, and R is the total resistance of the circuit. The following sections discuss specific definitions relating to each term.

The variable letters are representations of electrical quantities (explained in detail later in the chapter). For now, the letter I represents electrical flow or current. The letter I evolved from the notion that it was representative of the intensity of flow, measured in amperes (amps). The letter E represents the electrical pressure of the power supply, known as the voltage. The letter E evolved from the abbreviation of electromotive force, measured in volts (V). The letter V is sometimes used in formulas to indicate the voltage within the circuit. You will find that the letters E and V are used interchangeably from one reference to another, and in some cases both methods are used in one reference. The important concept to remember is that both E and V refer to electromotive force or voltage in the circuit. The letter R represents the opposition to the current flow in the circuit (the resistance), measured in ohms. Ohm's law is used for calculating circuit quantities with the following formula:

$$I = \frac{E}{R}$$

ELECTRICAL TERMS

In your study of electricity, there are many key terms you should learn. The first of these key terms are *current, voltage,* and *resistance.* In this chapter these and other terms are described in detail, along with an explanation of how they are used and how they relate mathematically to one another. This mathematical relationship among current, voltage, and resistance is known as **Ohm's law**. Many of you have probably heard of Ohm's law in high school, in technical school, or from some on-the-job experience.

FieldNote!

In 1827, a German named Georg Simon Ohm published a formula that expresses the single most important relationship in all of electricity. This formula, named Ohm's law in honor of its discoverer, finds application in virtually all aspects of the electrical and electronic industries.

Georg Simon Ohm was born on March 16, 1789 in Erlangen, Germany, and was educated at the University of Erlangen (Figure 1–2). From 1833 to 1849, he was director of the Polytechnic Institute of Nurnberg, and from 1852 until his death in 1854 he was professor of experimental physics at the University of Munich. Between 1825 and 1827, he developed a mathematical description of electrical current in circuits. What is now known as Ohm's law appeared in print in 1827. This work strongly influenced the electrical theory development of his day but was not well received by his peers.

Ohm

Figure 1–2 Georg Simon Ohm lived 1789–1854.

CURRENT

Your first challenge is to study the effects of DC (direct current) circuits and be able to predict what will happen on the basis of meter reading and observation of results of electricity. We will build the individual measurements and tie the individual definitions together through Ohm's law. We will make some assumptions at this point and then return to explain further as we delve deeper into the theory and verify the physical measurement. In other words, we will start with empirical evidence and go back to the theoretical concepts.

We will start with the basic premise of electricity. Electricity is the movement of electrons that have been activated or energized by an external energy. The external energy is imparted to the electrons, which "flow in a conductor."

The French physicist Charles A. de Coulomb (Figure 1–3) first studied the movement of electrons through materials. We call this flow of electrons through a material the current. Coulomb proved that like (positive to positive) charges repel and that unlike charges (positive to negative) attract. He also proved that this repelling force changes value as you change the distance between the charged materials. Coulomb studied the amount of charge on an electron and the amount of force between electrons, and on the basis of his research he defined the coulomb as equal to the total charge exhibited by 6.25×10^{18} electrons.

Coulomb

Figure 1–3 Charles de Coulomb, 1736–1806, is shown above.

Ampére

Figure 1–4 André Ampére, 1775–1836, is depicted above.

The electrons give up their energy at an electrical load as they produce heat, light, motion, and so on. As these electrons are moving in the conductor (wire), we have a way to measure the flow. The number of electrons is measured by the number of **coulombs** per second that move past a fixed point in the circuit. A coulomb is a fixed quantity of electrons.

If we could see electrons with the naked eye and we counted 6.25×10^{18} electrons moving past a point in 1 second, we would record the *flow* of electron **current** as 1 **ampere**. The flow of electron current is similar to the flow of water in a pipe. If we use the analogy of water flow, we could count the number of drops of water (electrons) and make a quantity of drops equal to 1 gallon (1 coulomb). If we measure 1 gallon of water moving past a point in 1 second, we would have the current of 1 gallon per second (i.e., 1 coulomb per second, or 1 ampere).

THE AMPERE

Now that we know what a coulomb is, how do we use the definition of the coulomb? Another French scientist, André M. Ampére (Figure 1–4), used the coulomb in his definition of the ampere. In the early 1800s, Ampére studied the effects of electrons flowing through wires. He defined an ampere as 1 coulomb of electrons flowing past a specific point in a wire in 1 second.

Although this definition remains accurate in the international system (SI) of weights and measurements, another definition is used in the meter-kilogram-second (MKS) measurement system adopted by the world in the 20th century. The newer definition in the MKS system defines *ampere* on the basis of the amount of force, in newtons per meter of length, created between two wires carrying equal amounts of current.

You will learn later that current flow creates a magnetic field. You also know through experience that two magnetic fields will either attract or repel each other, depending on their polarity. The new definition of an ampere is based on these facts and is stated as follows: When two long parallel wires, 1 meter apart, carry equal currents and the magnetic force between the two wires is .0000002 (2×10^{-7}) newtons per meter, the current flow in each wire is equal to 1 ampere.

DIRECTION OF CURRENT FLOW AND METERS

Current flow has a specific direction in a series DC circuit. In Figure 1–5, note that the battery has a positive and a negative side. The negative terminal is labeled negative because that is where there is a buildup or excess of electrons in the battery. Because electrons have a negative charge, this terminal is labeled negative.

The excess electrons leave the negative terminal and travel through each circuit component and back to the positive terminal. Thus, *outside* the battery the current flows from negative to positive. *Inside* the battery, the electrons move from the positive terminal toward the negative terminal.

DC Circuit Current Flow

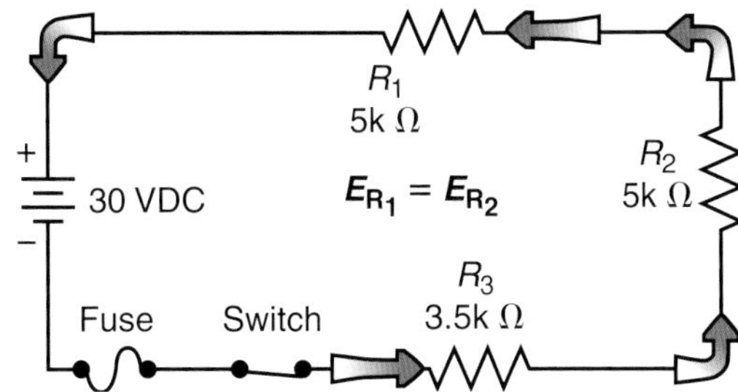

Figure 1–5 This DC circuit with a battery and three resistors shows current flow and electrical polarities.

Water Flow Meter

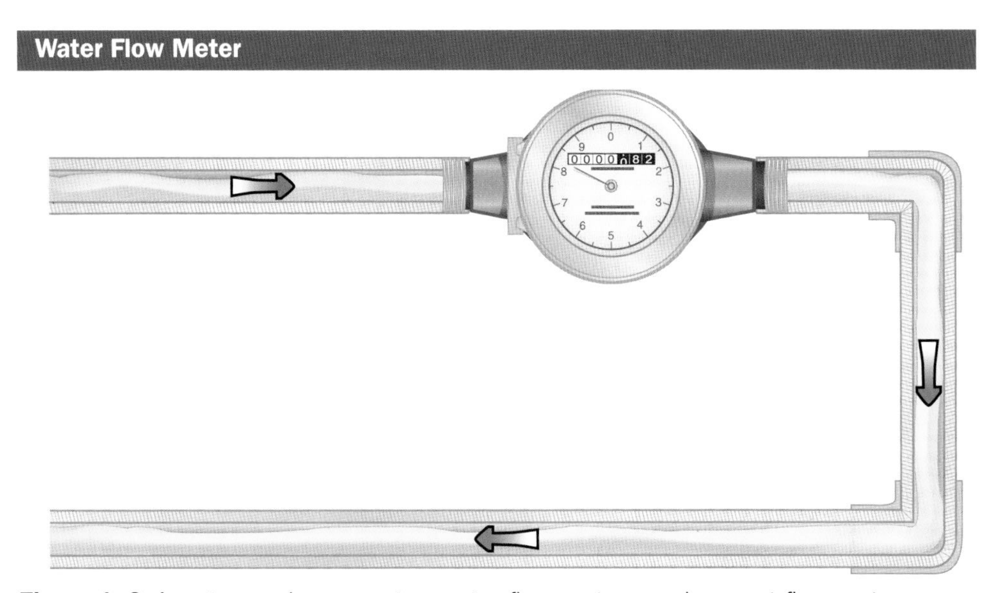

Figure 1–6 A water analogy equates water flow meters and current flow meters.

This concept is important in understanding circuits that are internal versus external to a power source.

MEASUREMENT OF CURRENT

We can measure the amount of current flow in a circuit by using a meter called an ammeter (not amp-meter). The ammeter uses various means for measuring current flow. Generally we need to measure the flow by inserting an ammeter into the flow, much like we would with a flow meter for water (Figure 1–6). As the current flows through the circuit it will also flow through the meter movement. The movement responds to the magnetic field created by the current flow and reacts by driving the meter pointer to the right, or upscale.

This type of meter is referred to as an analog meter, with a face marked as a scale and a pointer to point to the exact reading. Another popular meter is a digital meter, which automatically produces a digital readout of the current. Most of these meters are clamp-on meters that actually measure the magnetic field produced by the current and convert that quantity using electronics and a programmed chip to read in amperes (Figure 1–7).

Clamp-On Meter

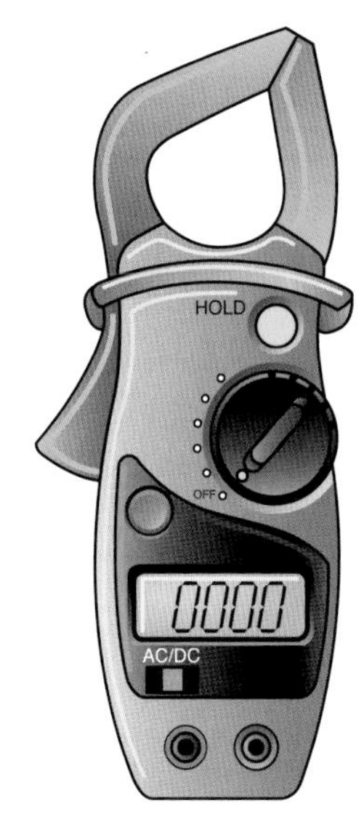

Figure 1–7 A clamp-on-style meter can be used to measure current.

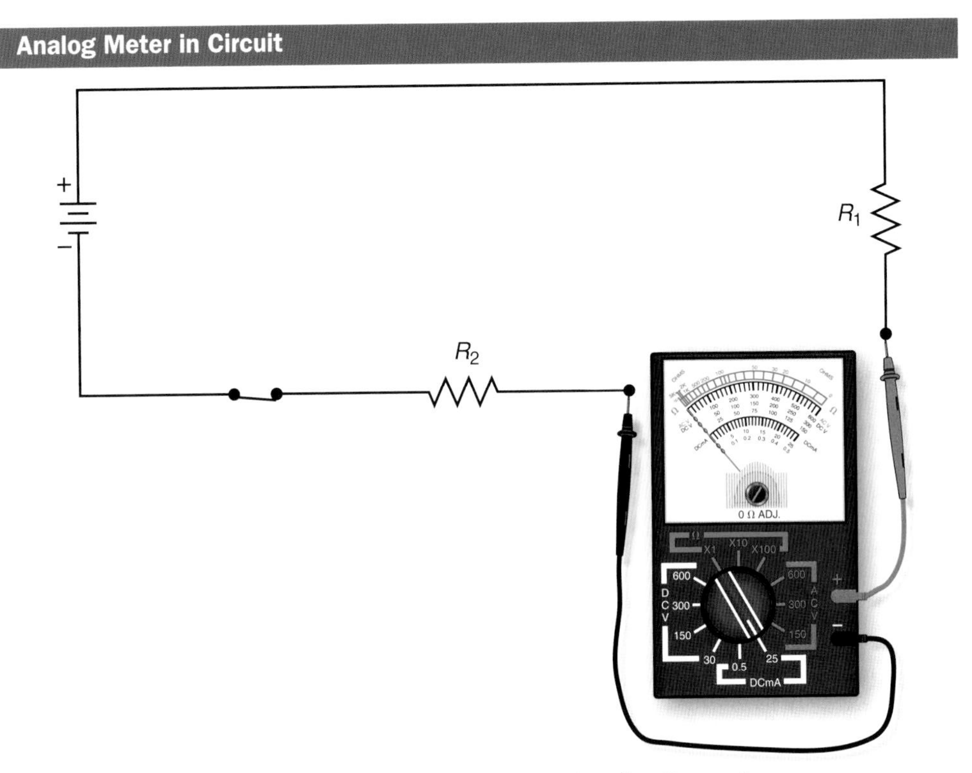

Figure 1–8 An analog meter can be inserted into the circuit, as above.

To measure current, the standard meter is inserted in the circuit current path and becomes part of the circuit (Figure 1–8). Although each type of meter may have its own internal power source, the operation of the meter is such that the actual current supplied by the circuit is used to drive the sensing element in the meter when the meter is used to measure current.

Because of the large variations available in both voltages and resistances in a circuit, current can vary over a large range of values, from a few microamps (one millionth of an amp) to thousands of amps. Ammeters and multimeters are not capable of measuring currents much above a few amps. For larger currents, it is impractical to build a series ammeter that can carry the full circuit current. This chapter is limited to a discussion of ammeters used for small DC currents.

Most multimeters also have multiple scales for use in current readings. Figure 1–9 shows an analog ammeter that has two DC current scales. The full-scale readings for these two scales are 0.5 DCmA and 25 DCmA.

DC Current Scales

Figure 1–9 The analog meter shown above has two different amperage scales for current measurement.

The maximum current that can be directly read using this meter is 25 mA. Currents greater than 25 mA can overdrive the meter element and damage it.

The scales for current on the analog meter are located on the meter face. The scale nearest the bottom corresponds to the 0 DCmA to 0.5 DCmA scale and is calibrated in 0.01 DCmA increments. The scales used for current measurements are much more linear than the scale used for resistance measurements. For this reason, it is much easier to read close to the ends of the scale. Maximum accuracy is still obtained, however, when the meter is read with the indicator near the center of the selected current scale.

When measuring current, you should always begin with the highest scale on the meter and then decrease the scale to obtain the proper reading if the first selected scale was too high. Most modern electronic multimeters also have fuse protection on current scales to prevent internal damage to the meter. The presence or absence of such a meter protection fuse must be verified for each meter used.

When using an analog meter to measure DC current, the polarity of the current is indicated by the direction of the meter needle's deflection. Normally, the meter is connected so that the red test lead (+) is connected to the more positive point in the circuit as the current is flowing toward the positive power supply terminal, and the black test lead is connected to the more negative point in the circuit that is supplied by the current source (Figure 1–10).

If the analog meter is connected incorrectly for more than a fraction of a second, the meter may be damaged because the current will try to drive the needle downscale and may in fact bend the delicate pointer. Damage to the meter movement (the part that moves the needle) may also affect the accuracy of future meter readings.

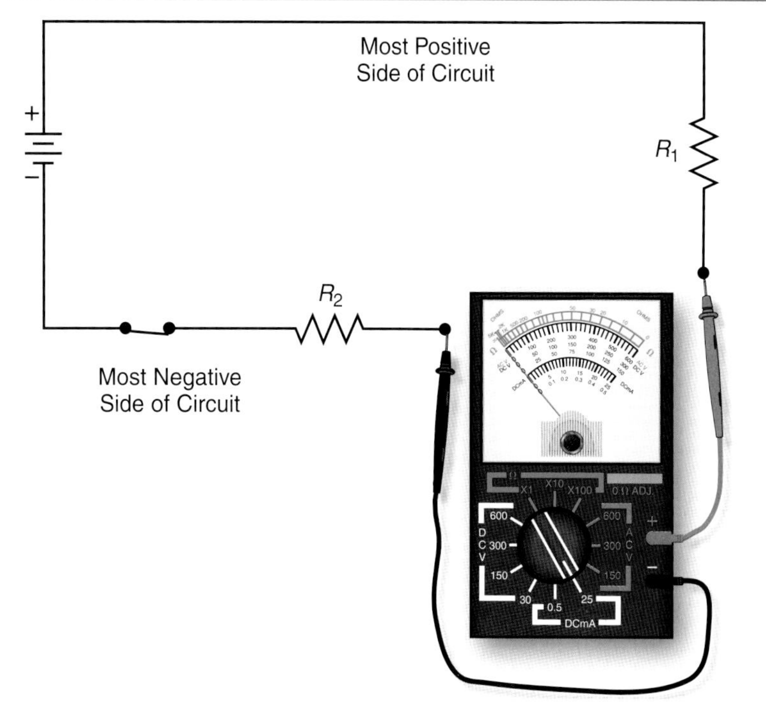

Figure 1–10 Analog meters need to have the DC circuit connected with the proper polarity.

CAUTION: When using any type of meter to measure electrical values in an energized circuit, be sure to follow the safety precautions recommended by the meter manufacturer. Do not attempt to use test equipment for purposes for which the meter was not designed. To do so will damage or destroy the meter and might injure the individual performing the test.

Figure 1–11 shows a digital meter, which can be used to measure current. The digital meter includes five different current scales, including 200 μA (microamp), 2 mA (milliamp), 20 mA, 200 mA, and 2 A. In addition, one position of the selector (20 m/10 A) serves a dual purpose. For all normal scales of the meter, the test leads are inserted into the COM and A jacks of the meter. However, when the selector switch is set to the 20 m/10 A position and the test leads are inserted into the COM and 10 A jacks, the meter will read 10 amps full scale.

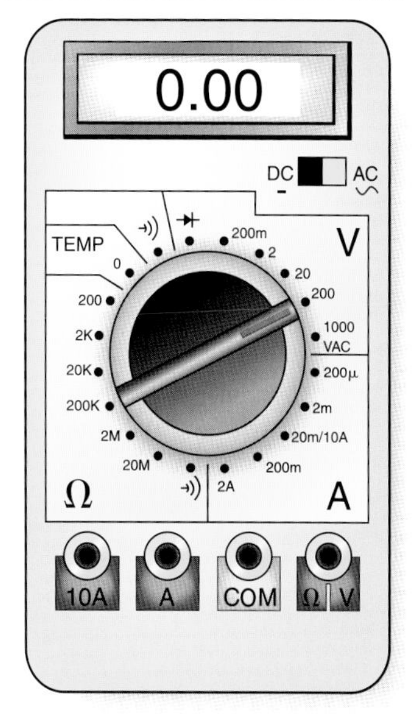

Figure 1–11 Above is an example of a digital multimeter used for current and other measurements.

For voltage sources up to 1,000 V, American National Standards Institute (ANSI), Canadian Standards Association (CSA), and International Electrotechnical Commission (IEC) have established four electrical work environment categories. They range from CAT I (least hazardous), to CAT IV (most hazardous). The category designation depends on potential surges, as well as circuit impedance. Multimeters selected for use must be certified by their manufacturer to meet CAT I, CAT II, CAT III, or CAT IV, as appropriate. Work should be performed only with appropriate personal protective equipment (PPE) as described in National Fire Protection Association's NFPA 70E. PPE ranges from natural cotton clothing for CAT I, to specialized face mask, gloves, and clothing furnished by the employer for CAT IV.

This is useful when measuring higher currents in a circuit. However, one additional precaution must be followed when using the 10 amp scale on the particular meter shown in Figure 1–11. All current scales for this particular digital multimeter are fuse protected, with the exception of the 10 amp scale. If the current exceeds 2 amps for any of the current scales except the 10 amp scale, the internal (replaceable) fuse will blow, creating an "open" to protect the circuit, multimeter, and possibly the user. If the current exceeds 10 amps when the meter is set up to read current on the 10 amp scale, the meter will be damaged or destroyed and the user could possibly be injured. It is important that you know if, and how, the particular meter you are using is protected against such occurrences.

When the digital meter is used to measure current, the scale is selected using the meter's selector switch. The position of the decimal point on the digital display will indicate the range of readings for the current scale selected. Table 1–1 outlines the scales for this meter. An additional switch, one that must be set for the digital meter shown, is the DC/AC switch. This switch is used to tell the meter if the current being measured is DC (direct current) or AC (alternating current). In this chapter, it is assumed that all currents are DC.

On some multimeters, an overcurrent (current larger than the selected scale) would be displayed with a single digit 1. Care must be taken not to confuse this type of overcurrent indication with a legitimate, in-scale reading such as 1.000, 10.00, or 100.0. Some digital multimeters indicate such an overload condition by displaying the letters OL. Read the manual supplied with any meter you might use to find out how that meter indicates overcurrent or overload conditions.

When the current reading on a digital meter is underscale, the reading will be shown as a series of zeros with the decimal point correctly located for the scale being used. In the 200 mA scale position, for example, when there is no current flowing through the meter or the current through the meter is less than 100 μA, the meter display will be 00.0. If you see this reading and you suspect there should be some current present in the circuit, adjust the selector switch to the next lower range and try to measure the current. Repeat this procedure until the selected scale allows the circuit current to be measured or until you get to the lowest scale. Currents below 0.1 μA are not detectable by this meter.

Table 1–1

The typical scales of the digital meter and what the display may indicate are listed.

Scale	Display at Maximum
200 μA	199.9
2 mA	1.999
20 mA	19.99
200 mA	199.9
2 A	1.999
10 A	10.00

Digital meters can measure current of either polarity. When the red test lead is connected to the more positive point in the circuit and the black test lead is connected to the more negative point, the digital meter will indicate the polarity by showing a plus sign (+) in front of the numerical reading or by not indicating a sign for the circuit current. When the red test lead is connected in the circuit to a point that is more negative than the black test lead, however, this reverse polarity will be indicated by a minus sign (–) in front of the numerical reading of the digital display. The meter's accuracy is not affected by the polarity of the applied signal.

TIP: It is a good practice to always identify lead polarity and connect accordingly. This helps prevent misconnections when using an analog meter.

When measuring current with either type of meter (analog or digital), the procedure to obtain accurate measurement of the current value is the same. Digital meters are often more precise and easier to read for steady-state currents or currents that are not changing rapidly, whereas analog meters often work better when currents are changing or varying and will not settle down to a precise value.

The normal power or voltage supply for the circuit must be applied in order to take current readings. However, the circuit must be turned off in order to safely connect the ammeter before taking the current readings. Caution must be used to ensure that the meter polarity is correct (even if using a digital meter) and that the scale selected for the reading is as great or greater than the highest anticipated circuit current. Generally, the meter's selector switch can be set to the highest scale and then adjusted downward if the scale is too high.

In the case of the digital meter, if the current is within the limits of the 10 amp range but greater than the 2 amp range, the 10 A test lead jack and 20 m/10 A switch selector position must be used. If later you wish to lower the range setting, power must again be removed from the circuit because removing the test lead will involve breaking the current path in the circuit.

Because there is a single current path through a series circuit, the circuit can be broken and the meter inserted at any place in the circuit. The current reading will be the same wherever the meter is inserted into the circuit. The following material explores circuits for which it is necessary to measure the current. Figure 1–12 shows the proper way to measure the current in this circuit. The circuit has been broken between resistors R_1 and R_2 and the meter inserted with the positive test lead toward the positive power supply lead. The meter could have been inserted at any place in the circuit by breaking the conductor at that location.

Ammeter: Proper Connection

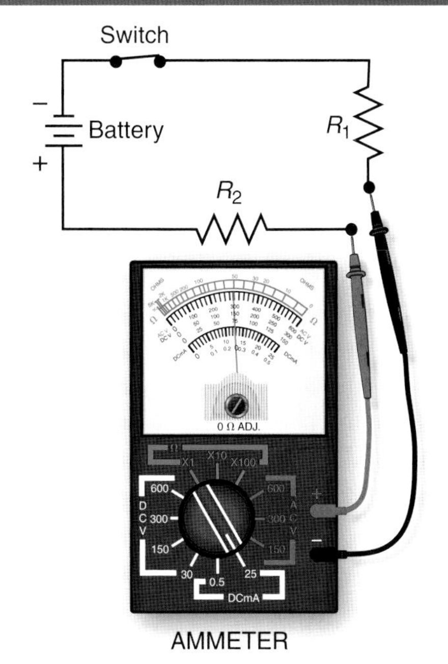

Figure 1–12 This in-line ammeter is connected correctly.

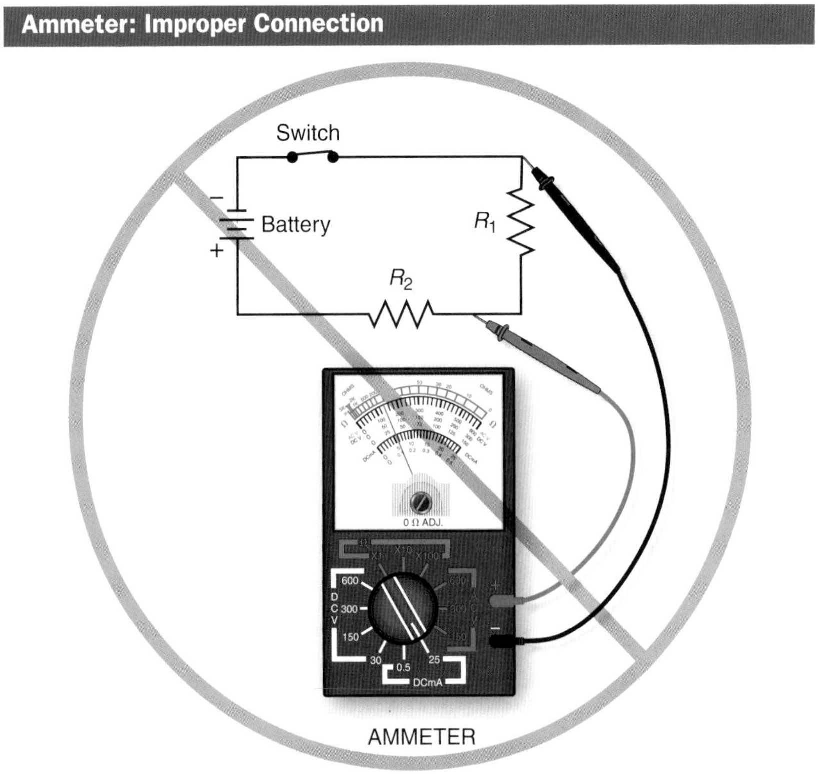

Ammeter: Improper Connection

Figure 1–13 This in-line analog ammeter is connected incorrectly across a component.

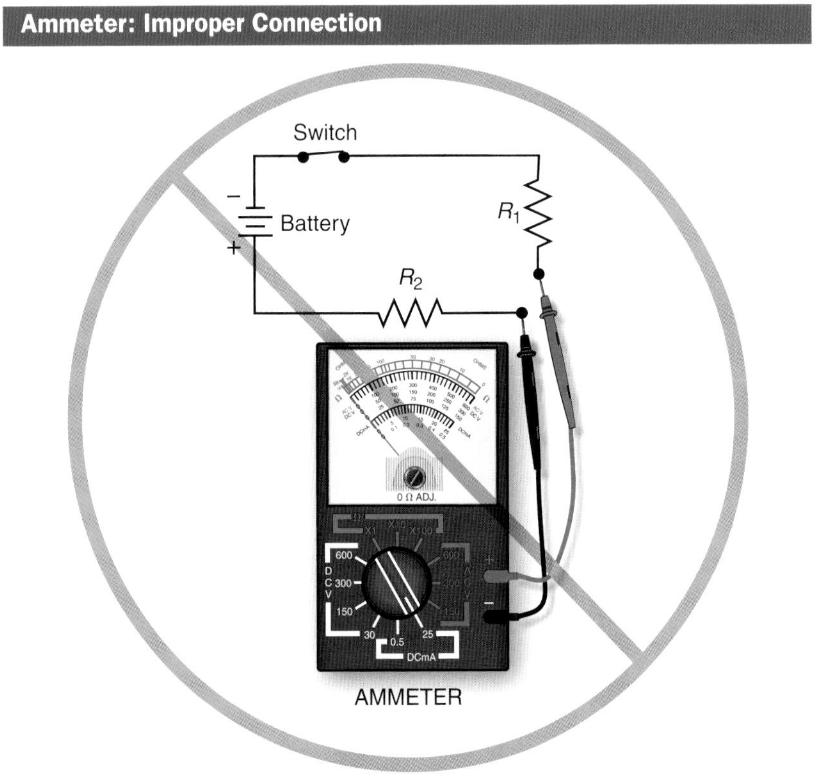

Ammeter: Improper Connection

Figure 1–14 An ammeter connected backwards in a circuit can cause damage.

Figure 1–13 shows a circuit in which the current measurement would be in error because the meter is improperly installed. The meter must be placed in series with other circuit components not in parallel, as shown. If the meter were connected as shown, the internal resistance of the meter would be in parallel with resistor R_1, creating an additional current path and changing the parameters for the circuit (thus giving an erroneous reading for circuit current).

Figure 1–14 shows a circuit in which an analog meter is incorrectly connected. The negative meter lead is connected to the more positive circuit point and the positive meter lead to the more negative circuit point. This could damage the analog meter's movement. This connection would not be a problem, however, if a digital meter were used for this measurement.

There are clamp-on ammeters that are used to measure circuit current without breaking the circuit. These work by sensing the magnetic field surrounding a conductor as a result of current flowing through that conductor. These meters are generally used on AC circuits but are also available for DC circuits.

CONVENTIONAL CURRENT FLOW

In the early days of electricity, before the development and understanding of electron theory, electricians did not know which way current flowed. They believed that *something* was flowing, and they needed to know which way it was flowing. On the basis of the magnetic field created and how it interacted with the earth's magnetic field, early experimenters decided that current flowed from the positive pole to the negative pole of a DC source. This concept is named **conventional current flow**, also known as hole flow or positive current flow.

In modern times, an explanation must be given for the concept of conventional current flow. When an electron leaves an atom, it creates a gap called a hole in the atom it left. The atom with a missing electron is now a positive ion and will want to attract another free electron. However, the new electron has to come from somewhere. This means it makes a hole somewhere else.

If you look at the holes' movements (hole flow), it looks as if they are moving in a direction opposite that of the electrons. Thus, we have a modern explanation for the concept of conventional current flow. Conventional current flow is still used in many engineering applications and in electrical engineering and physics classes in engineering colleges. Figure 1–15 illustrates the concept of conventional current flow. Thus, many people in the electrical industry learn by explanations involving conventional current flow.

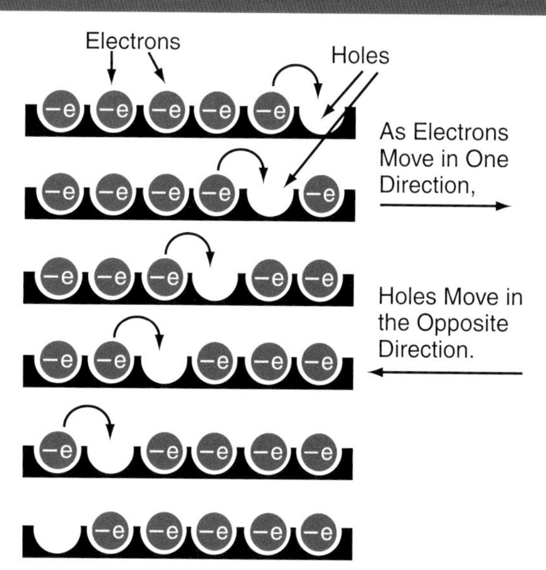

Figure 1–15 Electrons flow opposite to "hole" flow.

ELECTRON CURRENT FLOW

Now, of course, we know that the electrons are actually moving through the conductor (wire). Since the mid 1900s, electrician and technician training has used the more correct term **electron current flow** in all training and day-to-day applications. This book and others you will use as electricians use *electron current flow*. Figure 1–16 shows both conventions and their relative directions.

The maximum possible speed of the electrons is the speed of light in free space, which is equal to 299,792,458 meters per second (186,282 miles per second). You will learn that some materials allow electrons to move more freely than others. We will conclude that electrons will move more slowly in a conductor than through the vacuum of space. There are many factors to take into account, including material, voltage, and frequency of the current flow. (You will learn about frequency when you study AC theory.)

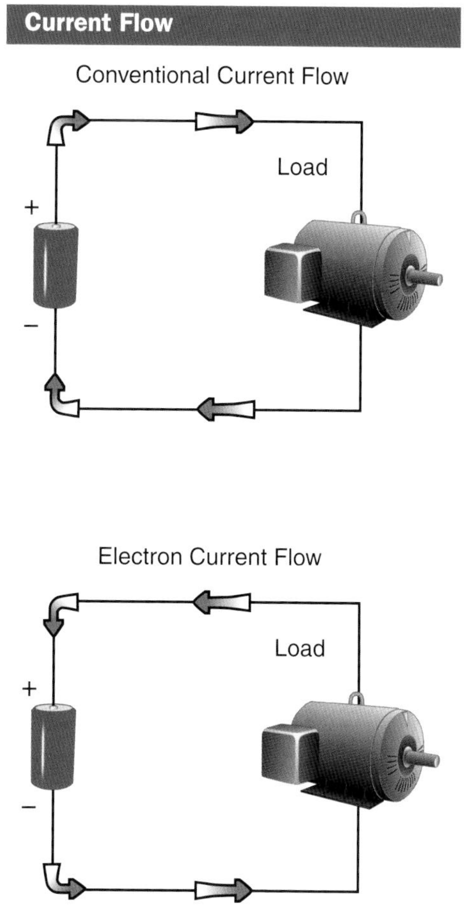

Current Flow

Conventional Current Flow

Load

Electron Current Flow

Load

Figure 1–16 Conventional current flow is opposite to electron current flow.

Generally, the current in a conductor (wire) will flow at something less than the speed of light, often 80% of light speed or less.

THE VOLT

Using the definition of the coulomb, we can now define the volt. Voltage is the **electromotive force (EMF)**, a potential difference in electron charges. A potential difference exists when one object has a greater or fewer number of electrons than another object. Because each electron has a fixed amount of charge, there is potential energy available because of the two different electrical charges. This potential difference causes electrons to be repelled or attracted from one material to another (see Figure 1–17). In a circuit, this is from the negative side to the positive side of a battery. The volt is defined in terms of the coulomb and joule. (A joule is the amount of energy equal to 0.737 foot-pounds.)

The volt is defined as the amount of potential that will cause 1 coulomb to do 1 joule of work. Keep in mind that it is the current doing the work, not the voltage. Voltage does not flow; it causes current flow, somewhat like a pump that causes water flow in a fluid system. The potential difference (EMF) shown in Figure 1–17 acts as a pump to force electrons to flow in the direction indicated. The analogy of the pump refers to the tendency of the electrons to move and to try to balance out. Refer to Figure 1–18 for an analogous situation with a water system.

A potential difference between the two points creates a pressure difference, and the water will move from where there is more water (excess of electrons) to a point where there is a shortage of water (not enough electrons). The EMF is the difference in potential energy between two points or difference in pressure. The positive polarity point has too few electrons, which creates a net positive charge. The negative polarity point has an excess of electrons and therefore has a negative charge. The difference is measured in voltage, also known as a difference in potential. The "potential" of the circuit is referred to as the EMF or voltage.

VOLTMETER

The means of measuring the potential difference, or difference in pressure, is a voltmeter. The meter actually provides a path for electrons to flow from negative to positive through the meter.

Potential Difference

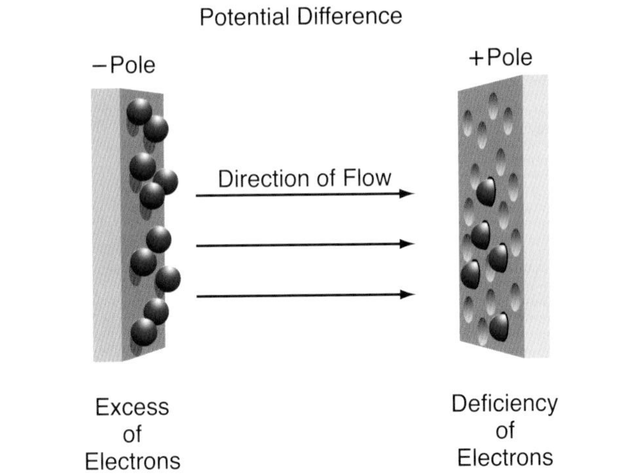

Potential Difference
−Pole
+Pole
Direction of Flow
Excess of Electrons
Deficiency of Electrons

Figure 1–17 Potential difference is caused by an imbalance of electrical charges.

Pressure Difference

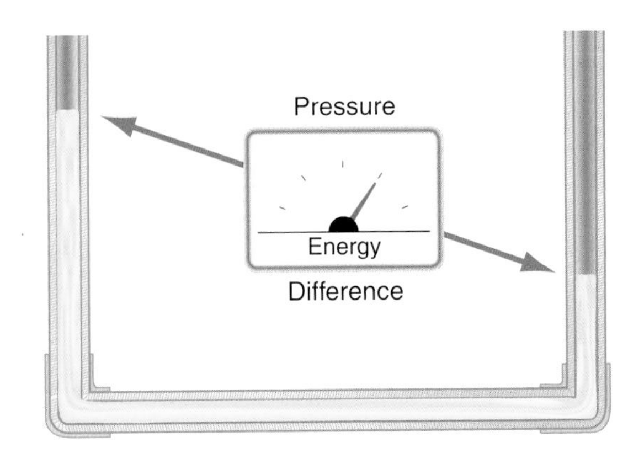

Figure 1–18 The difference in the amount of water creates a potential difference of energy.

Voltmeter Connection

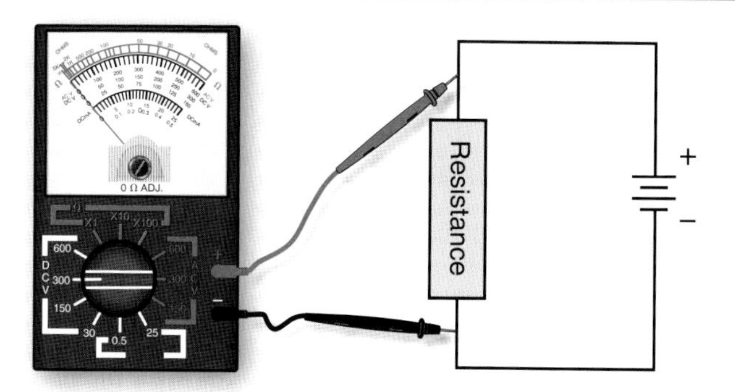

Figure 1–19 The correct connection of a voltmeter is across the component.

With a greater pressure difference, more current flows and the meter shows a higher reading. Conversely, with less potential difference in pressure there is less current flow and a smaller reading is obtained. The amount of current is very small and does not usually affect the circuit characteristics. See Figure 1–19 for connection of a voltmeter. We can connect the meter right across the positive and negative terminals of the power supply to see the amount of potential pressure of the source.

This is not to say that the internal resistance of the meter does not affect the circuit's operation, only that the voltmeter is not supposed to alter the operation of the circuit. The amount of current flowing through a good high-impedance (high resistance) meter is negligible and does not affect the operation of the circuit. Actually, a very small current does flow through the voltmeter. However, it is small enough that no significant change occurs in the circuit.

The voltmeter can be used to measure the voltage across any device in a circuit. Voltage is the driving force that produces movement of electrons in a circuit. Ohm's law relates all voltage, current, and resistance in a circuit. The amount of current in a circuit with a fixed resistance or load will be directly proportional to the amount of voltage present. Voltage may exist, however, even though no current is flowing in the circuit. A voltmeter can be used, for example, to measure the potential in a battery even though the battery is not connected in a circuit and no current is flowing from the battery. Voltage potential can exist either in or out of a circuit. Current is present only when both a voltage source and a current path are present.

As with current, voltages can vary from very small values (millivolts and microvolts) to very large values (megavolts). The meters described in this chapter can measure voltages as low as a few millivolts or (depending on the type of meter and the type of voltage being measured) as high as 1 kilovolt (1,000 V). For safety purposes, it is important to check the maximum voltage rating of a multimeter before using it. Many multimeters have a maximum rating of 600 V. To accommodate this range of voltages, different voltage scales are included with each meter.

Figure 1–20 shows the analog meter discussed previously. The DC voltage scales are located between the resistance and current scales on this meter.

Analog Meter

Figure 1–20 A typical analog meter with four switchable DC voltage selections is shown above.

Three different numerical scales are provided for the voltage ranges on the analog meter to provide four different DC voltage (DCV) ranges. The DCV ranges included on the analog meter include 30, 150, 300, and 600, indicating voltage ranges of 0 V to 30 V, 0 V to 150 V, 0 V to 300 V, and 0 V to 600 V. The same faceplate scales are used for both the DC and the AC measurements. However, different selector switch positions are used. The markings for the various scales are broken into 1 V, 5 V, 10 V, and 20 V increments, depending on the scale being used.

When measuring voltage, you should always begin with the highest scale. You can then reduce the scale, using the selector, to obtain the most accurate reading. Generally, voltmeters are temporarily connected to a circuit by means of test probes, but voltmeters of a nonmultimeter type can be installed where continuous monitoring of a circuit voltage is required.

Voltages are polarized just like currents. When using the analog meter, you must connect the meter to the correct polarity. The red test lead is inserted into the positive (+) test-lead jack and is connected to the more positive point in the circuit. The black test lead is inserted into the negative (–) test-lead jack and is connected to the more negative point in the circuit. As in performing current tests, the meter may be damaged if left connected to the wrong polarity for more than a fraction of a second.

Figure 1–21 shows the digital meter discussed previously. The digital meter has five different voltage ranges, as well as a special voltage scale indicated by the diode symbol (→+). Voltage ranges on the digital meter include 200 m, 2, 20, 200, and 1,000, indicating DC voltage ranges of 0 V to 200 millivolts, 0 V to 2 V, 0 V to 200 V, and 0 V to 1,000 V. The scale designated by the diode symbol is a special voltage scale used for testing semiconductor devices and is slightly different from a normal voltage scale. This special scale is common but is not important to the discussion at this time.

To measure voltage with the digital meter, the test leads are inserted in the test-lead jacks marked COM and Ω/V. When used to measure voltage, the scale selected on the meter's selector switch will determine the position of the decimal point on the display. Scales for this meter are outlined in Table 1–2. As in current measurements, the DC/AC switch must be set to select DC or AC voltage measurements.

Digital Multimeter

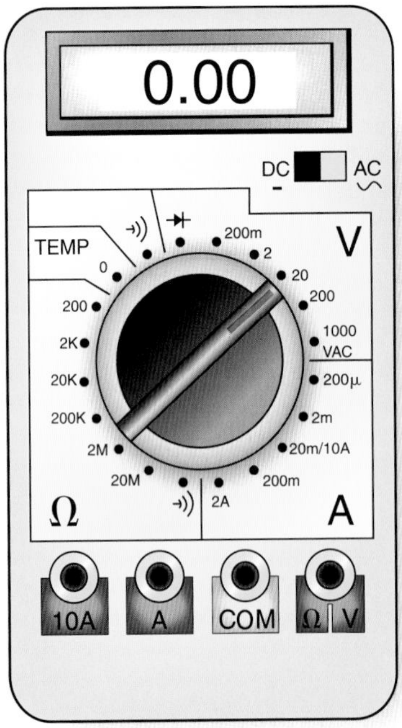

Figure 1–21 A digital multimeter with multiple voltage ranges is shown above.

Table 1–2

Voltage scales on a digital meter are indicated here, with maximum readings on each scale.

Scale	Display at Maximum
200 mV	199.9
2 V	1.999
20 V	19.99
200 V	199.9
1,000 V	1.000

Overvoltages would be displayed as a 1 and should not be confused with voltage displays in which all digits would be shown. Some meters display the symbol OL (overload) to show voltage values beyond the range of the meter. Be sure to read the manual supplied with the meter you might use to find out how it indicates overvoltage or overload conditions.

When the voltage reading on a digital meter is underscale, the display will show a series of zeros with a decimal point correctly located for the scale being used. In the 200 mV, position, for example, when there is no voltage present at the test leads or the voltage is less than 0.1 mV, the meter display will be 00.0. If you get such a display on any scale other than the 200 mV scale (already the lowest scale available) and you suspect there should be some voltage present at the test leads, adjust the selector switch to the next lowest range to attempt to measure the voltage. Repeat this procedure until the selected scale allows the voltage to be measured or until you reach the lowest scale. Voltages below 0.1 mV cannot be measured with this multimeter.

Digital meters can measure voltage of either polarity. When the red test lead is inserted into the Ω/V test-lead jack and connected to the more positive point and the black test lead is inserted in the COM test-lead jack and connected to the more negative point, the meter will indicate a positive voltage by displaying a plus sign (+) or by displaying no sign in front of the numerical value. If the polarity of the test leads is reversed, the numeric display will be preceded by a minus sign (–). The digital meter's accuracy is not affected by the polarity of the applied voltage.

MEASURING VOLTAGE

With either type of meter, the procedures for measuring voltage are the same. As with resistance and current, voltage values are more easily read with digital meters when the voltages are steady-state (unchanging). Generally, analog meters work better for voltage values that fluctuate rapidly or that will not settle down to a precise value.

Most voltmeters measure voltage by drawing a very small amount of current from the circuit being tested. The voltage to be measured is used to produce a small current through a known resistance. The meter is then calibrated to show the voltage rather than the current. All voltmeters produce some error when connected to a circuit, but some designs produce less error than others. Generally, the higher the input impedance or load (resistance) provided by the meter, the more accurate the voltage reading will be. Meters with very high input impedances (greater than 10 million ohms) generally do not affect the circuit to which they are connected because the current through the meter is extremely small (Figure 1–22). Each range creates a preset voltage drop across the meter movement to read the voltage being measured.

Analog meters are polarity-sensitive and may be damaged if connected incorrectly. Be sure the scale selected for the voltage measurement is equal to or greater than the highest anticipated circuit voltage. Generally, the meter's switch should be set to the highest voltage scale and then adjusted downward if the scale is too high.

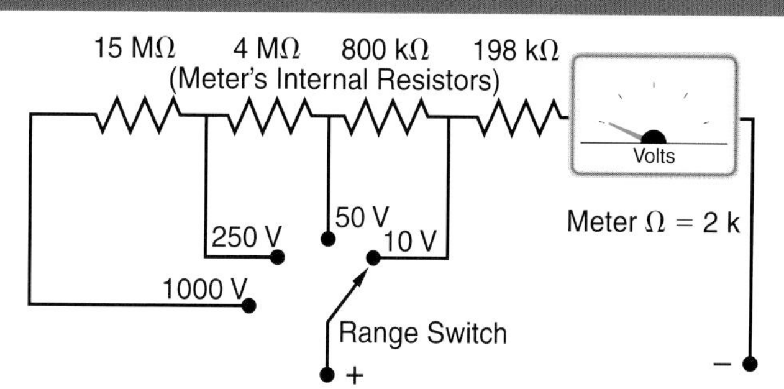

Figure 1–22 A voltmeter has internal resistance to adjust for each voltage range.

In electrical and electronic circuits, voltages are developed or are present across each component in the circuit. Each of these voltages can be measured independently, or the voltage across two or more components can be measured at the same time. Voltage measurements are measurements of the difference in potential between any two points in a circuit.

Figure 1–23 shows the proper way to measure the voltage across resistor R_1. The voltmeter is connected across the resistor and will show only the voltage developed as current flows through that resistor. The positive meter lead is connected to the more positive end of the resistor (the end closest to the positive terminal of the power supply or battery), and the negative test lead is connected to the more negative end of the resistor (the end closest to the negative terminal of the power supply or battery). Figure 1–24 shows a voltmeter connected across resistors R_1 and R_2. The voltage indicated would be the total voltage developed across both of these resistors, which would be equal to the power supply voltage.

Figure 1–25 shows a voltmeter inserted in series, with the circuit between resistors R_1 and R_2 as it might be connected to measure current. Ammeters have a very low internal resistance and do not significantly affect the circuit when installed in series. Voltmeters, as discussed previously, have a very large internal resistance. As a result, the current path would be greatly altered if the voltmeter were installed in this manner. The voltmeter reading would not reflect the voltage across R_1 or R_2 but would indicate a voltage very close to the power supply potential, because the very small current flowing through the meter would not produce a significant voltage drop across resistors R_1 and R_2.

Voltmeter: Proper Connection

Switch

Battery 12 VDC

R_1

R_2

0 Ω ADJ.

VOLTMETER

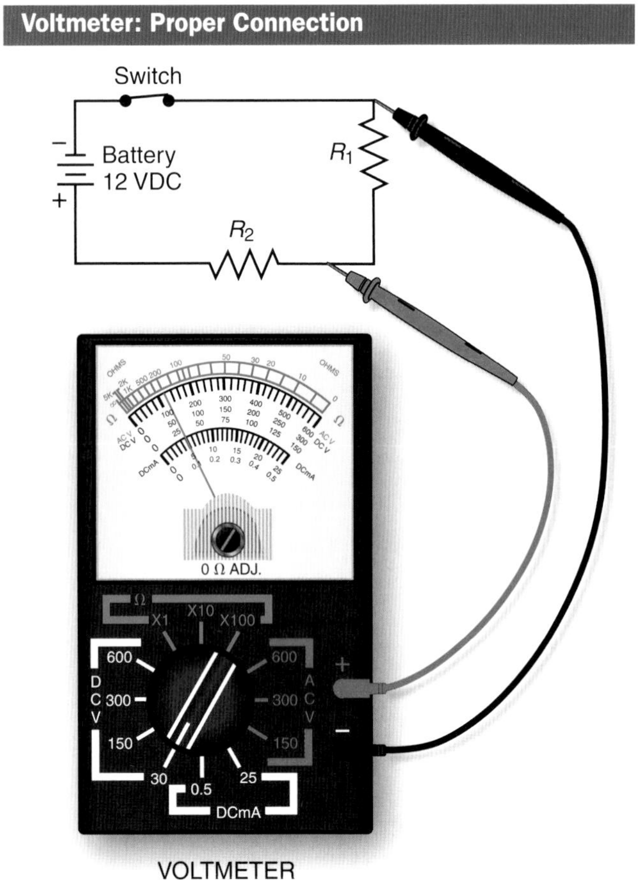

Figure 1–23 This analog meter is correctly connected across a component.

Voltmeter: Proper Connection

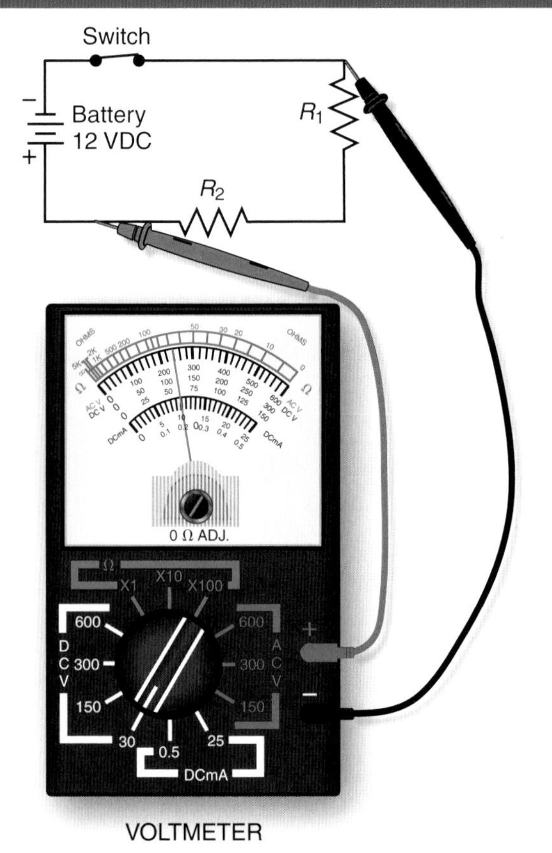

Figure 1–24 This voltmeter is correctly connected across two components.

THE OHM

Now that we have defined the ampere and the volt, let's look at the **ohm**. The ohm is a measurement of the amount of opposition to current flow that exists in a circuit object. We can go to the water analogy again and see the effects of opposition. As the water pipe is restricted, it takes more pressure to force the same amount of water flow through the pipe. If we constrict the pipe too much or add a great deal of resistance to the flow, the water flow will diminish until it stops with maximum restriction. The resistance to electron flow in a conductor has the same effect. More resistance means less flow if the pressure stays the same.

By definition, 1 ohm is the amount of **resistance** in a circuit that allows 1 ampere of current to flow when a potential of 1 volt is applied. This is a statement of Ohm's law. The unit of measure of resistance is the ohm, and the symbol used for the ohm is the Greek symbol omega (Ω). Resistance is a property of all materials through which current can flow. Some materials allow current to flow more easily than others. A material that has many free electrons is a conductor. Likewise, a material with few free electrons is an insulator. A conductor, therefore, has a low resistance and an insulator has a high resistance.

Voltmeter: Improper Connection

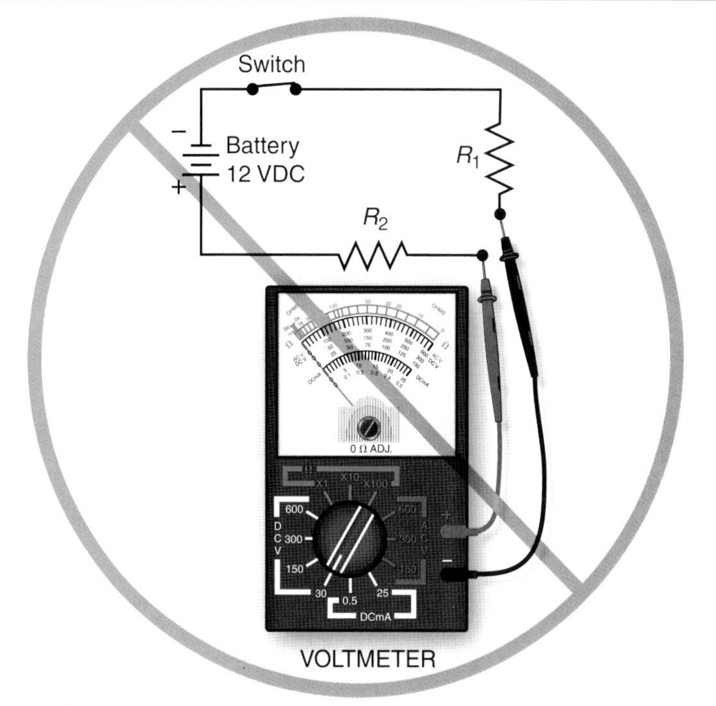

Figure 1–25 This voltmeter is incorrectly connected in series with the components.

Ohmmeter

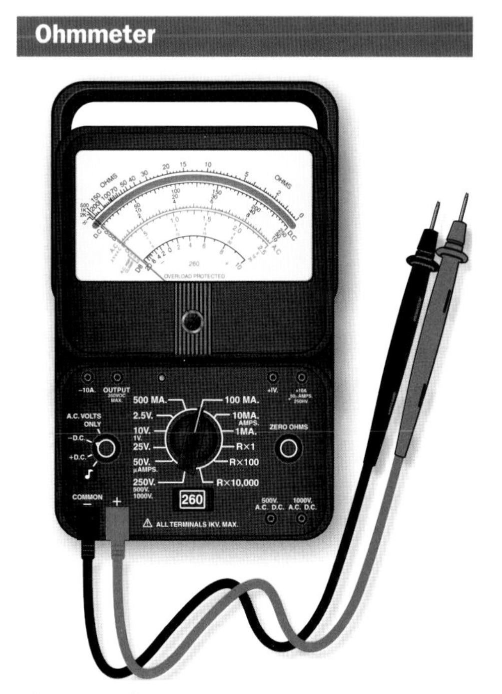

Figure 1–26 The ohmmeter has its own power supply that provides a source of current.

MEASURING RESISTANCE

The way to measure the resistance that a circuit or component has to the flow of electrons is to use an ohmmeter. This meter typically has its own source of current, such as a battery. As you use the test leads (Figure 1–26), the battery supplies a small current to the test situation and measures how much current flows with a specified amount of voltage or pressure. The meter is set to convert this small current to a scale marked in ohms.

If the test situation has a small resistance, more current will flow and the meter reads upscale and therefore indicates a low ohm reading. Conversely, a high-resistance component will not allow much current to flow and the meter reads low on the scale (high in ohms).

In Figure 1–27, note that the scale is reversed, compared to normal scales, with low resistance at the far right and high resistance at the far left of this analog meter. A digital meter uses the same scenario but converts the small current flow through the test item directly to ohms on the readout.

CAUTION: Never read ohms of a circuit with the power on. The only power must be that supplied from the meter.

USING AN OHMMETER

A meter designed to measure resistance is called an ohmmeter. Meters designed to measure several circuit characteristics are called multimeters. A multimeter generally contains an ohm scale for reading resistance. Resistance is a physical characteristic of a conductor or anything conductive. Ohmmeters and multimeters that measure resistance do so by applying a known voltage across the resistor and measuring the resultant current or by supplying a known current to a resistor and then measuring the resultant voltage.

These are both applications of Ohm's law. Scales on ohmmeters are calibrated to read resistance rather than voltage or current. Multimeters have multiple scales, each having a maximum resistance value (calibrated in ohms) used for resistance measurements. Resistances in circuit components vary from very small to very large values. For different applications, it is sometimes necessary to measure resistance values that vary from micro-ohms (0.000001) to megohms (1,000,000). It is not practical to design a meter that could read such a large range of resistance values on a single scale. Most ohmmeters available today have a range selector switch that allows the user to select the appropriate range for reading the resistance values being measured.

Ohmmeter Scale

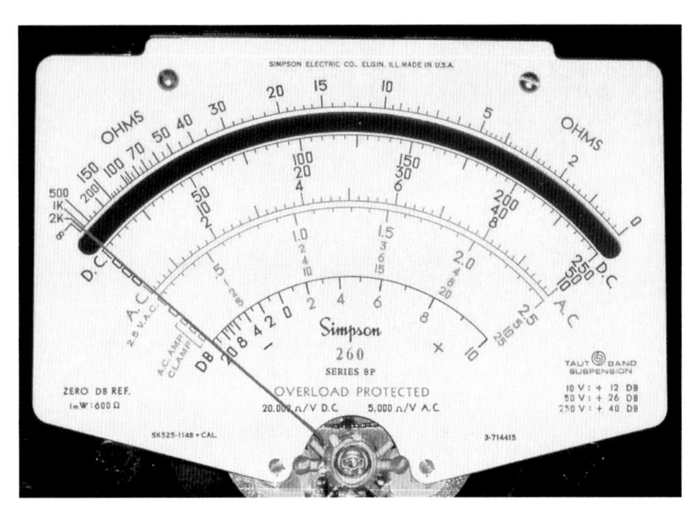

Figure 1–27 The ohmmeter scale on an analog multimeter is reversed, compared to the other scales.

ANALOG METERS

Figure 1–28 shows an analog multimeter. The analog multimeter uses a moving indicator and a fixed set of graduated scales. As the value being measured changes, the meter needle or pointer will move to some location between the meter scale limits. The value being measured can then be read from the scale being used. Multimeters are capable of measuring voltage, current, and resistance. The top scale on the meter shown in Figure 1–28 is the ohms scale used for resistance measurements. A selector switch allows the user to select three different ohm-scale ranges.

All ohm ranges use the same (top) scale. Multiple resistance scales are necessary because of the nonlinearity of the ohm scale. Even though the scale goes from 0 ohms to ∞ (infinity, or an open circuit), the scale is very difficult to read accurately at the upper end. Note that the numbers are closer together at the left, or high end, of the scale.

The analog meter includes a selector switch that can be used to select scales X1, X10, and X100 (Figure 1–29). This makes reading the correct value much easier because a scale can be selected to match the value of the resistance to be measured. If the pointer reads close to either end of the scale, selecting a different scale may allow the user to move the needle toward the center of the meter's scale.

Analog Multimeter

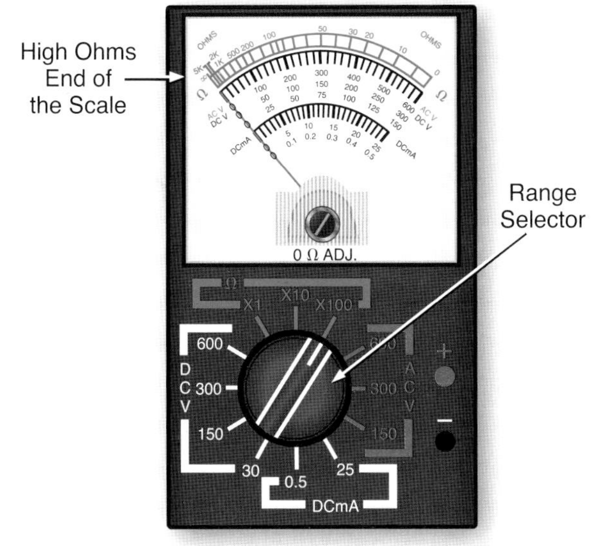

High Ohms End of the Scale

Range Selector

0 Ω ADJ.

Figure 1–28 A typical analog multimeter with multiple scales is shown above.

Ohmmeter Ranges

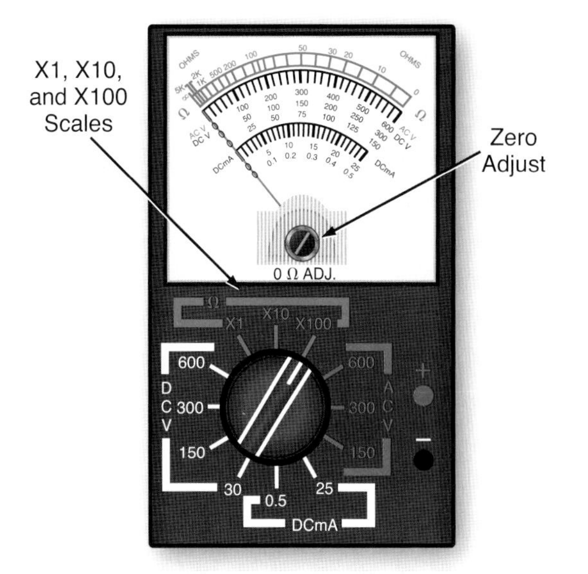

Figure 1–29 Ohmmeter ranges often use the same reading scale but use multipliers to calculate the actual ohm measurement.

This provides for increased accuracy in reading the resistance value. When the selector switch on the face of the meter is in the X1 position, the meter's scale is read directly, and simply reading the value indicates the correct resistance.

When the selector is in the X10 position, multiply the value on the scale faceplate indicated by the pointer by a factor of 10. In the X10 position, for instance, if the pointer indicated a resistance of 30 the value would actually represent a resistance of 10 × 30 (300 ohms). Similarly, when the selector switch is in the X100 position, the value on the faceplate is multiplied by 100. In the X100 position, with the pointer indicating 2 K (or 2,000 ohms), the actual reading would be 2 K × 100 (200 kΩ).

A 0 Ω ADJ (adjust) control knob allows the meter to be adjusted to compensate for test-lead resistance and battery voltage level in resistance measurements. To adjust this control, the user would "short" the leads by touching them together and then adjust the control to give a 0 Ω reading on the faceplate. This adjustment eliminates the resistance of the test leads from any resistance reading taken with the meter and must be done every time the meter is used and whenever the scale is changed from one range to another. The zero on the meter should be checked periodically, as the zero value may drift as the batteries weaken.

DIGITAL METERS

Figure 1–30 shows a digital multimeter that can be used to measure resistance. Digital meters are read differently than analog meters. As you can see, the moving pointer and multiscale faceplate are no longer present. In the digital multimeter, they have been replaced by a liquid crystal display (LCD).

On a digital meter, a single value is digitally displayed that represents the resistance being measured. A selector switch is used to select the maximum resistance to be read on the selected scale. The amount of information shown depends on the type of display on the meter. Meters may display two and one-half digits (up to 199), three digits (999), three and one-half digits (1999), or four digits (9999).

Digital Multimeter

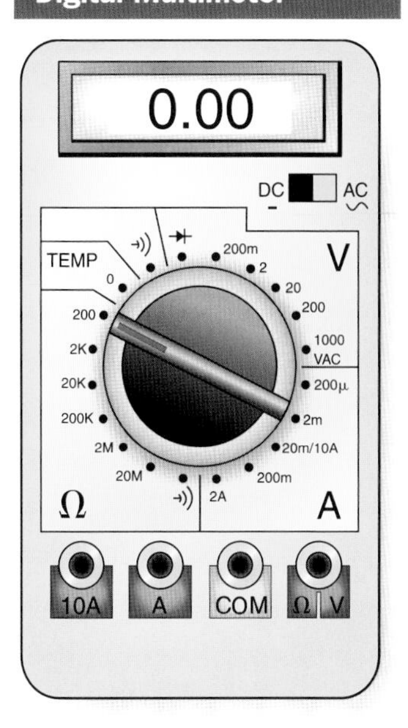

Figure 1–30 A typical digital multimeter can be used to measure ohms as well.

The meter shown is classified as a three-and-one-half-digit meter. This means that the display has three digits that can show any value from 0 to 9 and a fourth digit that can indicate either a 0 or a 1. The maximum reading for this meter is 1999. A decimal point is used to scale this reading. Scales shown for this meter include 200, 2 K, 20 K, 200 K, 2 M, and 20 M. These scales represent the maximum resistance readable for that scale and would be displayed as indicated in Table 1–3.

On the digital meter, when a value exceeds the maximum value that can be displayed for that scale, the meter indicates the overscale condition by displaying a 1 or a 9 or by flashing 9999. When the value of resistance being measured represents a resistance value below the scale of the meter (as might occur if a 200 Ω resistor were to be checked while the meter was set on the 2 M scale), the meter will indicate the underscale condition by displaying a 0.00.

When measuring resistance on the digital meter shown in Figure 1–30, if you encounter a 1 you should increase the scale reading (turn the selector knob counterclockwise on the meter shown) to bring the meter within scale. If you encounter a 0.00 on the display, you should decrease the scale reading (turn the selector knob clockwise) until you have the correct reading. Other meters may differ. Be sure to understand how the meter you are using operates.

One additional resistance scale, indicated by the symbol ·))), is used for continuity tests. A continuity test is used to indicate a current path. Continuity exists between two points. This type of test is used when the user is concerned not about the specific resistance in the current path but only that a continuous current path exists in the circuit. When set to this scale, the meter will indicate a 1 for an open circuit and a 000 for a closed or shorted circuit. In addition, with some meters an audible signal is heard when a closed continuous circuit is encountered.

The digital multimeter does not have a 0 Ω ADJ control, as zero adjustment is normally not required for digital multimeters. Some digital multimeters have autoranging capabilities that automatically adjust the meter's range to suit the value being measured.

Table 1–3

Ohmmeter scales and the maximum reading available with each scale are shown.

Scale	Maximum Reading (Ω)	Display at Maximum
200	200	199.9
2 K	2,000	1.999
20 K	20,000	19.99
200 K	200,000	199.9
2 M	2,000,000	1.999
20 M	20,000,000	19.99

MEASURING RESISTANCE IN A CIRCUIT

When measuring resistance with either type of meter, the procedure to obtain accurate measurement of the value of resistance is the same. The digital meter is easier to read when the resistance of a specific component is required. Digital meters are less desirable, however, when reading the resistance of circuit components that have a quickly changing resistance (such as a resistive sensor in which the resistance depends on some physical variable, such as light level or temperature). The main reason for this is that the variable being measured is rapidly changing and the process required within the digital meter to read, interpret, and display the measured value often exceeds the time between the changes in that variable.

This does not mean that the digital meter is slow. Rather, it may be inappropriate for the type of variable being measured. For these types of quickly changing values, the analog meter is often a better choice. Note that many modern digital meters are equipped with an LCD analog bar that can be used for tracking changes. Because meters that measure resistance have their own power sources, all power must be removed from circuits before attempting to measure resistance. Circuit power sources will not only affect the reading on resistance measuring devices but potentially pose a safety hazard if left connected.

CAUTION: Never attempt to measure the resistance of a component in a circuit while power is applied to that circuit. To do so not only will damage or destroy the meter used for the test but might injure the individual performing the test.

Other components connected in a circuit might also affect the accuracy of resistance measurements. When measuring resistance, you must verify that the only current path available is the one through the component whose resistance you wish to measure. Generally, when measuring the resistance of components in a circuit, the polarity of the ohmmeter is not important. Most resistive components are not sensitive to the polarity of the voltage applied to those components during a resistance test.

Electronics devices such as diodes and transistors are polarized. Therefore, caution must be applied to ensure that the polarity of the meter's internal source is correct for the test to be accurate. Figure 1–31 shows the proper way to measure the resistance of resistor R_1. The power source and all other current paths have been isolated from the component whose resistance is to be measured.

Figure 1–32 shows a circuit in which the resistance measurement would be in error because the power is still connected to the circuit.

Ohmmeter: Proper Connection

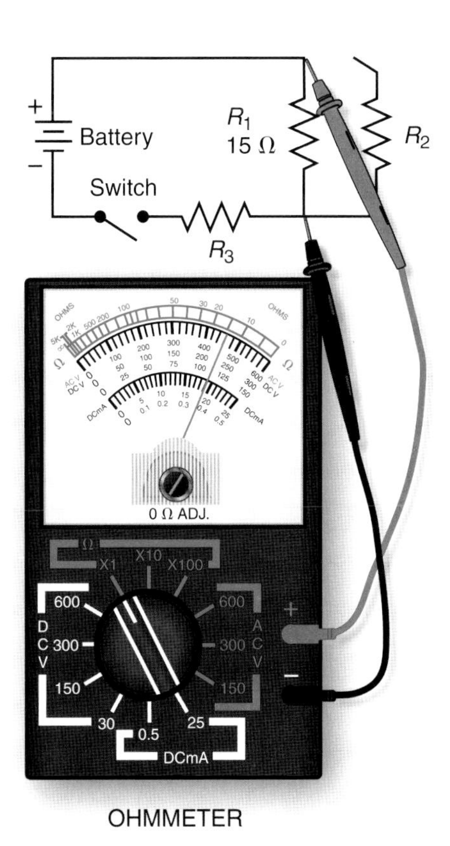

Figure 1–31 Correct ohmmeter connection on a component is illustrated above. When the switch is open, the power source will not affect the measurement.

Ohmmeter: Improper Connection

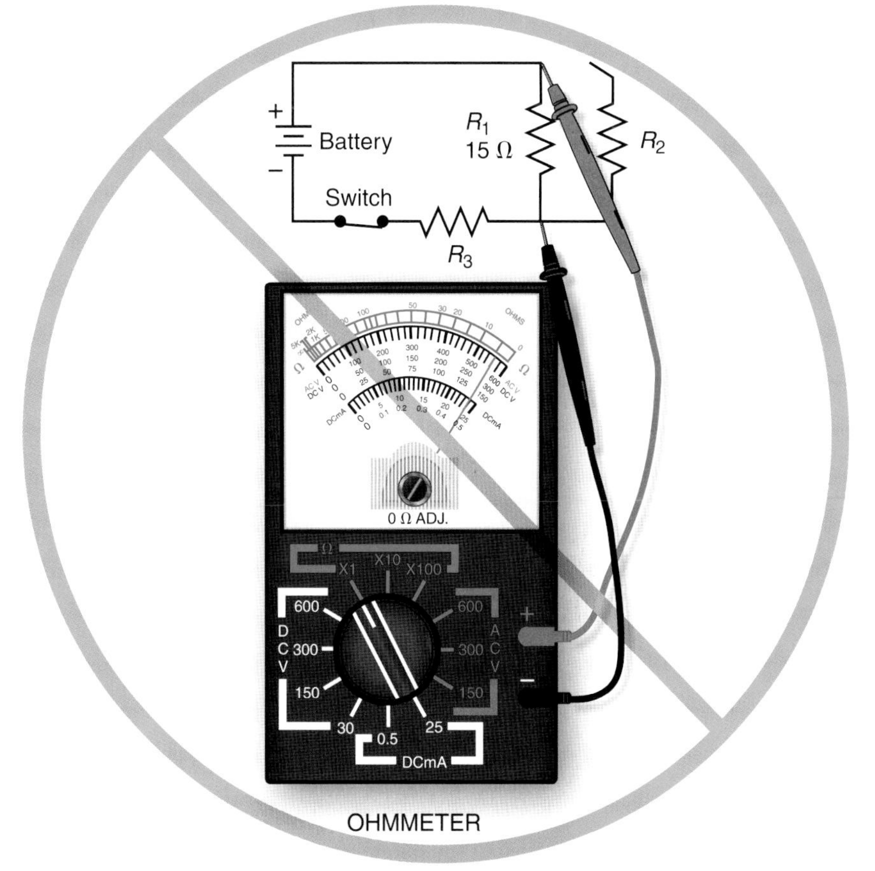

Figure 1–32 An improper connection with ohmmeter and power source still connected is shown. When the switch is closed, the power source will affect the measurement.

The reading would show that the resistance of R_1 was higher (or lower) than the actual value because of the presence of the circuit power source, which would subtract from (or add to) the ohmmeter's own internal power source.

Ohmmeter: Improper Connection

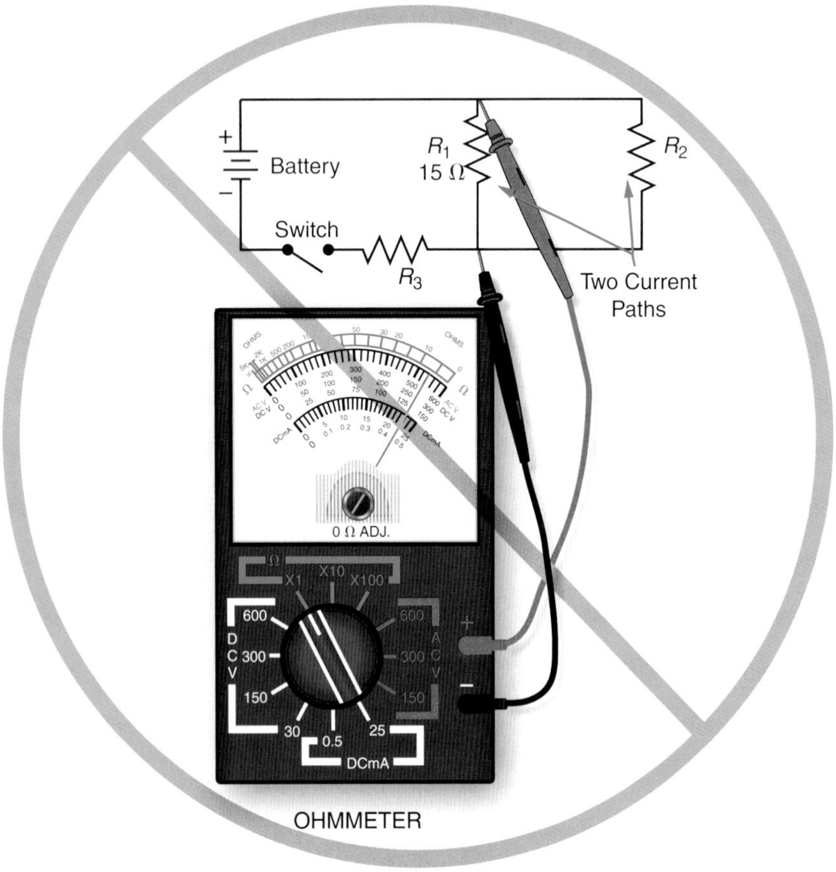

Figure 1–33 An improper ohmmeter connection with more than one resistance in the path is shown above.

Simple Complete Circuit

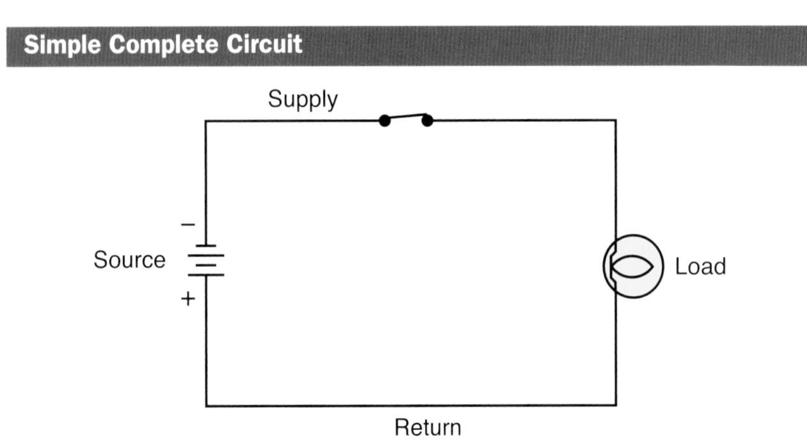

Figure 1–34 A simple complete circuit with voltage source, load resistance, and current flow path is depicted above.

CAUTION: This connection could also be dangerous. Do not do this.

Figure 1–33 shows a circuit in which the resistance measurement would be in error because more than one current path exists for current to flow in the circuit. The internal current source from the ohmmeter would flow through both R_1 and R_2, and the resistance reading would represent the equivalent resistance of resistors R_1 and R_2 combined—which would be lower than the resistance of R_1 alone.

A CIRCUIT

As we have been discussing the three essential quantities of electrical theory, we have referred to a **circuit** (Figure 1–34). For electricity to provide useful work, there must be a path for current to flow from a source of pressure (pump) through conductors (pipes) to a device (water wheel), and then the electrons must be able to get back to the source to reach the positive terminal of the source.

For a circuit to be complete, it must have a source (which will provide the force needed for electrons to flow), a supply path, and a return path for electrons to flow from the source, through the load, and then back to the source. In other words, the circuit must provide a path back to the *original* source which caused the electrons to flow. Once a path back to the source is connected, the circuit is complete, and as long as the "pump" has power supplied, it will continue to force electrons to flow in a completed system.

CLOSED CIRCUIT

All of the knowledge you have can now be combined to create a circuit. You know that a potential difference (voltage) must exist to cause electrons to flow (current) within a conductor. And you know that all conductors will have some resistance (ohms) to the current flow. There are many types of circuits you may encounter in your career as an electrical worker.

Because of the ever-changing designs for new electrical systems and circuits, not all are alike. You must become familiar with these differences to be successful. This section does not attempt to cover these circuits in detail but does give you an overview.

For current to flow, there must be a complete path (closed circuit, Figure 1–35). Note that the electrons flow from the battery through the motor and back to the battery. The battery causes the potential difference for the electron flow. Another way to think of it is that the electrons make a solid chain through the circuit. If one electron moves, they all move in a circular path. The circuit is a "closed loop."

OPEN CIRCUIT

Circuits are designed to deliver current to a component that consumes electricity, known as an electrical load. Remember that the component does not use up electrons, because they must be returned. It does not consume voltage but does create a change in pressure. This circuit allows electrons to flow through the component and back to the source. If we create a break in the circuit (as in Figure 1–36), the electrons cannot return to the source and no electrons can "flow" in the circuit. As many nonelectrical people may say: There must be a short! Contrary to this misconception, the situation when no current flows is an "open" circuit. As you can see from the diagram, the circuit is open and the ammeter reads zero amps.

If there is a break (open) somewhere in the circuit, the electrons cannot continue to flow (Figure 1–36). This is called an open circuit. The potential difference (voltage) still exists, but the electrons cannot move (there can be no current). This is another way of saying that R is infinite ($R = \infty$).

SHORT CIRCUIT

The correct definition of a short circuit is represented by that shown in Figure 1–37. In this case, the current has found a shorter path back to the source than going through the resistance of the full circuit.

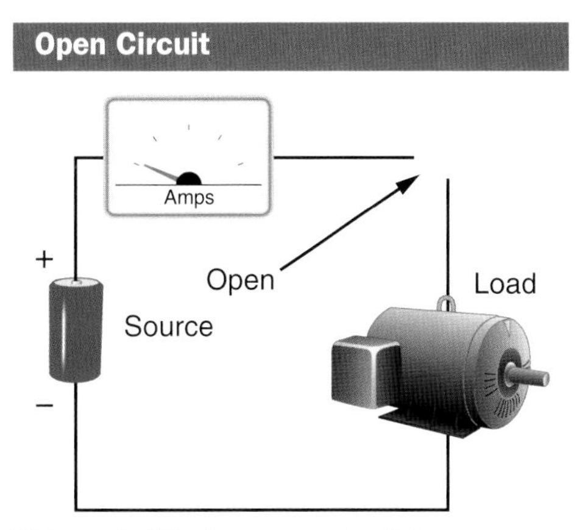

Figure 1–36 An open circuit has no current flow.

Closed Circuit Path

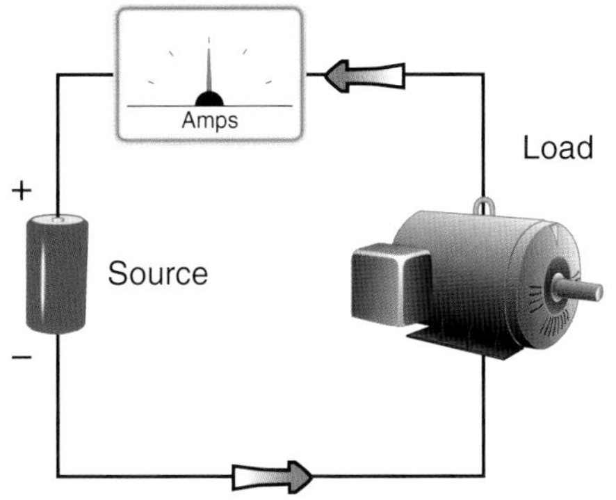

Figure 1–35 Above is an illustration of a closed circuit path.

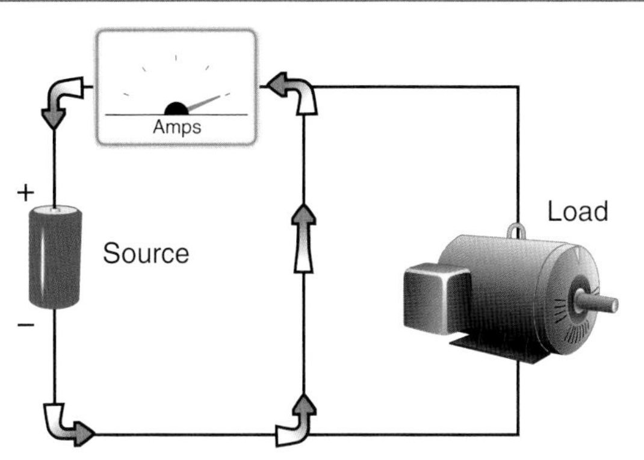

Figure 1–37 A short circuit has almost no resistance and a high-current flow.

When the circuit does not exert much resistance, the current increases, often beyond the capability of the supply to provide enough electrons. The source therefore shuts down, or a fuse opens the circuit. In the case of a short circuit that has had a fuse open, the circuit may first appear as an open circuit. The circuit with an open link in the fuse is an open circuit, but it was caused by a short circuit. The short circuit is characterized by maximum current, whereas an open circuit is characterized by zero current.

OHM'S LAW OF PROPORTIONALITY

Ohm's law is a law of proportionality, stating that it takes 1 volt to push 1 amp through 1 ohm of resistance. Another way to look at it is that the current *(I)* in a circuit is directly proportional to the voltage *(E)* applied to the circuit and inversely proportional to the resistance *(R)*.

Figure 1–38 can help you remember the relationship. To find the value you are looking for, cover it with your finger to see the relationship that remains. From Figure 1–38, you can derive the three equations you need to make any calculation using Ohm's law. Recall from earlier in this chapter that *V* is sometimes used in place of *E*.

$$E = I \times R$$

$$I = \frac{E}{R}$$

$$R = \frac{E}{I}$$

Ohm's Law Circle

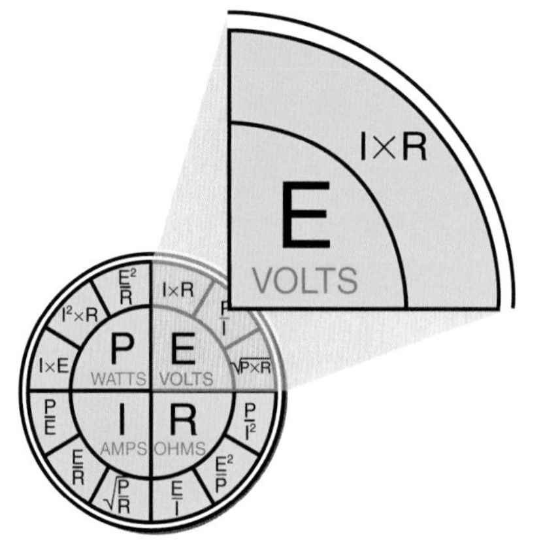

Figure 1–38 Ohm's law circle is used to create a formula to find the unknown values in a circuit.

CALCULATIONS INVOLVING OHM'S LAW

Example

If a circuit has a total resistance of 20 ohms (Ω) and a current of 3 amps, what is the circuit voltage?

Solution:

Using Figure 1–38, cover the *E* with your finger and you have $I \times R$ left over. Thus, the equation is:

$$E = 3\text{ A} \times 20\ \Omega$$

$$E = 60\text{ V}$$

Example

If a circuit has an applied voltage of 24 volts and has a current of 2 amps, what is the resistance of the circuit?

Solution:

Again, looking at Figure 1–38, cover the *R* and the equation becomes:

$$R = \frac{E}{I}$$

$$R = \frac{24\text{ V}}{2\text{ A}}$$

$$R = 12\ \Omega$$

Example

Solve for current in a circuit that has 120 ohms (resistance) and 240 volts of EMF.

Solution:

Referring to Figure 1–38, cover the I part of the diagram and the remaining variables are E and R, forming the equation:

$$I = \frac{E}{R}$$

$$I = \frac{240\ \text{V}}{120\ \Omega}$$

$$I = 2\ \text{A}$$

PREFIX MULTIPLIERS

As you have noticed in the preceding information, prefixes are often used to aid in dealing with small or large quantities of measure. These prefixes are used in many applications and are universally accepted because we use the powers of 10 in our decimal system. Electrical and electronic component and circuit values can be extremely small or very large.

Prefix multipliers are used to simplify the printing and use of small or large figures. For example, 0.001 is expressed using the prefix *milli.* Thus, 0.001 amperes may be expressed as 1 milliampere. The method of applying such prefixes is straightforward. Divide the original value by the number value of the prefix and add the prefix. For example:

2,000 amperes = 2,000/1,000 amperes = 2 kiloamperes

Note that prefix multipliers are sometimes abbreviated even further. The value 2 kiloamperes is usually expressed as 2 kA. Table 1–4 outlines the values, names, and powers of 10 for the most commonly used prefixes. When using prefixes, remember to convert the units to full unit values when performing calculations. You cannot disregard the prefixes because there are significant figures involved with the computation.

Table 1–4

This table shows prefixes and numeric equivalents.

Number Value	Metric Prefix	Power of 10
1 000 000 000 000	Tera—T	10^{12}
1 000 000 000	Giga—G	10^{9}
1 000 000	Mega—M	10^{6}
1 000	Kilo—k	10^{3}
100	Hecto—h	10^{2}
10	Deka—da	10^{1}
1	Units or 1	10^{0}
0.1	Deci—d	10^{-1}
0.01	Centi—c	10^{-2}
0.001	Milli—m	10^{-3}
0.000 001	Micro—µ	10^{-6}
0.000 000 001	Nano—n	10^{-9}
0.000 000 000 001	Pico—p	10^{-12}

Example

A circuit has 12 kΩ and 120 V. Find the circuit current.

Solution:

$$I = \frac{E}{R}$$

$$I = \frac{120\ \text{V}}{12{,}000\ \Omega}$$

$I = 0.01$ A (also denoted 10 mA)

Example

There is a complete circuit with 50 kV and a current of 50 A. What is the circuit resistance?

Solution:

$$R = \frac{E}{I}$$

$$R = \frac{50{,}000\text{V}}{50\ \text{A}}$$

$R = 1{,}000\ \Omega$ (also denoted 1 kΩ)

Example

Find the voltage of a circuit when 30 mA flows through 60 kΩ.

Solution:

$E = I \times R$

$E = .030 \text{ A} \times 60{,}000 \text{ ohms}$

$E = 1{,}800 \text{ V}$ (also denoted 1.8 KV)

ELECTRICAL POWER

Another value that is helpful in designing and working on electrical circuits is the power requirement. In electricity, the unit of power is the **watt**. Power in a circuit is the amount of work being done per unit of time, represented as:

$$\text{Power} = \frac{\text{Work}}{\text{Time}}$$

Power is what is consumed when a voltage (volts) is applied to a circuit and current (amps) flows through the load. The relationship can be seen between the applied voltage and the current consumed by the load as power is consumed. This is represented as:

$$P = E \times I$$

The amount of power consumed by a load is directly related to the amount of voltage applied to the load and the amount of current flowing through the load. Using the water analogy, we can visualize a water wheel (Figure 1–39). As water flows to the water wheel, it causes the wheel to turn, producing mechanical work. If we increase the water flow (current), the wheel will turn faster or create more effort at the mechanical load. If we increase the pressure (voltage), the wheel will also do more work. The power (work / time) is dependent on the pressure and the flow (voltage and current).

The load in the circuit uses the power to change or convert electrical energy into some other form of energy. For power to exist in an electrical circuit, this change must take place. The change may be in the form of current flow causing a heater element to heat up (change from electrical energy to heat) or a voltage supplied to a motor causing the motor to rotate (electrical energy to mechanical energy). The key point to remember is that the change or conversion from one form to another must take place.

FieldNote!

The electrical unit of measurement for power is named after James Watt for his work on the power delivered by the steam engine.

Water Wheel Analogy

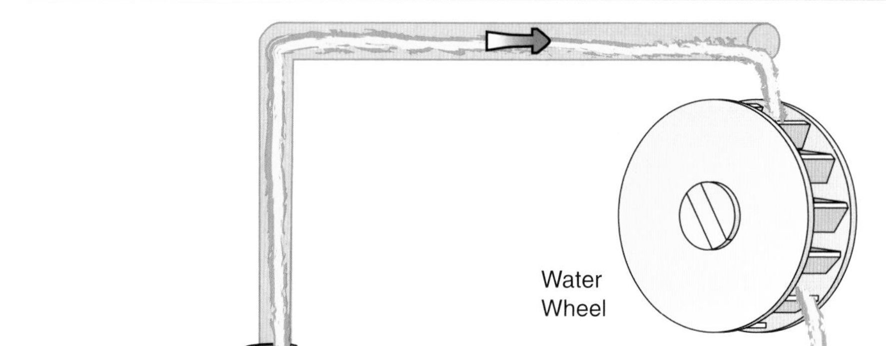

Figure 1–39 A water wheel analogy explains the concept of work: energy per time.

The more current that is allowed to flow through a circuit, the hotter the circuit becomes. This heat is produced by the collision of the flowing free electrons with the fixed atoms. But that is not the strange part. If the current is doubled, the heat produced will go up by four times, and if the current is tripled the heat produced goes up by nine times!

This showed early scientists that the heat produced, and hence the work being done in the circuit, was proportional to the force applied but not to the current. The proportion was of an unexpected nonlinear nature.

The heat produced by the current through the resistance in the circuit is called I^2R losses because it is heat lost in the system, also known as watt losses. The formula is found by combining the formula $P = I \times E$ with the formula $E = I \times R$. By substituting $(I \times R)$ for E in the $P = I \times (E)$ formula, i.e.,

$$(P = I \times [I \times R]),$$

you get

$$P = I^2R$$

Because of Joule's work on electrical heat, he is credited with proving the relationship between electrical power and current and was given the honor of having a unit named after him (the joule). The **joule** is an amount or unit of energy used to measure the heat or work produced or consumed in a system. In an electrical circuit, this is the heat produced or the work done in the circuit due to the amount of current flowing through it. The work used in heating a circuit is called lost work because this work cannot be used to move anything.

Because of the heat generated in electrical components, specifically resistors, a rating system has been developed so that components will not overheat during normal operation. This power rating is calculated with the following equations:

$$P = \frac{E^2}{R}$$

$$P = I \times E$$

Further, in solving for each of the values I, E, and R, a relationship diagram can be derived. Using Figure 1–40, you can solve most equations involving power, current, voltage, and resistance. The chart shown in Figure 1–40 is derived by combining the formula $E = I \times R$ with the formula $P = I \times E$. Other variations of the relationship between these variables can be found, such as $I = \frac{P}{E}$ and $E = \frac{P}{I}$.

English physicist James Prescott Joule (1818–1889) dealt with this problem of nonlinear proportionality. He wrote a paper in 1840 called "On the Production of Heat by Voltaic Electricity." In that paper, he explained his experiments—which we will not go into here. His conclusions are called Joule's law.

Joule's law states that "the total amount of heat produced in a conductor is directly proportional to the resistance times the square of the current" (see TechTip below). Mathematically, the heat (power) produced can be represented as:

$$P = I^2 \times R$$

Heat and light are produced with current and voltage. Heat is produced when current flows through the element in this light. This heat is lost work.

Formula Chart

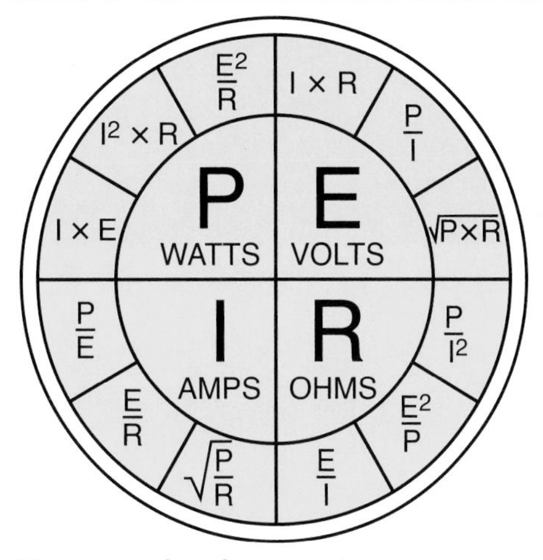

Figure 1–40 A formula chart can be used to track watts and Ohm's law.

Example

What is the voltage requirement for a circuit with resistance 150 Ω and a current of 12 A, as shown in Figure 1–41?

Solution:

Using Figure 1–40, you can see that in the voltage quadrant there are three formulas to choose from. Because you know resistance and current, the equation becomes:

$$E = I \times R$$

$$E = 12\text{ A} \times 150\ \Omega$$

$$E = 1{,}800\text{ V}$$

Example

A room heater has a rating of 2,000 W and uses 120 V to power it. What is the resistance of the heater?

Solution:

In Figure 1–40, the resistance quadrant shows that because you know the power and the voltage, the formula to use is:

$$P = \frac{E^2}{R}$$

or

$$R = \frac{E^2}{P}$$

$$R = \frac{120^2}{2{,}000}$$

$$R = 7.2\ \Omega$$

Example

What is the power lost due to heat in a motor that draws 20 A of current and has a resistance or 5 Ω?

Solution:

In Figure 1–40, the power quadrant shows that because you know current and resistance, the formula to use is:

$$P = I^2 \times R$$

$$P = 20^2 \times 5$$

$$P = 2{,}000\text{ W}$$

Example

Using Figure 1–42, what is the power consumed by the lamp?

Solution:

In Figure 1–40, the power quadrant shows that because you know the voltage and the current, the formula you use to solve for the circuit power is:

$$P = I \times E$$

$$P = 0.5\text{A} \times 10\text{ V}$$

$$P = 5\text{ W}$$

Example Calculation

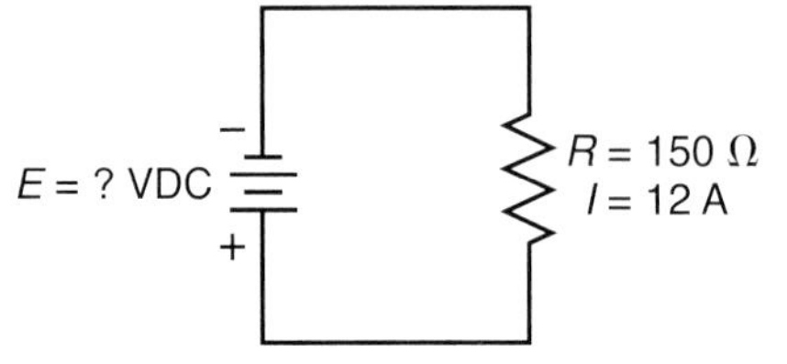

Figure 1–41 Use this circuit for the example calculations.

Example Calculation

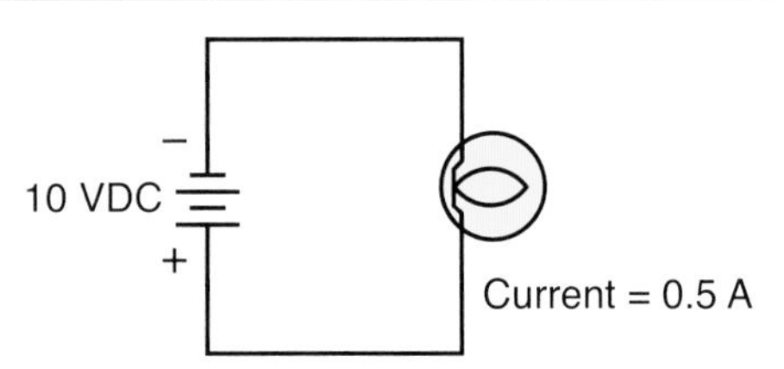

Figure 1–42 Use this simple DC circuit with lamp load for example calculations.

Example

Assume, based on the circuit in Figure 1–42, that the circuit power was 75 W and the circuit voltage was 3 V. Calculate the circuit current.

Solution:

In Figure 1–40, the current quadrant shows that because you know the circuit power and circuit voltage, the formula you use to solve for the circuit is:

$$I = \frac{P}{E}$$

$$I = \frac{75\text{ W}}{3\text{ V}}$$

$$I = 25\text{ A}$$

Example

Assume that the circuit in Figure 1–42 has a circuit current of 7 amps and consumes 42 W of power. Calculate the circuit voltage.

Solution:

In Figure 1–40, the voltage quadrant shows that because you know the circuit current and the circuit power, the formula you use to solve for the circuit voltage would be:

$$E = \frac{P}{I}$$

$$E = \frac{42\text{ W}}{7\text{ A}}$$

$$E = 6\text{ V}$$

SAFETY AND ARCS

Electricity is a versatile form of energy and is also lethal. The electrical power can create electrocutions, create arc blasts and arc burns, ignite combustible vapors, and heat materials to a point that creates combustion. As an electrical worker, it is your job to keep the public safe as you deliver this power for their use. It is also your job, and maybe your life or livelihood, to know how to work safely when around electrical circuits. According to the United States Occupational Safety and Health Administration (OSHA), there is one electrocution every day in the United States.

There are only two outcomes to an electrical shock. Either the victim survives and has some discomfort for a short duration (or lasting effects) or the victim dies from the shock. OSHA places the threshold voltage of electric shock at 50 V. This means that at 50 V of electrical pressure there is enough pressure to push current through your bare skin and into a circuit created by your body. Once a circuit is created through your body, the current through your body's blood and tissue creates severe damage. The electrical current that has entered your body also disrupts the electrical signals from your brain to your muscles, including your heart and lungs.

See Table 1–5 for the physiological damage to your body as current flows.

Table 1–5

Physiological effects of current on the human body are listed.

Amount of Current (mA)	Physical Reaction	Possible Results
1–4	Tingling sensation	Little danger from brief contact
5–10	Muscular contractions (can vary)	Difficult or impossible to let go/break away
11–39	Extreme–Severe contractions (throughout body)	No voluntary control while current continues
40–99	Respiratory arrest—until flow ceases Treatment—1st need to remove victim Secondary cardiac arrest can result	Short duration—breathing may resume Longer—may need resuscitation (CPR/ALST)
100–200	Ventricular fibrillation (possible seizures), burns, and tissue damage	Trained medical support must begin immediately, even if brief contact
201 and above	Respiration and cardiac arrest, severe burns, internal bleeding	Death

Note that the amount of current that creates muscular paralysis is quite small, meaning that you may not be able to free yourself from a live circuit. As the current flow increases, especially if it flows through your chest cavity and your vital organs, the heart may go into ventricular fibrillation with as little as 100 milliamps. **CAUTION:** Do not create a path for current to flow through your body.

Never work on a live circuit unless you must. There will be times when you need to work on live circuits, but OSHA and your employers do not want you to take chances if it is not necessary. The rules you need to follow are covered in National Fire Protection Association (NFPA) Document 70E, regarding safe work practices for electrical work. This document is a system for following the OSHA guidelines when working around electrical equipment. One of the main directives is to work deenergized whenever possible.

The systems should be turned off and tested to be sure there is no electrical power. They should then be locked out with your own lock and tagged to identify that you are working on the system; this is referred to as lockout/tag-out (LOTO). Only you are supposed to remove your own lock when you finish working. You must not remove another worker's lock. There are specific procedures to follow to guarantee safe work practices for you and your co-workers and to maintain compliance with OSHA rules.

OSHA guidelines and the related NFPA 70E have established rules for working on live equipment. As the rules are updated, new editions of the regulations are published. You and your employer are responsible for being up to date on the regulations as they apply to your safety practices. Specifically, you need to be aware of the distance requirements designed to provide safety from contact with live parts of a system. There are charts that tell you what the approach distances are for equipment operating at different voltages.

Three main categories of approach distances are the limited approach, restricted approach, and prohibited approach (Figure 1–43).

OSHA Approach Distances

Nominal System Voltage Range	Flash Protection Boundary	Limited Approach Boundary		Restricted Approach Boundary	Prohibited Approach Boundary
Phase to Phase		Exposed Moveable Conductor	Exposed Fixed Circuit Part	Includes Standard Inadvertent Movement Adder	Includes Reduced Inadvertent Movement Adder
Energized Part to Employee (Distance in Feet—Inches)					
300 V & less	3 ft 0 in.	10 ft 0 in.	3 ft 6 in.	Avoid Contact	
Over 300 V, not over 750 V	3 ft 0 in.	10 ft 0 in.	3 ft 6 in.	1 ft 0 in.	0 ft 1 in.
Over 750 V, not over 2 kV	4 ft 0 in.	10 ft 0 in.	4 ft 0 in.	2 ft 0 in.	0 ft 3 in.
Over 2 kV, not over 15 kV	16 ft 0 in.	10 ft 0 in.	5 ft 0 in.	2 ft 2 in.	0 ft 7 in.
Over 15 kV, not over 36 kV	19 ft 0 in.	10 ft 0 in.	6 ft 0 in.	2 ft 7 in.	0 ft 10 in.

Figure 1–43 Examples of OSHA approach distances near live power are listed above.

Each distance has special conditions, with the "prohibited" being the category that assumes you may be in contact with the live electrical conductor. This means that you need specific "voltage-rated" tools (Figure 1–44), including voltage-rated gloves (Figure 1–45) and safety glasses.

Another hazard of working on live parts is the risk of arc flash and arc blast. As you already know, when electricity moves through the air it creates an arc (such as lightning). It creates extreme heat—up to 35,000 degrees Fahrenheit at the arc. This extreme heat melts copper conductors and just about everything else near the arc. The extreme heat and light are known as the arc flash. This flash can burn unprotected skin, and its heat can set clothes on fire. The flash can cause temporary blindness or long-term eye problems. You must be protected from this extreme heat and light.

Use fire-rated (FR) clothing designed for the amount of heat that could be present. This heat is expressed as calories per square centimeter of "incident energy." Different categories of FR clothes create the needed protection. To protect yourself adequately, a "flash hazard analysis" is calculated on the basis of the amount of current present if a short circuit occurs, the time the arc is allowed to burn, and the distance the person is from the arc.

The distance you may see on a rating is called the "flash protection boundary." This distance refers to how far you must be from the flash to sustain a burn that is "just curable" according to OSHA. "Just curable" means that it might be a second-degree burn but not a third-degree burn. Ask your safety representative about clothing and gear that protects you from arc flash injuries.

The other dimension of an arc is the blast or the explosion that occurs as the gases expand around the superheated arc. The blast propels materials and molten metal toward you. It may blow you out of the equipment or pelt you with molten material, creating bodily injury. The hot gases that are expanding may knock the air from your lungs. Your natural reaction is to try to breathe again, possibly inhaling hot gases and molten metal.

Voltage-Rated Tools

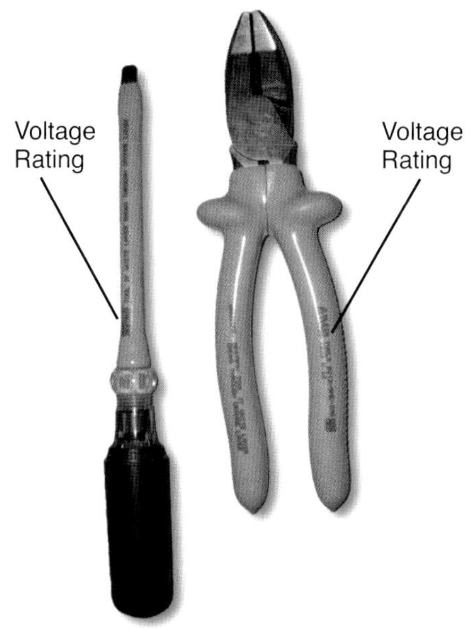

Voltage Rating

Voltage Rating

Figure 1–44 Voltage-rated tools must have a voltage rating marked on them.

Personal Protective Equipment

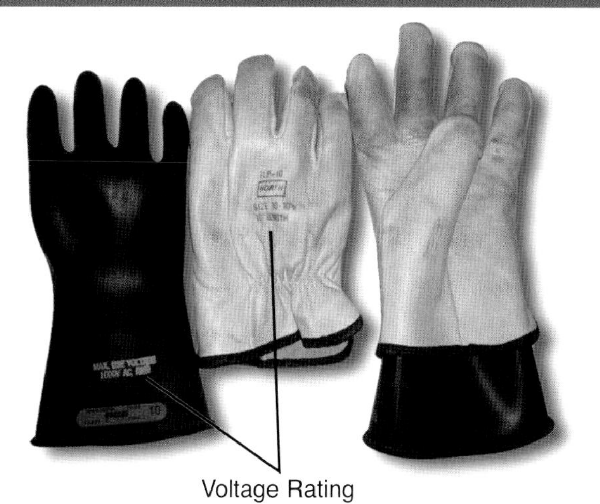

Figure 1–45 Gloves for working on live equipment must have voltage rating marks and must be tested.

Ground Fault Circuit Interrupter

Figure 1–46 Ground fault circuit interrupter (GFCI) receptacle style.

We need to follow all safety rules as we learn to work efficiently and effectively. A basic knowledge of fire extinguishing and the types of extinguisher to use on electrical fires is essential. Look around your surroundings and try to determine the fastest escape route if there were a fire or another emergency. Assess the type of equipment and the construction materials around you. Is the building construction wood or concrete and steel? Are there flammable materials or gases used for welding or for temporary heating stored nearby? Are the housekeeping and cleanup adequate to prevent injury or fires? Are there fire extinguishers nearby?

Look to be sure you are using the right type of extinguisher on a fire. Type A fires are combustibles such as wood, paper, and cloth (on which type A extinguishers are used). Type B fires are flammable liquids and gases. Type C fires are electrical fires where the source may still be energized. These fires are smothered by type C extinguishers. Never use water on a live electrical fire. Type D fires are combustible metals.

There are several types of circuit protections designed to prevent electrical circuits from starting fires or electrocuting people. Fuses or circuit breakers are intended to protect the premises from fire caused by overheating wiring or from ignition due to short circuits. The fuse is an in-line device that will melt open if the current through the circuit conductor (wire) becomes too great based on the accepted limits of the conductor (wire) size. As the current exceeds the specified value, the fuse melts open (or the circuit breaker trips), severing the circuit path to the conductor (wire). Therefore, the circuit is allowed to cool before it can generate enough heat to cause a fire. The protective device can open in time with a slight overload, or may open quickly with a large overload of current (as in a short circuit).

Another type of protection is for personnel. A device called a ground fault circuit interrupter (GFCI) is used to monitor not only the amount of current leaving the source (as a circuit breaker does) but the current returning to the source. In a class A GFCI, the amount of current going and coming must be within 5 mA of each other. If there is a larger discrepancy than that, the protection will shut off the circuit. The assumption is that the current left the source and may be returning through a person to the source and not through the intended wiring. See Figure 1–46 for an example of a home receptacle type of GFCI.

Figure 1–47 illustrates the case where current passes through a person rather than the wiring. If you refer to Table 1–5 you will notice that 5 mA is the threshold where a person may not be able to let go of a live circuit and the current is passing through their body. If the circuit shuts off at this point, little

Possible Current Paths

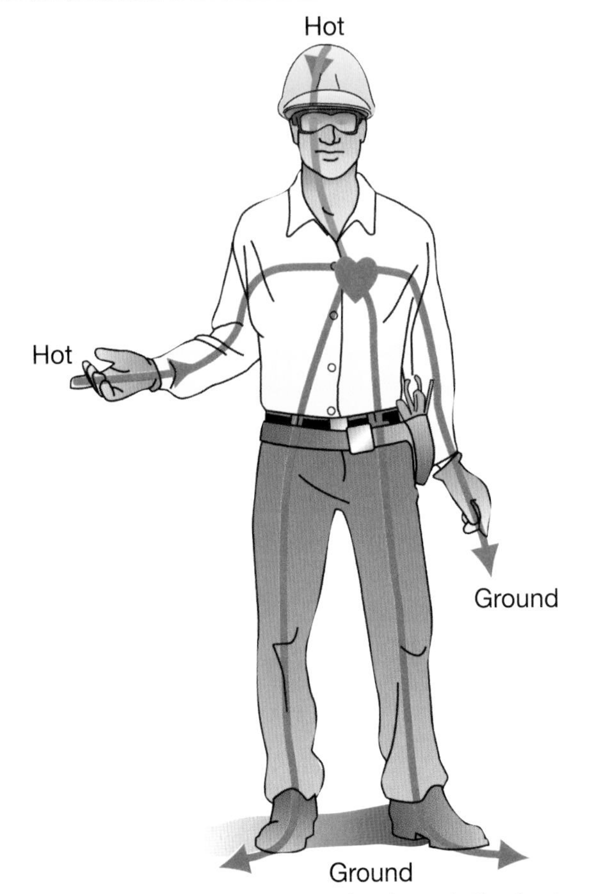

Figure 1–47 These possible current paths through the body can cause severe damage.

permanent damage results. The person feels the shock but will be okay. This is considered personnel protection.

Another type of protection now mandated in the *National Electrical Code*® is called an arc fault circuit interrupter (AFCI, Figure 1–48). The AFCI is designed to determine if there is an arc in the protected circuitry that could start a fire. If a cord-and-plug-connected device has a faulty cord and is actually arcing across the broken cord, it may not draw enough current to trip a circuit breaker but may cause enough heat to start a fire. This is a particularly common occurrence in bedrooms where cords get pinched behind beds or dressers. The AFCI circuit detects an arcing fault and shuts off the circuit.

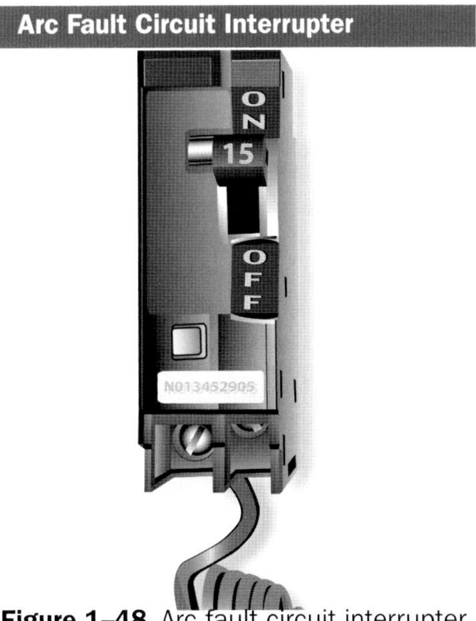

Figure 1–48 Arc fault circuit interrupter (AFCI).

SUMMARY

What is electricity and how do we measure it? You have not looked at the electrical systems from the subatomic view yet, but you have gained a little insight into how electricity is actually measured and an understanding of what the essential components of an electrical circuit are. You have been introduced to the ideas of current flow, electrical pressure or voltage, the resistance to current flow measured in ohms, and electrical work measured in watts.

You have been able to see how meters are connected to measure circuit quantities and use these measured quantities with Ohm's law to determine circuit parameters. With these measurements, we found that some quantities are very large and some are very small, leading to the "shorthand" method of using prefixes to describe the multipliers used in notation. Prefixes such as milli (m) and micro (μ) as well as kilo (K) and mega (M) are in everyday use in the electrical field, and it will take some time to get used to them.

As you start making measurements and working with live power, it is essential that you stay alert and have a solid understanding of the dangers involved with electrical work. A short explanation of electrical safety was included to make you aware of some of the safety systems available to you. You are ultimately responsible for your own safety. Do not take chances and do not assume you can just try electrical connections you are not sure of. As you continue to learn about electrical systems, you will become more capable and be able to do more complex jobs, but always think about the safest way of accomplishing the task.

CHAPTER GLOSSARY

Ampere The unit of electrical current flow, often abbreviated as amp or just A. It is a measure of the movement of electrons as 1 coulomb per second past a fixed point.

Circuit A completed path for current to flow from a source of current through a load and back to the source of current.

Conventional current flow The theory in which current flows from a positive charge to a negative charge; also known as hole flow or positive current flow.

Coulomb The unit of electrical charge equal to the total charge possessed by 6.25×10^{18} electrons. Abbreviated C.

Current The flow of electrons through a material. Measured in amperes.

Electricity A class of phenomena that results from the interaction of objects that exhibit a charge (electrons and protons). In its static form, electricity exhibits many similarities with another naturally occurring force—magnetism.

Electromotive force (EMF) Electrical pressure created between a region of positive charge (fewer electrons) and a region of negative charge (more electrons), measured in volts and represented in formulas with "*E*".

Electron current flow The theory in which current flows from a negative charge to a positive charge.

Joule A joule refers to the amount of energy. One joule of energy used each second is equal to 1 W of work. One joule is the amount of energy used when 0.737 pounds is lifted a distance of 1 foot.

Ohm's law The mathematical relationship among current, electrical potential, and electrical resistance, measured in amperes (amps), volts, and ohms, respectively.

Ohm The unit of electrical resistance, often shown as the Greek letter omega (Ω).

Resistance The physical opposition to electrical current. Resistance (measured in ohms) is caused by the energy loss that occurs when an electron displaces other electrons in a valence ring.

Watt (W) The unit of electrical power. One horsepower is equal to 745.7 W or approximately 746 W.

REVIEW QUESTIONS

1. Define current flow in terms of amperes and coulombs.
2. Write Ohm's law in its three forms.
3. Describe the concept of electron current flow.
4. What is conventional current flow (also called hole flow)?
5. What is a coulomb, and how does it relate to electric current?
6. Calculate the resistance of a circuit with 10 mA and 100 V.
7. Calculate the current of a circuit with 50 k Ω and 500 mV.
8. Describe the differences among closed circuits, open circuits, and short circuits.
9. If a circuit has 50 V and 10 A of current, find the watts dissipated.
10. Explain the different ways of measuring current in a circuit.

For questions 11 through 15, choose the correct term in parentheses.

11. When connecting a voltmeter, connect in (series, parallel) with the component.
12. Electrocution is caused by (current, voltage) as it passes through your body.
13. An essential precaution while using an ohmmeter when measuring resistance in a circuit is to have power (connected, disconnected).
14. When connecting an analog ammeter, the negative meter lead goes to the (positive, negative) side of the circuit.
15. At the flash protection boundary, you could receive a burn that is (just curable, incurable) on exposed skin.

2

Elemental Electricity

TOPICS

- INTRODUCTION
- MATTER AND ATOMS
- ELECTRICAL PROPERTIES OF MATERIALS
- MOLECULES AND COMPOUNDS
- SOURCES OF ELECTRICAL ENERGY
- VOLTAGE POLARITY
- CONDUCTOR RESISTANCE

OVERVIEW

As you learned in the last chapter, electricity has many component parts that involve the fundamentals of electron flow or current, pressure or voltage, and opposition or resistance. This chapter explains the basic molecular systems that allow us to know how electricity reacts as we apply the theories, create the circuits, and measure the results. The basic causes and effects of elemental electrical actions are described so that the process is more easily understood and so that you can make better predictions of behavior and even decide if an innovative new product could work. You will learn how and why electricity works from the molecular level and how that applies to our everyday work.

OBJECTIVES

After completing this chapter, you will be able to:

- Describe the basic structure of the atom
- Name the three main particles that are part of all but the simplest atom
- Describe the electrical characteristics of an atom
- Describe the relationship between the valence electrons and electron movement (current flow)
- Describe different means of producing electrical potential
- Determine voltage polarity of sources and loads

INTRODUCTION

As you learned in the first chapter, **electricity** in action is a movement of electrons from a point of excess electrons (the negative power supply) to the positive power supply point to complete a circuit. This invisible energy provides light, sound, motion, heat, and many other applications. The transfer of this energy from the source to the desired point of use is what the study of electricity is all about. If we know the characteristics of electricity and the results of electricity, we just have to be able to control it between the source and the application.

The common link among all of these applications is the electrical charge. All of the materials we know—gases, liquids, and solids—contain two basic particles of electric charge: the electron and the proton. The electron has an electric charge with a negative polarity. The proton has an electric charge with a positive polarity. The interaction of these subatomic particles is what makes electricity work, and the understanding of this phenomenon allows us to use and control electrical systems.

MATTER AND ATOMS

As you look around you, everything you can see, feel, or touch is made of **matter**. All things in the universe are "assembled" or composed of matter. Matter is the "stuff" that makes up the universe. Currently there are five forms of matter in the universe: solid, liquid, gas, plasma, and Einstein–bose. The most common types of matter that exist in the universe are in the form of a solid, a liquid, or a gas, although only one form at a time.

In 1869, Dmitri Ivanovich Mendeleev and Julius Lothar Meyer assembled and published all elements known at that time into a structured table known as a Periodic Table of the Elements (Figure 2–1). The Periodic Table shows recurring ("periodic") chemical properties of all the natural and some of the man-made elements. The smallest piece of an element that still has the characteristics of the element is called an atom.

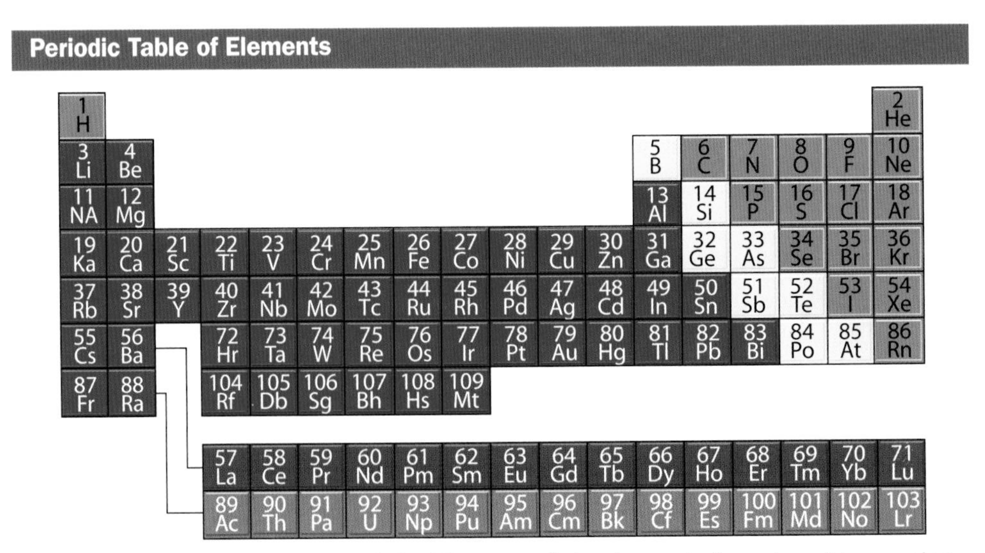

Figure 2–1 The Periodic Table of the Elements lists elements in order of increasing atomic number (i.e., the number of protons in the atomic nucleus). Rows are arranged so that elements with similar properties fall into the same vertical columns (groups).

A good example of this is water. Water is naturally a liquid. When cooled to freezing, it becomes ice (a solid), and when heated to boiling it becomes steam (a gas). The simplest form of any matter is called an **element**. The element is the smallest form of matter in which the unique characteristics of the substance can still be identified.

The smallest piece of an element that has all the characteristics of that element is the **atom**. Atoms have three main parts: **electrons**, **protons**, and **neutrons**. The proton and neutron combine to form the atom's nucleus. Smaller subparticles of the proton and neutron are called **quarks**. Figure 2–2 illustrates a simple form of an atom: one electron, one proton, and one neutron. Recall that the electron has a negative charge and the proton a positive charge. Note that the neutron is neutral; it has no charge. This means that the nucleus, which is made up of the proton and neutron, is positively charged.

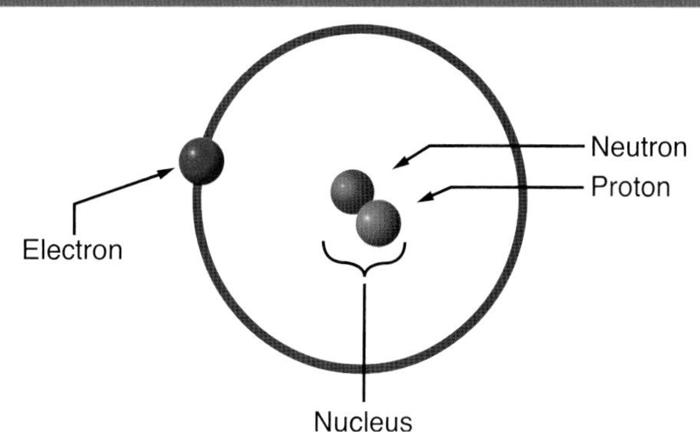

Figure 2–2 A hydrogen atom has one proton, one neutron, and one electron.

You can tell the type of element by the number of protons or positive charges in the atom's nucleus. For example, silver has 47 protons in its nucleus, iron 26, and oxygen 8. The number of protons also equals the element's atomic number on the periodic table of the elements (Figure 2–1). Not all elements have a nucleus that has the same number of protons and neutrons. For example, carbon has a nucleus that contains 6 protons and 6 neutrons. Copper, however, has a nucleus that has 35 neutrons and only 29 protons (Figure 2–3).

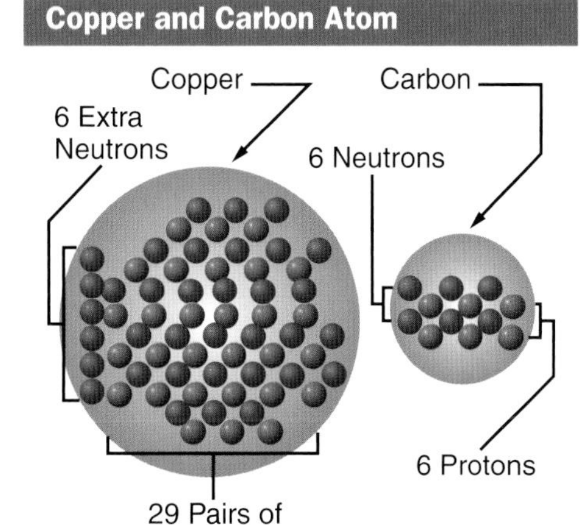

Figure 2–3 Different elements (such as carbon and copper) have differing numbers of protons, neutrons, and electrons.

Some materials may come in several different forms, with the same number of protons but different numbers of neutrons. These materials are called **isotopes**. One of the best-known examples of these is carbon. As stated previously, carbon usually has 6 protons and 6 neutrons. This common isotope of carbon is called carbon-12 and is abbreviated as C^{12}. Another very useful isotope of carbon has 6 protons and 8 neutrons—carbon-14 or C^{14}. Fortunately, all of the various isotopes of elements behave in the same way electrically.

ATOMS AND LINES OF FORCE

The electron and proton have two characteristics in addition to charge: size and mass (heavy or light). The electron is about three times larger than the proton but is much lighter. Even though it is only one-third the size of the electron, the proton has a mass that is more than 1,836 times greater.

A practical example would be comparing a 15 inch–diameter balloon to a 5 inch–diameter lead cannonball.

Atoms that have the same number of electrons as they do protons are considered electrically balanced or electrically neutral in nature. This neutral condition means that opposing forces or lines of force are exactly balanced, without any effect either way. Refer to Figure 2–4. When you look at the proton and the electron, you see that each of them has lines of force. The difference between the proton and the electron is the direction of those lines of force. The electron's lines of force "flow" inward toward the electron. This inward flow is caused by the negative charge of the electron. The proton is just the opposite. Because the proton has a positive charge, the lines of force of the proton "flow" outward in all directions.

When you need to use these electrical lines of force, work must be done to separate the electrons and protons. Changing the balance of forces produces evidence of electricity. Recall Benjamin Franklin's convention: Unlike charges attract; like charges repel. This means that electrons are attracted or pulled toward protons and that protons repel or push away from other protons (Figure 2–5).

NUCLEAR FORCES

So what keeps all of the protons and neutrons together in the center of the atom? Because the protons all repel each other, why don't they simply fly apart? The accepted theory states that a **strong nuclear force** holds the nucleus together. This force is carried by another particle called the **gluon**, which is also responsible for binding quarks together to form protons and neutrons.

CENTRIFUGAL FORCE

The next logical question is: What keeps the electrons from falling into the nucleus, in that they are attracted by the protons? As the electrons rotate about the nucleus, there is a force that tries to cause them to fly off into space. This force is called the centrifugal force, which exactly balances the force trying to pull them into the nucleus. The centrifugal force is also the force that keeps the string tight on a model airplane as it whirls about the operator. Figure 2–6 illustrates the concept of centrifugal force.

ELECTRONS AND THEIR ORBITS

Although there are many possible ways in which electrons and protons might be grouped, they come together in very specific combinations that produce stable arrangements (atoms).

Lines of Force

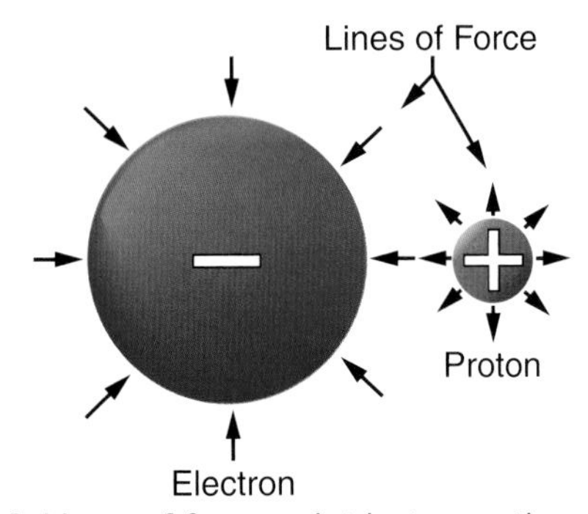

Figure 2–4 Lines of force exist between the proton and the electron.

Charge Behavior

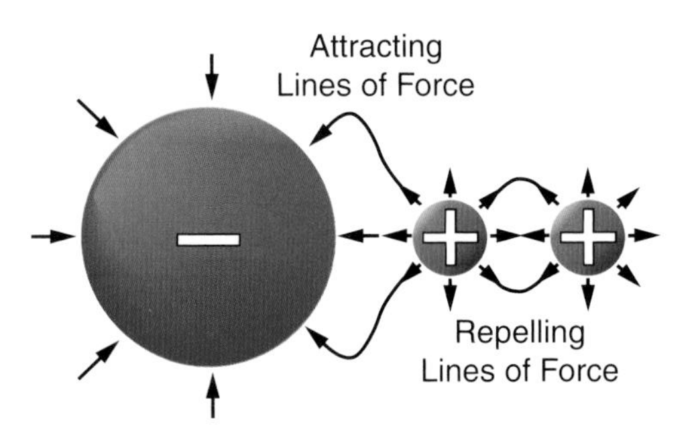

Figure 2–5 Like charges repel each other and unlike charges attract each other.

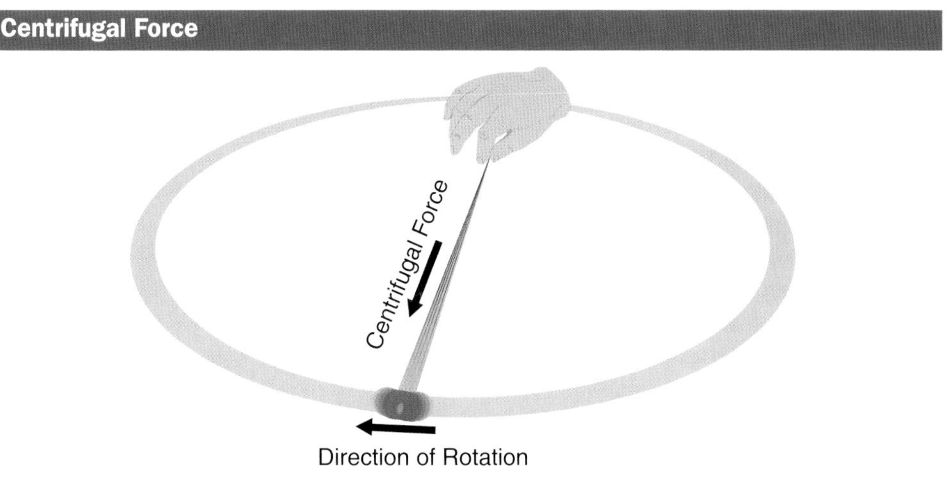

Figure 2–6 Centrifugal force tends to push an item outward as it revolves around a center.

The simplest form is the hydrogen atom (Figure 2–2), which contains one proton and one electron, the electron spinning in "orbit" around the nucleus. There are a number of orbital rings (shells) around a nucleus that can hold electrons. Figure 2–7 shows an atom with a number of rings filled with electrons in orbit around the atom's nucleus.

The number of electrons each shell or orbital ring can hold is determined by the formula $2N^2$. N stands for the number of the shell or ring. Follow the examples given here to calculate the number of electrons each shell or orbital ring can hold. For the first shell or orbital ring (the number in parentheses stands for N, or the number of the shell or orbital ring):

$$2N^2 \text{ or } 2 \times (1)^2 \text{ or } 2 \times 1^2 = 2$$

For the second shell or orbital ring:

$$2N^2 \text{ or } 2 \times (2)^2 \text{ or } 2 \times 4 = 8$$

For the third shell or orbital ring:

$$2N^2 \text{ or } 2 \times (3)^2 \text{ or } 2 \times 9 = 18$$

For the fourth and fifth shell or orbital ring (these rings can hold a maximum of 32 electrons):

$$2N^2 \text{ or } 2 \times (4)^2 \text{ or } 2 \times 16 = 32$$

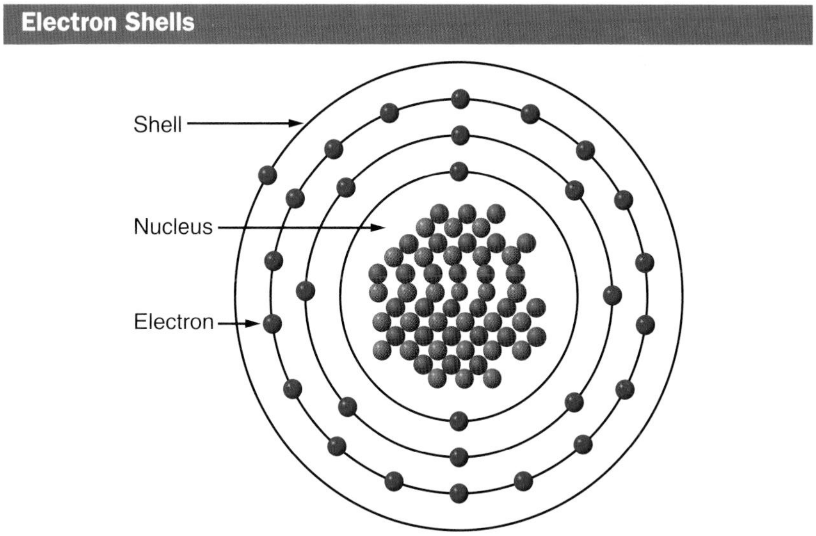

Electron Shells

Shell
Nucleus
Electron

Figure 2–7 Electrons orbit the nucleus in standard shells of electrons.

VALENCE SHELL AND ELECTRONS

Of particular interest in the study of electricity are the electrons in the outer ring, called the **valence ring** of the atom. The valence ring can contain a maximum of eight electrons. These electrons are called **valence electrons**, which can be easily freed in those materials that have few electrons in this outer orbit. Materials with a partially filled valence ring are chemically unstable. Because of this unstable condition, less energy is required to remove electrons from their orbits.

The concept of a valence ring and electron is shown in Figure 2–8. With few electrons in the outer ring (valence ring), there is less attraction to the nucleus. Generally, the farther electrons are from the protons in the nucleus, the weaker the proton attraction on them. Therefore, if the valence ring has only one electron and the electron is far from the nucleus, the easier it is for the electron to pull away from the nucleus.

ELECTRICAL PROPERTIES OF MATERIALS

All materials will allow electricity to flow to some degree. However, some materials conduct quite readily and some do not conduct well at all. A material that allows electrical current to flow easily is called a **conductor**, a material that strongly resists the flow of electricity is called an **insulator**, and a material that falls between a conductor and an insulator in terms of the conductivity scale is called a **semiconductor**. Where a material falls on this scale is determined in part by the number of electrons in their valence ring.

CONDUCTORS

Conductors are good examples of materials whose atoms have only one or two electrons in the valence ring. Examples of good conductors are silver, gold, platinum, copper, and aluminum. Silver, gold, platinum, and copper all have one valence electron, but silver conducts the best of these metals. Aluminum has three valence electrons (three electrons in the valence shell) and is a better conductor than platinum, which has only one valence electron. Remember that the easier it is for the electron to be pulled away from its valence orbit the better the atom "conducts." Figure 2–9 shows a copper atom, one of the most common conducting metals.

A material is said to "conduct electricity" when one electron of an atom is forced from its orbit by another atom's electron. When an atom has only one valence electron, it can be easily bumped or forced from orbit by another electron. When an electron impacts (strikes) another electron, the electron being hit takes energy from the striking electron. See Figure 2–10 (a billiard ball). As the cue ball strikes the next ball, the energy is transferred to the new moving ball.

Valence Ring

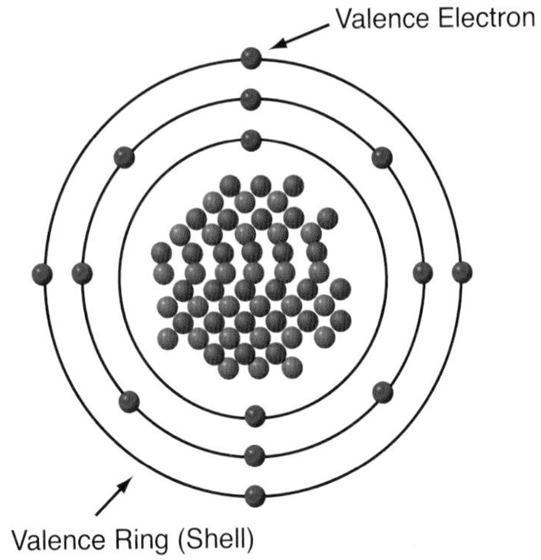

Valence Electron

Valence Ring (Shell)

Figure 2–8 The outermost ring (or shell) of electrons is referred to as the valence ring.

Copper Atom

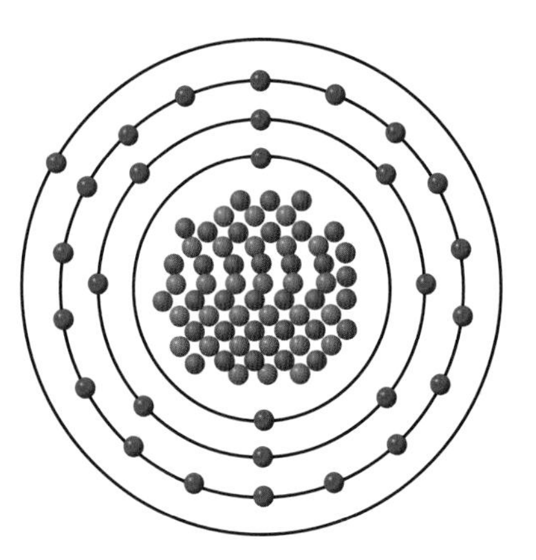

Figure 2–9 A copper atom's fourth shell has one valence electron.

This extra energy forces the "hit" electron to leave orbit and jump to a neighboring electron's orbit, and the process is repeated. The electron that impacted the first valence electron now takes the leaving electron's place as the new valence electron. Figures 2–11 and 2–12 detail this "electron flow."

An important fact to note here is that not all of the energy is transferred when the electrons collide. Some of the energy is lost in the form of resistance. This resistance comes from the original valence electron's attraction to its own nucleus. The energy expended in moving electrons is released as heat (Figure 2–11). That is why conductors (wires) that conduct electricity get warm. Too much electron flow (electrical current) in too small of a conductor (wire) or with too few electrons can overheat the conductors (wires) and the electrical insulating covering and may melt or ignite them.

When one electron strikes a valence shell with two valence electrons, the energy of the incoming electron is divided between the two valence electrons.

Energy Transfer

Figure 2–10 As one ball strikes the next, the energy is transferred to the next ball.

Electrons Imparting Energy

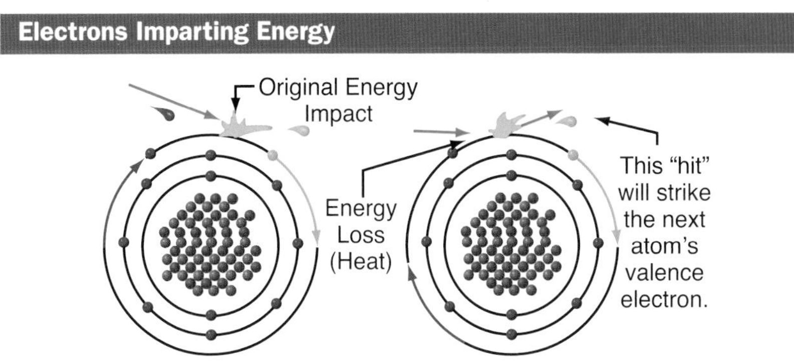

Figure 2–11 Electrons are moved from one atom's shell to another atom's shell, imparting energy.

Divided Energy

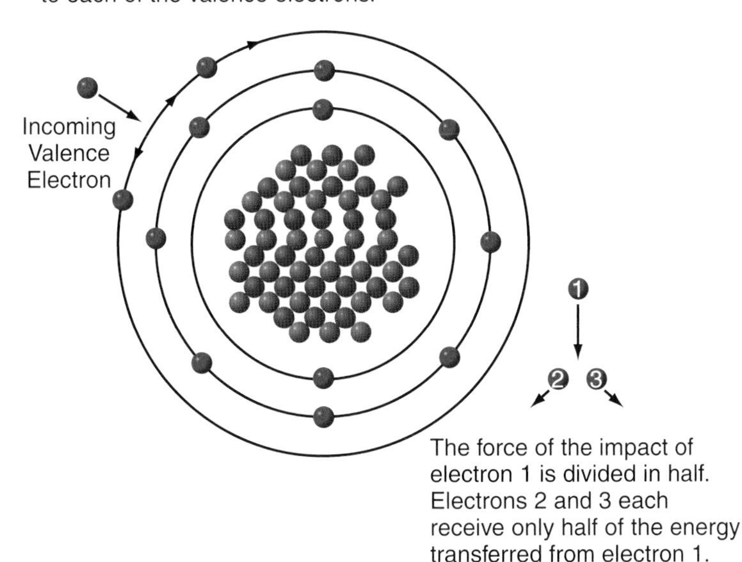

Figure 2–12 Two valence electrons each receive half the imparted energy.

FieldNote!

Since 1947, when the transistor (a semiconductor device) was invented, semiconductors have been used in almost all phases of our life. Solid-state or semiconductor devices are used in everything from radios, televisions, toasters, and pagers to motor controls and garage door openers. Some of the most common semiconductor materials are carbon, silicon, and germanium. Silicon is most often used in diodes, transistors, and integrated circuits. Before a "pure" semiconductor material such as silicon can be used in a device such as a transistor, it must be "doped" (mixed) with an impurity. You will learn more about semiconductors in your study of electronics.

This is illustrated in Figure 2–12. In electrical circuits, the electrons travel through the electrical circuit. The force of attraction or repulsion of the electrons in a circuit is fundamentally a measure of the quantity of the electron's charge. This ability or "potential" to attract or repel other electrons refers to the possibility of doing work. Any charge has the potential to do the work of repelling or attracting other charges. When you talk about two unlike charges (a positive and a negative), you have a potential difference.

The flow of these electrons is always from the pole of an electrical energy source that has an excess of electrons to the pole that has a deficiency of electrons. This is in compliance with the atom attempting to reestablish electrical balance. The direction of current flow, negative to positive, is defined as the polarity of the circuit. This is an important concept in the study and understanding of many electrical principles. Figure 2–13 shows how this electron flow works.

Circuit Polarity

Potential Difference

−Pole

+Pole

Direction of Flow

Excess of Electrons

Deficiency of Electrons

Figure 2–13 Current flows from excess electrons (negative polarity) to deficiency of electrons (positive polarity).

INSULATORS

Materials with atoms in which the electrons have a very high resistance to leaving their own valence rings are known as insulators. These materials cannot conduct very well because their electrons do not readily move from atom to atom. The valence shell or ring that is almost full (seven or eight valence electrons) is very stable. These electrons require a tremendous amount of energy to break free of their orbits.

Any incoming electron gives up its energy to all electrons in the valence ring. These electrons do not break free but do store the incoming energy, as shown in Figure 2–14. Examples of good insulator materials are rubber, glass, plastic, paper, wood, air, and mica. Insulators are useful when it is necessary to prevent current flow. They can also be used to store an electrical charge.

SEMICONDUCTORS

Semiconductors are materials that partially conduct. They conduct electricity less readily than good conductors, but better than good insulators. Semiconductors have four electrons in the outermost ring (the valence shell). This means that they neither gain nor lose electrons but share them with other similar atoms. The reason for this is that four is exactly half the stable condition of eight electrons in the valence ring. Figure 2–15 shows a diagram of a semiconductor.

Electrical Insulators

Energy from incoming valence electron is absorbed by each of the 8 electrons. None of the "hit" electrons leave their orbit rings.

Incoming Valence Electron

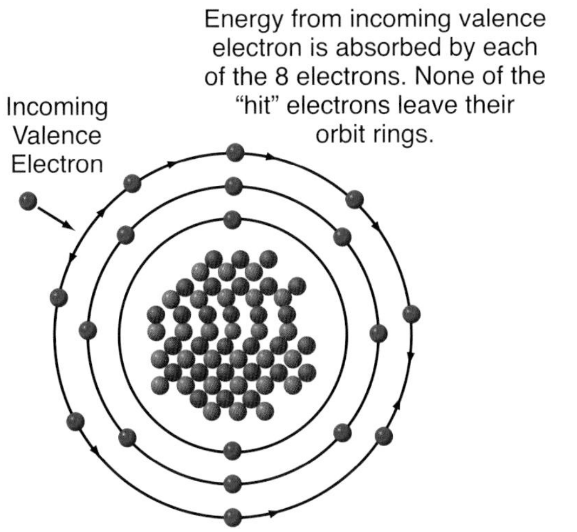

Figure 2–14 Electrical insulators have many valence electrons and do not transfer the energy to another atom.

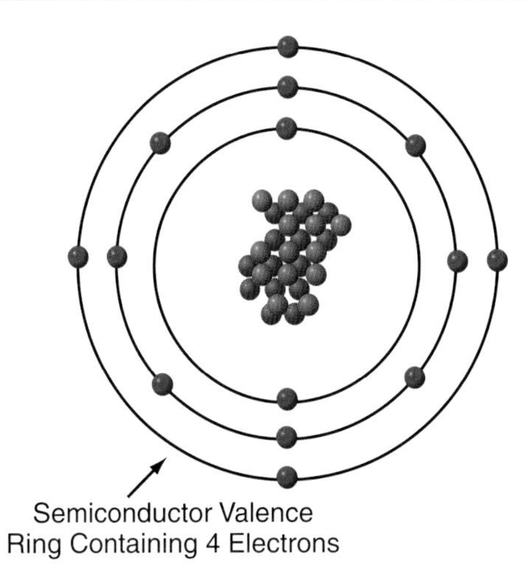

Figure 2–15 An atom structure that would be used for a semiconductor is illustrated.

The addition of heat causes changes in the conductivity of all materials. When an insulator or a semiconductor is heated, its resistance tends to go down. Heat is energy, and as it adds to the total amount of energy a smaller amount of external energy is needed to break the electrons free of their orbit and allow them to move. Consequently, they become a better conductor. When a conductor is heated, its resistance goes up. This characteristic of many conductors is referred to as a positive temperature coefficient. You will study the effects of this later. This information will be useful in understanding conductor sizes and current flow.

MOLECULES AND COMPOUNDS

A group of two or more atoms forms a **molecule**. For example, two hydrogen (H) atoms make a hydrogen molecule (H_2). When this hydrogen molecule combines chemically with an oxygen (O) atom, you have a water (H_2 + O, or H_2O) molecule. **Compounds** are made up of two or more elements (such as oxygen, hydrogen, copper, or silicon). A molecule is the smallest unit of a compound with the same chemical characteristics. You can have molecules for elements [as in the example of two hydrogen atoms combining (H_2)] or for compounds, such as H_2O. See the examples shown in Figure 2–16.

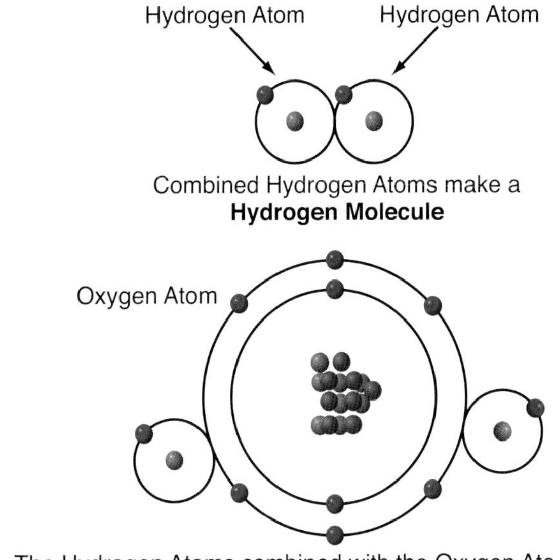

Figure 2–16 Two hydrogen atoms join with oxygen to form a water molecule (H_2O).

IONS AND ELECTRICAL USES

Ions are atoms that have either gained or lost electrons. An atom that has lost electrons is a positive ion. An atom that has gained electrons is a negative ion. The process of an atom losing or gaining electrons is called ionization. Negative ions (atoms) have more electrons than protons. Positive ions (atoms) have fewer electrons than protons.

Remember that electrical current (electron) flow is caused by a potential difference between two poles. One pole has an excess of electrons (the negative pole) and the other has a deficiency of electrons (the positive pole, Figure 2–13). Normally you think of this current flow as moving through a conductor. However, depending on the atoms, element structure, and molecules, this electrical current could be through a gas or a liquid. In gases or liquids, the electrical flow is caused by individual ions rather than individual electrons.

An example of two different elements' atoms (chlorine and magnesium) will show the effects of ionization. The element chlorine is not considered a metal because it has seven valence electrons (electrons in the outermost valence shell). Magnesium is considered a metal because it has two valence electrons in its outermost valence shell (Figure 2–17.)

When magnesium is mixed with chlorine gas and heated, the two elements (one atom of magnesium and two atoms of chlorine) combine to form a molecule of metallic salt called magnesium chloride. Figure 2–18 shows a diagram of the combination results. These atoms are now called ions. The magnesium atom is a positive ion because it has lost two electrons, and the chlorine atoms are negative ions because they have each gained an electron from the magnesium atom. Another molecule that combines a metal element atom (sodium) with chlorine in this same way is sodium chloride, or salt.

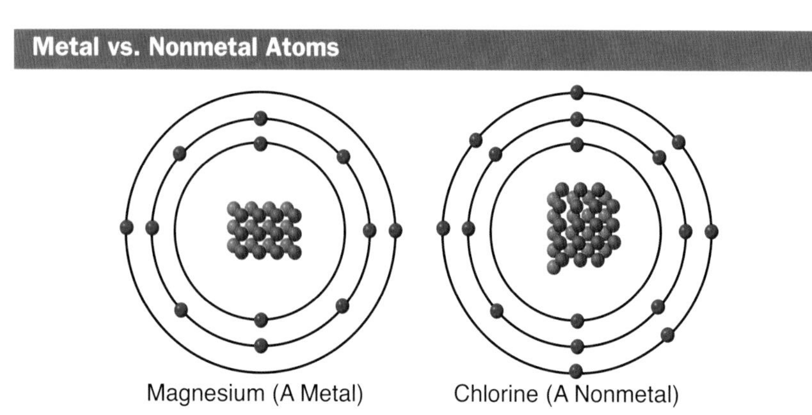

Figure 2–17 Magnesium and chlorine atoms show the effects of ionization.

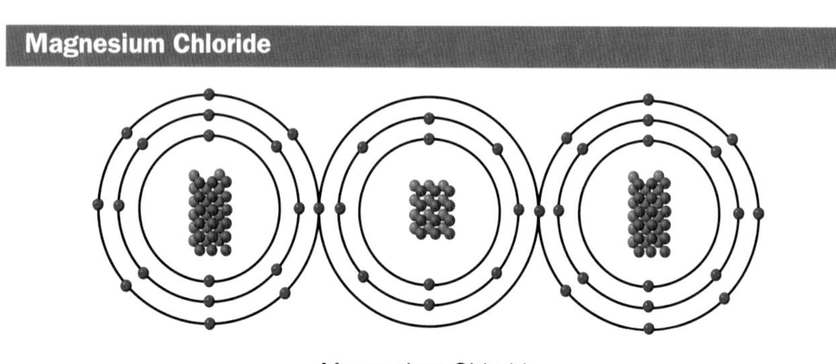

Figure 2–18 A molecule of the compound magnesium chloride, with positive and negative ions, is depicted above.

ELECTROLYTES

When a salt such as sodium chloride (table salt) is dissolved in water, the ionized particles will allow the passage of electricity. Figure 2–19 shows how a glass of pure water (a nonconductor) can be transformed into a conductor by dissolving salt in the water. Note that the electrodes shown in Figure 2–19 are made of the same material, copper, for example. Acids and alkalis can also be used to cause electrical conduction in liquids. These types of solutions are called **electrolytes**. Refer to Figure 2–19.

When two unlike metals are placed in an electrolyte solution, an electrochemistry process takes place that causes the transfer of electrons, through the electrolyte, from one of the metals to the other.

The metal that accumulates electrons is the negative pole, and the metal that gives up electrons is the positive pole. In this manner, an electrical potential is developed between the two poles. This action is discussed again when you learn about batteries.

OTHER USEFUL IONS

Other types of materials also have useful electrical functions. One of the more useful is copper sulfate, which is used in the copper electroplating process. Its symbol is $CuSO_4$. When mixed with water, two ions are created: a negative sulfate ion (SO_4), which has two extra electrons from the copper atom, and Cu (a cupric ion), which is positive because it has lost two electrons to the sulfate ion.

Sulfuric acid is also a compound made of two ions. Its main use is the electrolyte liquid used in lead-acid car batteries. Sulfuric acid's symbol is H_2SO_4. This indicates that sulfuric acid is made up of two hydrogen atoms, one sulfur atom, and four oxygen atoms. When H_2SO_4 is combined with water, its molecules separate into three ions: SO_4, H, and H. The two hydrogen (H) atoms are positive because they have lost their valence electron to the SO_4 ion. The SO_4 ion is negative because it now has two extra electrons. The lead and the sulfuric acid and water react to create the voltage potential in the battery cells.

SOURCES OF ELECTRICAL ENERGY

We have been describing the makeup of atoms and how electrons can flow or be forced from one atom to another. This causes different effects and combinations to happen. Many of these effects have useful electrical purposes. What causes the electrons to leave their valence shells and "hit" or take up orbit in another atom's valence ring? The answer is in what causes the initial electron to be forced from its shell. There are six methods that are known to force electrons from the valence ring of an atom to bump another atom and become an electrical current flow. These external forces are:

- Friction
- Chemical
- Heat
- Pressure
- Light
- Magnetism

With the exception of friction, electrical energy can produce the same effects as those used to produce it. That is, if the proper load is connected across a source, any one (or more) of the following effects will occur when electricity is used. It can create:

- Chemical action
- Heat
- Pressure
- Light
- Magnetic fields (you will study electromechanical generation in detail in Chapter 10)

As you can tell, electricity is created by other energies. It is transported through conductors by colliding electrons which give up energy in another form of energy at the load.

Metals in Electrolyte

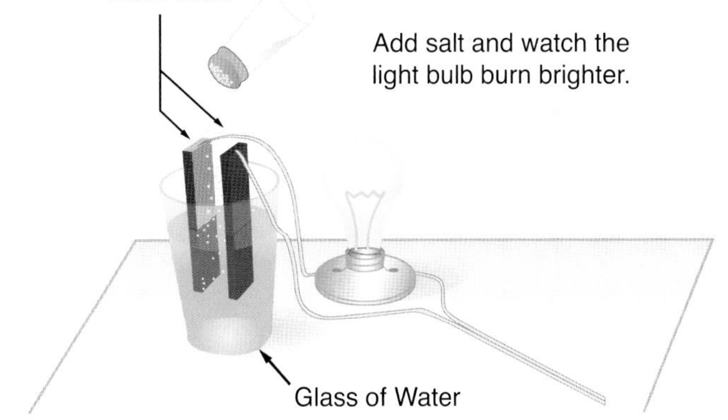

Figure 2–19 Water can be turned into an electrical conductor by adding salt.

FieldNote!

The **battery** (electricity produced by chemical action) was first discovered in 1791 by Luigi Galvani. He was conducting anatomy experiments using frog legs preserved in a salt solution. He discovered that when the legs were suspended by copper wires and touched by an iron scalpel they would "twitch." He knew this twitch was caused by electricity, but he mistakenly thought the electricity was caused by the frog's contracting muscles.

Alessandro Volta repeated Galvani's experiment in 1800. He correctly identified the cause of the electricity as a combination of the copper wire, iron scalpel, and salt solution. Volta continued experimenting until he developed the first battery from zinc and silver disks held apart by pieces of cardboard saturated with salt water. This single structure with one silver disk, one zinc disk, and one piece of saturated cardboard is called a **cell.** When many of these cells are connected in series or parallel, the resulting structure is called a battery. Volta called his "battery" a voltaic pile because it was a series of connected stacks.

The symbol on the left in Figure 2–20 represents a single-cell battery, which is represented by a long and a short line. The short line is the negative side of the power source, and the long line is the positive side. In Figure 2–20, the symbol on the right represents a multi-cell battery. If each cell produces 1.5 VDC, the entire battery will produce 3 VDC.

Battery Symbology

A Cell A Battery

− + − +

Figure 2–20 A single-cell battery and a multi-cell battery are represented by these diagrams.

FRICTION

Friction was the first known method of generating electricity. Its discovery is usually credited to the Greeks. This method consists of rubbing two pieces of material together, such as a rubber rod and wool or silk and a glass rod. The friction created by rubbing the objects together caused electrons to be transferred from one of the materials to the other. The change that is created is called **triboelectricity**. Friction is a source of **static electricity**.

Static means that the electricity is merely stored and is at rest. There is no movement of electrons after the transfer is complete. The static charge is potential electrical energy that will cause the excessive electrons on one body to flow to a body with a deficiency of electrons until they are electrically balanced. In nature, the tendency is always to establish an electrical balance. The body with excessive electrons is said to be negative, and the body with a deficiency of electrons is said to be positive.

As you know, static charges accumulate as you walk across wool carpet in stocking feet. The friction of your feet and the carpet create a difference of charge and you will discharge (get a shock) as you touch another contact point. Lightning is also a static charge that discharges from ground to cloud, or cloud to ground, as it tries to neutralize the total change imbalance.

CHEMICALS AND CHEMICAL ACTION

The chemical source of electricity is best represented in our everyday lives by the battery. These devices are used to start our cars, power portable systems, and operate emergency lighting and other systems when our primary source of electrical power fails. Batteries can be large (a car or truck battery), or they can be small enough to fit in the back of a wristwatch. The battery is a primary source of the DC charge.

BATTERY CONSTRUCTION

An easy-to-construct example of a cell is shown in Figure 2–21. Connect a piece of aluminum wire to one side of a voltmeter and a copper wire to the other side. Insert the other ends of the aluminum and copper wire into each end of a raw potato. The acid in the potato acts as the electrolyte, and current will flow between the two unlike metals (aluminum and copper). You should be able to see a small voltage register on the voltmeter. An even stronger electrical cell can be constructed using a copper wire, a galvanized roofing nail, and a grapefruit.

BATTERY METALS

Individual cell voltage depends on what metals are used for each plate. A list of special metals called the electromotive series of metals is shown in Table 2–1.

Example of a Cell

0.020

Copper Wire and Electrode

Aluminum Wire and Electrode

Potato

Figure 2–21 A chemical reaction with the electrodes and a potato creates electricity.

This list is not complete, but it provides good examples listed in the order in which they can receive or give up electrons. The first metal, lithium, receives electrons more easily than the rest. Note the difference on the chart between aluminum and silver. Because aluminum will receive electrons more easily than silver, when the two are paired, aluminum will be the negative electrode and silver the positive electrode.

As a practical matter, not all of these materials will make good metal for battery cells. Many will corrode, and others will cause chemical reactions in the electrolyte. Either of these problems will cause resistance to electron flow and reduce the cell's ability to produce electricity.

Table 2–1

An electromotive series of metals shows the relationships of electrodes that could be used for batteries.

Metals
Lithium
Potassium
Sodium
Magnesium
Aluminum
Manganese
Zinc
Iron
Cobalt
Nickel
Tin
Lead
Copper
Mercury
Silver
Platinum
Gold

BATTERY CATEGORIES

Batteries are usually divided into two categories: **primary cells** and **secondary cells**. Primary cells, such as most batteries used in portable radios and flashlights, cannot be recharged.

Once they have depleted their chemical action, they must be thrown away. Secondary cells, such as those used in automobiles, can be recharged several times. Table 2–2 presents a chart of primary and secondary cells.

When a secondary cell type battery is recharged, the chemical action that allows electric current to be drawn from the battery is reversed. Passing electric current through a solution of water and sulfuric acid will separate the water molecules into two hydrogen ions and one oxygen ion. The positive hydrogen ions will attract negative sulfate ions and produce sulfuric acid.

The negative oxygen ions are attracted by the positive electrode. Therefore, as the battery charges, more and more of the electrolyte is changed from water to sulfuric acid, increasing the battery's charge. Other chemical actions (such as electrolysis) are caused by an electrical current passing through certain materials. Electroplating gold jewelry is a good example of this chemical action.

HEAT AND ELECTRICITY

Heat is generated any time current flows through a conductor. This happens because energy is used to move electrons. Usually, the effect of heating a wire is undesirable. For this reason, metals such as copper and aluminum are normally used as conductors because they need less energy to move the electrons. For certain applications, heat can be a desirable outcome. Products such as toasters, irons, electric clothes dryers, and space heaters are designed to produce heat for practical purposes. These items use special wire that is designed to heat when current flows. Typically, this type of "heater element wire" is nichrome.

A transfer of electrons can also take place when two dissimilar metals are joined at a junction and then heated. This is known as the **thermoelectricity** process. For some active metals, the ambient room temperature (at normal comfort level) is sufficient to cause the transfer of electrons from one metal to the other. Increases in the temperature cause a greater transfer of electrons. This type of junction of dissimilar metals is called a **thermocouple**.

A thermocouple is very common in gas appliances such as water heaters and some furnaces. The thermocouple is placed directly in the pilot flame. As it is heated, it generates a small voltage that keeps the pilot gas valve open and supplies voltage to the main valve.

Table 2–2

Electrode–electrolyte combinations lead to a chemical production of electricity.

Cell	Positive Plate	Negative Plate	Electrolyte	Volts per Cell
		Secondary Cells		
Lead-acid	Lead dioxide	Lead	Diluted sulfuric acid	2.2
Nickel-iron	Nickel dioxide	Iron	Potassium hydroxide	1.4
Nickel-cadmium	Nickel hydroxide	Cadmium	Potassium hydroxide	1.2
Silver-cadmium	Silver oxide	Cadmium	Potassium hydroxide	1.1
		Primary Cells		
Carbon-zinc	Carbon manganese dioxide	Zinc	Ammonium chloride	1.5
Alkaline	Manganese dioxide	Zinc	Potassium hydroxide	1.5
Mercury	Mercuric oxide	Zinc	Potassium hydroxide	1.35
Zinc-air	Oxygen	Zinc	Potassium hydroxide	1.4

As the thermostat calls for more heat, a main gas valve will open if the pilot light is burning. This prevents a large amount of gas from entering the combustion chamber if there is no pilot flame.

PRESSURE AND ELECTRICITY

When placed under pressure, certain crystalline substances generate minute potentials (potential differences). The force of the pressure causes the electrons to be driven out of orbit in the direction of the force. The electrons leave one side of the material and collect on the other side. Depending on how the crystal is cut, some respond best to bending pressures and some to a twisting action.

Electricity derived from pressure is known as the **piezoelectric** effect. Figure 2–22 shows a strain gauge or pressure transducer. It is possible to reverse the process and produce pressure with electrical current. The pressure is produced by the physical displacement of the ions in the crystal. This reaction to producing pressure with electrical energy is the concept behind producing sound waves for speakers and small headphones.

Strain Gauge

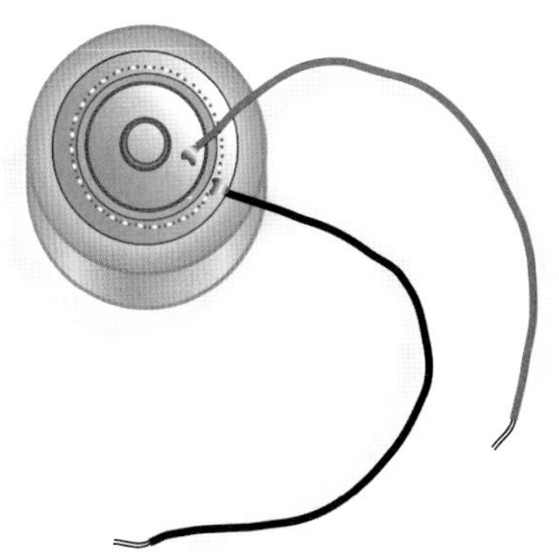

Figure 2–22 A strain gauge is used to transform pressure into an electric supply.

ELECTRICITY PRODUCED BY LIGHT

Light has small particles of energy called photons. When photons strike certain types of photosensitive material, they release energy into the material. There are three types of photoelectric effects of interest in the study of electricity.

1. *Photoemission:* When photons strike the surface of the material, electrons are released in a vacuum tube. A positive plate placed in the tube will cause the electrons to flow to it. More electrons will flow with greater light energy.
2. *Photovoltaic:* When photons strike one of two photosensitive plates that are joined, electrons move from the plate that is bombarded by the light energy to the adjacent plate. An electrical potential then exists between the two plates. This action is similar to that of a battery, except that light energy is used rather than chemical energy. This component is called a solar cell.
3. *Photoconductive (photoresistive):* When photons bombard some types of materials that are normally poor conductors, electrons are freed from the valence shell to participate in electrical current flow. This in turn lowers the resistance of the material. This component is called a photocell.

LIGHT PRODUCED BY ELECTRICITY

When enough electrical current is passed through a poor conductor, not only is heat generated but also some materials begin to glow red or even white-hot. Everyone has seen the red glow of a burner on an electrical stove. The common electrical incandescent lamp is designed to give off white light when it is heated to incandescence. The heat given off with this process is usually considered an energy loss or "heat loss." There are four other methods of producing light with electricity that do not result in as much heat loss as in the incandescent lamp.

1. *Electroluminescence:* Some materials (such as neon gases, argon, and mercury vapor) give off light when current is passed through them.

Because the usable amount of light is small, electroluminescence methods are used mostly for displays such as neon signs.

2. *Phosphorescence:* Light occurs when an electron beam strikes a surface covered with phosphors. This is how light is provided on the screen of a cathode ray tube used in television sets and oscilloscopes.
3. *Fluorescence:* Fluorescent lamps make use of both electroluminescence and phosphorescence to provide a higher level of light output than that of an incandescent lamp while using an equal amount of electrical power. These lamps have a gas, such as mercury vapor, that becomes an ionized carrier of electrical current. This ionization process causes ultraviolet radiation that strikes the phosphorescent coating on the inside of the fluorescent tube, causing "white light."
4. *PN junction luminescence:* This type of luminescence occurs when DC is applied to a PN junction. The PN junction must be specially doped with other materials, such as the semiconductors discussed earlier in this text. When electricity passes through the doped PN junction, electrical energy is absorbed by the junction. This electrical energy is then released in the form of light and heat energy. These special PN junctions are called light-emitting diodes (LEDs). LEDs can be manufactured that will produce any color of light, from infrared to nearly ultraviolet. Individual LEDs do not generate a large amount of light but are used extensively as indicator lights, in numeric displays, and in specialty equipment such as signs and long-life emergency exit light lamps.

Single-Filament Light

Figure 2–23 A single-filament light has only one level of light.

Incandescent Lamps

Incandescent lamps are designed to produce light, but the process requires us to heat the filament until it incandesces, or glows white with heat. Tungsten filaments have many different shapes, as required by the applications of lamps, but all use heated elements to produce light, with relatively poor efficiency of transfer from electric to light energy. Single-element lamps (Figure 2–23) use one element, and the circuit conducts current from the center point on the bottom of the base through the filament and back to the screw shell.

A three-way lamp uses two elements. For example, the lamp shown in Figure 2–24 is rated for 50 W, 100 W, and 150 W. As you view the bottom of the base, you will note the center tab, a ring connection point, and the screw shell. A first connection allows current to flow through the 50 W incandescent element. The second position of a three-light switch allows current through only the 100 W element and not the 50 W element. The third light level is the combination of the two elements for 150 W total. This is an example of how to cause different light levels by heating (burning) different filament structures to produce light.

Two-Filament Light

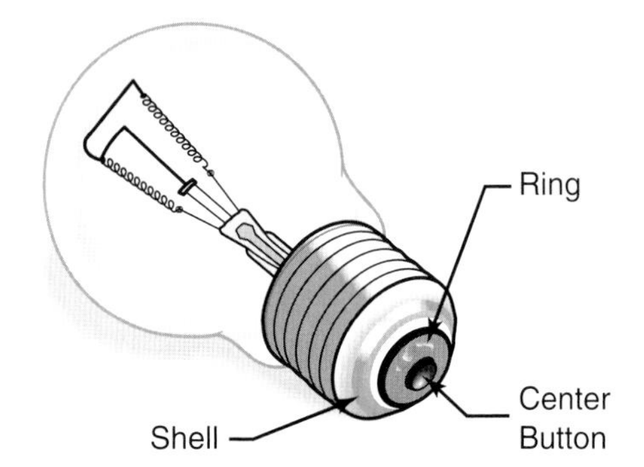

Figure 2–24 A two-filament light has three light levels.

VOLTAGE POLARITY

When you use a battery or a DC instrument such as a voltmeter or an ammeter, you will note that the terminals are marked with a plus sign (+) and a minus sign (–).

These symbols indicate the polarity of the terminals, and they must be understood to avoid damage to the equipment. Polarity is the property of a component that has two points with opposite characteristics. For example, a battery has two terminals (plus and minus), and a potential exists between those two terminals. The measure of that potential difference is called voltage.

Connecting some types of electrical or electronic equipment with their indicated polarities reversed can result in damage to or destruction of the equipment being used. For this reason, equipment of this type will normally have polarity marks indicating which terminal is positive or negative. Always check before you actually make a measurement with such equipment.

CAUTION: Failure to observe polarity during measurements may result in equipment damage and/or severe injury.

DETERMINING VOLTAGE POLARITY

When current flows through a load or resistor, it takes on a polarity. Because electrons flow from negative to positive, it is obvious that when electrons flow through a resistor, the voltage drop ($I \times R$), as determined by Ohm's law, is developed across the resistor. One end of the resistor becomes negative and the other positive. A diagram of a load, with polarity markings indicated, shows which way electrons will flow. Analyzing electrical circuits correctly often requires the use of laws and theorems that depend on the principle of polarity.

You must know how to determine the polarity across each component to solve these types of problems. In Figure 2–25, for example, the current through the resistor is flowing from left to right. The left side of the resistor, where the current enters, is more negative than the right side. An easy way of remembering this polarity of a load is to think the electrons are always moving toward the positive point in the circuit. Therefore, as they enter a component and move through a component, they move from the entry point (negative) to the exit point (positive).

In a circuit, the polarity of components is determined by tracing the current path, starting at the negative side of the power source. When a component is encountered, a negative sign is placed at the point where the current enters, and a positive sign at the point where the current exits. Continue doing this until all components are marked with polarity signs. A completed circuit is shown in Figure 2–26.

Note that where the current exits, R_1 is positive, and where it enters, R_2 is negative. This does not mean that the negative side of R_2 is more negative than the positive side of R_1. They are actually the same electrical potential. The polarity signs show ends of the same component *compared* with each other. If a voltmeter was used to measure the difference between points A and B, it would read zero because there is no difference in potential. A voltmeter will measure the difference between point B and point C. This is the voltage drop across the resistive component, or drop in electrical pressure.

In the same respect, a bird sitting on a high-voltage transmission line is not electrocuted because there is effectively no difference in potential between its two feet. Both feet are at the same electrical potential, and therefore no current tries to flow from one point to the other.

Current Flow

Figure 2–25 A current through a load moves from the negative point toward the positive point.

Establishing Polarities

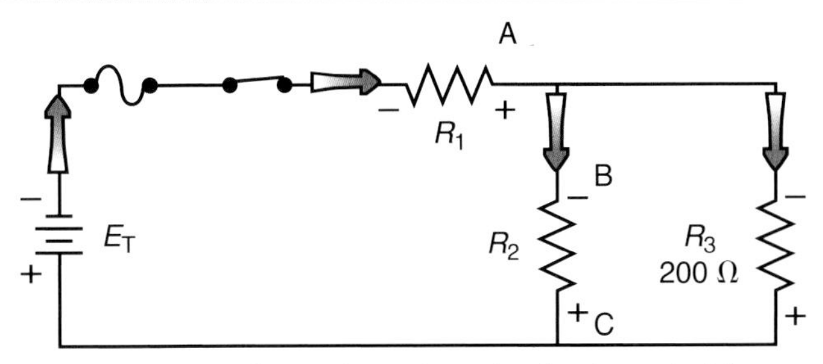

Figure 2–26 As current flows through a circuit, the polarities are established as shown.

Because every wire has some resistance, there would be minute voltage, depending on the wire size and circuit current. The bird is sitting between two points that have very little potential difference (note that the positive side of R_1 and the negative side of R_2 in Figure 2–26 are the same potential). Later, when you learn about AC circuits, you will find that polarity and polarity markings are used there as well.

CONDUCTOR RESISTANCE

Many factors must be taken into account to prevent cables from being underrated when wiring buildings or other facilities that use electrical cabling. Underrating a cable can cause the insulation for the cable to deteriorate, even to the point of it becoming a fire hazard. The *National Electrical Code®* (*NEC®*) specifies the ratings for conductors and cabling based on application of the physical law governing electrical circuits.

The code must always be followed when installing the cables to ensure that electrical fires do not start because of overheating conductors or devices. In addition, the conductors must be chosen so that rated equipment can operate at the correct voltage and current. Systems with too much inherent resistance will reduce both the voltage to the load and the efficiency of current delivery. The calculation for actual ohmic resistance of conductors is covered in Chapter 3.

AMPACITY

The **ampacity** of a cable or conductors is the maximum current the cable can carry safely without exceeding its temperature rating. The *NEC®* requires that branch circuit conductors have an ampacity greater than or equal to the noncontinuous load plus 125% of the continuous load. In other words, the wiring must be able to conduct enough current to the load without overheating. Heat builds up with time, and if a load is connected for 3 hours or more it is considered continuous, and we need to increase the physical size of the conductor to avoid the heating of the conductors.

SUMMARY

The study of electricity is interesting because you cannot see the actions taking place. The transfer of energy from different sources (such as friction, pressure, and light) is completed at the subatomic level. Your understanding of this transfer allows you to make educated assumptions about conducting materials, insulators, and semiconductors.

You now have knowledge about the molecular level of electrical actions, including the means of creating a potential difference by chemical action. You know the importance of the valence shell and how it affects conductivity. In most cases the cause of the electric potential difference (such as light, heat, and chemical action) can again be reversed at the load. This means that the electrical energy can produce light, heat, or chemical actions at the load point. We reviewed how polarity of the power supply means one thing, but the polarity of the voltage at the circuit component is different. Finally, we saw how to create different light levels in incandescent lamps as an example of light production in everyday use.

CHAPTER GLOSSARY

Ampacity The amp-carrying capacity of an electrical conductor. The contraction of amp and capacity create an electrical term for how much current a conductor can carry safely.

Atom The smallest particle still characterizing a chemical element.

Battery Series and/or parallel combination of cells. A group of cells connected in series (more voltage) or parallel (more current).

Cell A single chemical structure composed of an electrolytic solution (sulfuric acid) and two different metallic electrodes (lead and lead peroxide).

Compound A material made from the chemical combination of two or more elements.

Conductor A material that easily passes electrical current. Examples include silver, copper, and aluminum.

Electricity A class of phenomena that results from the interaction of objects that exhibit a charge (electrons and protons). In its static form, electricity exhibits many similarities to another naturally occurring force—magnetism.

Electrolyte Any material that will dissolve into ions when immersed in a liquid. The liquid thus becomes an electrical conductor.

Electron One of the three main components of an atom. The electron is a fundamental particle, and by definition has a negative electrical charge (from the Greek word *elektron*, meaning "to be like amber").

Element The simplest form of matter. There are more than 103 known elements, 92 of which occur naturally. All matter is made from chemical combinations of elements.

Gluon The particle that mediates or transmits the strong nuclear force between quarks. The fundamental particle responsible for the strong nuclear force.

Insulator A material that does not allow electrical current to flow easily. Examples include rubber, plastic, and mica.

Ion An atom or molecule that has gained or lost one or more electrons. A positive ion has lost electrons, and a negative ion has gained one or more electrons.

Isotope One of two or more atoms with the same number of protons but different number of neutrons.

Matter The material from which all known physical objects are made.

Molecule The chemical combination of two or more atoms. The smallest particle of a compound that has the same chemical characteristics of the compound.

Neutron One of the three main components of an atom. The neutron has no electrical charge and is therefore classed as electrically neutral. The neutron has been shown to be composed of even smaller particles, called quarks.

Piezoelectric Electricity created by stress or pressure in a material—especially a crystalline material.

Primary cell A cell that cannot be recharged after it has depleted all of its stored chemical energy in the form of electricity.

Proton One of the three main components of an atom. By definition, the proton has a positive electrical charge. The proton has been shown to be composed of even smaller particles, called quarks.

Quark One of the fundamental particles of matter. There are six different types of quarks that are assembled in different combinations to create larger particles, such as protons and neutrons.

Secondary cell A cell whose chemical energy can be restored by forcing electrical energy into it.

Semiconductor A material that falls between conductors and insulators in terms of electrical conductivity.

Static electricity An electrical charge that is stationary or nonmoving. Sometimes called triboelectricity.

Strong nuclear force The force that holds quarks together to make up neutrons and protons. The residual strong nuclear force is also responsible for holding protons and neutrons together in the nucleus despite the electrical repulsion trying to force the protons apart.

Thermocouple A junction of two dissimilar metals that creates an electrical potential when heated.

Thermoelectricity Electricity created by heat.

Triboelectricity An electrical charge created by rubbing two materials together. Sometimes called static electricity.

Valence electrons The electrons that make up the valence shell. Valence electrons are free to participate in current flow.

Valence ring or shell The outermost shell of electrons in an atom.

REVIEW QUESTIONS

1. What are the three particles that make up atoms?
2. Which particle has a negative charge? Which has a positive charge?
3. What is the valence shell of an atom?
4. Which shell of electrons is the primary determinant of the electrical behavior of a material?
5. How many electrons are found in the valence shell of a sodium (Na) atom?
6. Electron current flow is always from a __________ concentration of electrons to a _________ concentration of electrons.
7. How does a battery store electrical energy?
8. Explain what a molecule is.
9. How can energy be transferred by means of electrons?

Use Figure RQ2–1 for questions 10 through 14.

10. Identify the polarities of the power supply.

11. Identify the polarity of the voltage drop across component 1.

12. What is the difference in potential from point B to point C, based on a 4 V power supply?

13. To measure the voltage drop across component 1, the negative voltmeter leads should go to point _____.

14. If each cell of the battery produces 2 VDC, what is the circuit supply voltage?

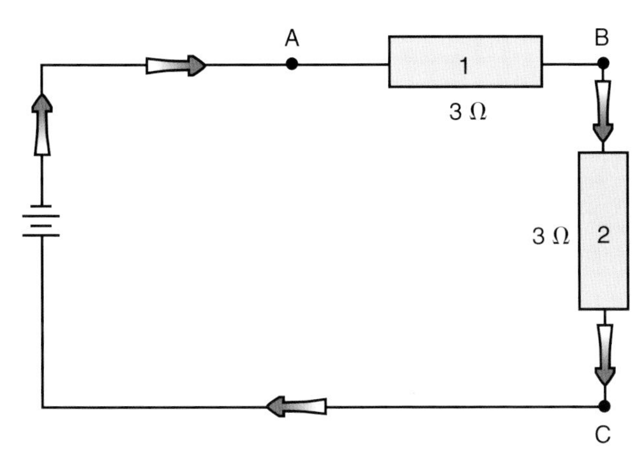

Figure RQ2–1 This circuit drawing shows current direction and two components.

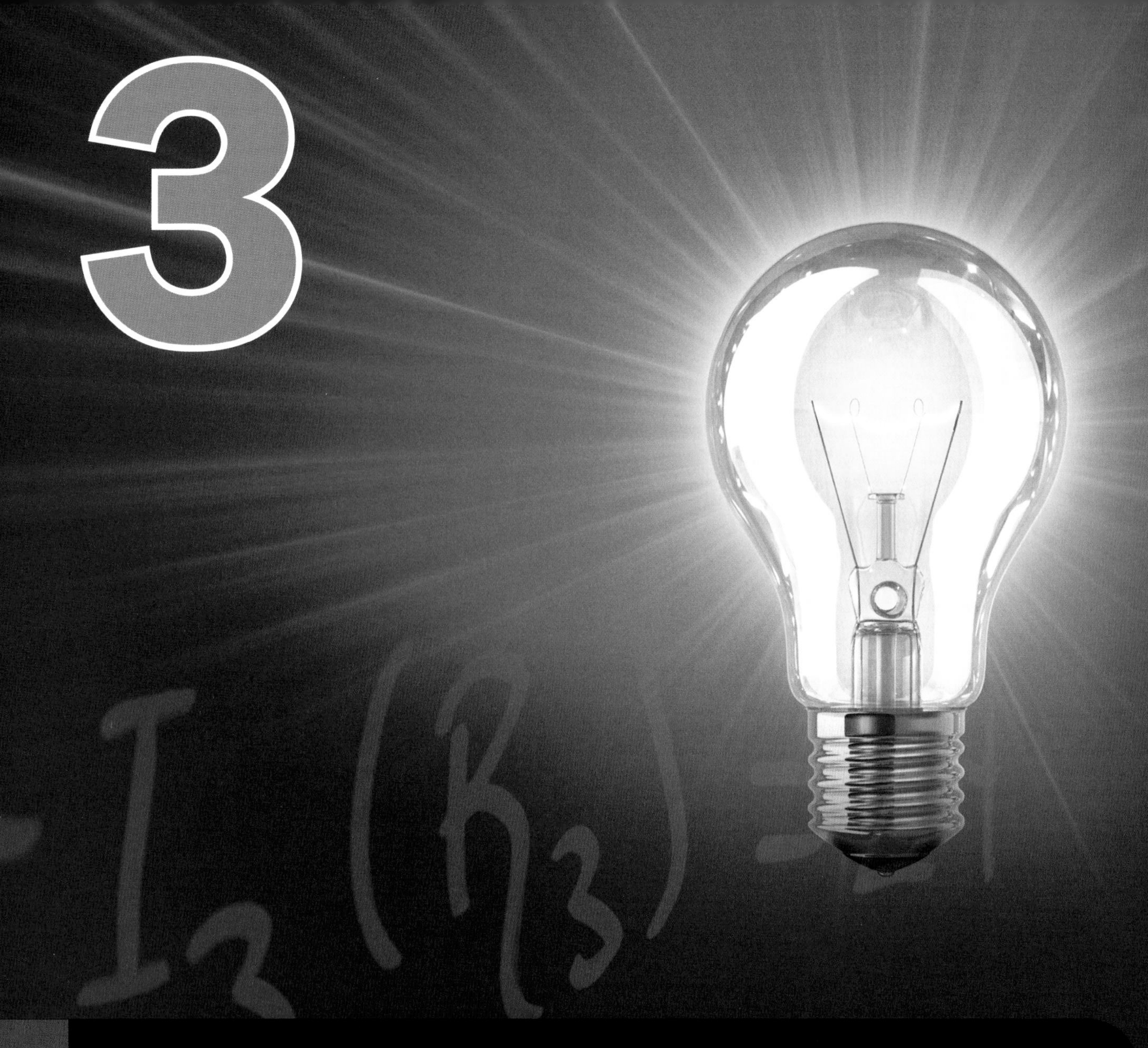

3

Circuit Theory and Switches

TOPICS

- INTRODUCTION
- WIRE VS. CONDUCTOR
- CIRCUIT CONCEPTS
- SWITCHES
- CONDUCTOR SIZE
- VARIATION OF RESISTANCE WITH AREA
- MATERIAL TYPE EFFECTS ON CONDUCTOR RESISTANCE
- DETERMINING THE TOTAL RESISTANCE OF A CONDUCTOR
- CIRCULAR MILS AND SQUARE MILS
- CONDUCTOR RESISTANCE, VOLTAGE DROP, AND WATTAGE LOSS
- INCANDESCENT LAMPS

OVERVIEW

You learned in previous chapters the basic measurements of electrical circuitry and the subatomic theory of what makes electricity work. Now we will start putting the theory and measurement into practical application. You will construct circuits to do work. In order to have the ability to control the electrical components of a circuit, we have to safely open, close, and conduct the correct current to and from the load. Switches are used in various styles and combinations to safely switch the current and voltage of the circuit. You will learn the mechanics of determining conductor size to accommodate current and calculate the results of conductors made of different materials, in various diameters, and of various lengths.

OBJECTIVES

After completing this chapter, you will be able to:

- Describe circuit concepts of open, closed, and short circuits
- Determine the current flow scenarios for open, closed, and short circuits
- Describe switch ratings and styles to accommodate various applications
- Use switch patterns in schematic form to connect actual switch circuits
- Determine the conductor gauge; determine circular mil and square mil cross-sectional area of conductors
- Determine voltage drop on conductors in circuit conditions
- Connect and analyze incandescent lamps as loads

INTRODUCTION

In Chapter 1 you learned the fundamentals of a circuit and what was required to make electricity work. We will spend more time analyzing these concepts and then find ways to control the current flow using switches. As we find ways to control the circuit operations, we will learn that the connecting conductors also have characteristics that will determine circuit operational limits. Using meters to troubleshoot simple lamp circuits will provide experience with circuit operations, meter operations, and conductor characteristics.

WIRE VS. CONDUCTOR

Before we study actual circuit concepts, it is important that you first understand a few terms that are often used interchangeably in the electrical industry. These terms are wire and conductor. The first term we will discuss is *wire*; this term is often used as a slang term throughout the industry as a replacement for the word *conductor*. As an example, you may hear a phrase such as, "Go grab me a spool of THHN wire" instead of the phrase, "Go grab me a spool of THHN insulated conductor." You may also hear a phrase such as, "We have to finish this wire pull before we can leave for the day" as opposed to, "We have to finish this conductor pull before we can leave for the day."

The term *conductor* is also used frequently. Some of the common methods where the term *conductor* is more appropriately used are: "What is the resistance of a #12 copper conductor per 1000 feet?" or "How many #14 conductors can be installed into a ½-inch EMT conduit?" The term *conductor* is the term that is more commonly used throughout the *National Electrical Code®* (*NEC®*). However, if you look up the term *conductor* in the *NEC®* it is defined as one of the following three:

1. Conductor, Bare
2. Conductor, Covered
3. Conductor, Insulated

Each of the three conductor terms will be explained in further detail later in your studies; it is important that you simply understand that the term *conductor* is the recognized term used by the *NEC®* in most cases. As an example, when you are required to find the resistance of a single conductor, you will find the resistance chart under conductor properties in the code book, not under wire properties.

When both terms are used in the same statement, you may hear something like this: "Your job for the day is to wire the control cabinet, but before you start terminating the wires make absolutely sure that you don't exceed the amount of conductors allowed by the *NEC®* in each conduit."

As you can see by the last statement, these terms are loosely used in normal conversation. If you look closely at the way these terms were used you will see that the term *wire* was more appropriately used (as a verb in this case) in the first part of the statement. It could be replaced by the word *conductor* in the middle part of the statement, and in many cases it would replace the word *conductor* in the last part of the statement.

When you actually install wire to make the real circuit, the concept is the same. Start at one terminal of the power source, move through the conductor to the load, and back to the power source to complete the circuit. The process is the same as when you first learned to read a map. Start at home, trace the path to your destination, and return home.

The schematic is your map to the electrical system. As you physically do this wiring installation, you may want to check off the wire on the schematic diagram. This is a simple task at this point, but as circuits become more complex the process of installation of actual wiring compared to schematic wiring becomes more difficult. This is complicated even more if you have several electricians working to complete a project and you need to know who did what. The check-off process helps avoid confusion.

You may be asking the question "Why wouldn't you simply use the word *wire* in all three locations?" The answer is quite simple; recall that the word *conductor* is the term more commonly used throughout the *NEC*®. The word *conductor* is used in the last part of the statement to instruct the person to check the *NEC*® to find out how many conductors are allowed in each conduit. When the term is used in this context, it is accurately describing the method used in the *NEC*®.

Regardless of how these terms are used, the important thing to know as you study electricity is many textbooks use these terms interchangeably.

CIRCUIT CONCEPTS

Studying circuit concepts requires us to see a circuit on paper as a representation of an actual wiring diagram. The circuit you see on paper that depicts the systematic way current would flow through the wires, switches, and loads in a logical progression is referred to as a schematic diagram. Electricians must take the schematic diagram and transfer that concept to a physical wiring of the circuit. These two diagrams may not look much like each other. The reverse of this process is used when troubleshooting a wiring system in a building. That means you will have to look at a hardwired circuit and be able to translate back to a schematic diagram. This takes practice.

The first circuit concept to investigate is that of a closed (complete) circuit (Figure 3–1). As you wire a simple circuit, start at the negative terminal of the power source and try to find a way to get back to the positive terminal of the power source. (Pretend that this is your job as an *electron*). Try to find the path of least resistance back to the source. Because the circuit is connected with conductors to the load, follow the conductor to the load. Go through the load, always moving toward the positive terminal, and then through the conductor back to the positive terminal.

When there is an "open" in the conductive path, you must stop. In reality, no current flows when there is an open in the circuit. However, it is easier to visualize the effects if you are an electron that has to stop (Figure 3–2). Now compare the current flow in an open circuit with no current flow. The drawbridge for traffic flow is open and there is no traffic moving. An open circuit can be created by a switch in the open position, a conductor that is broken and not complete, or a missing connection at a terminal. In any case, there is no complete path for current to flow.

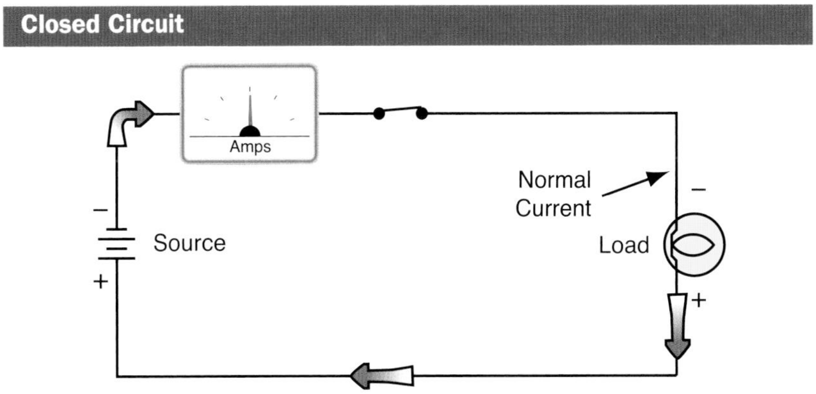

Figure 3–1 A complete or closed circuit has a path for current from the source and back.

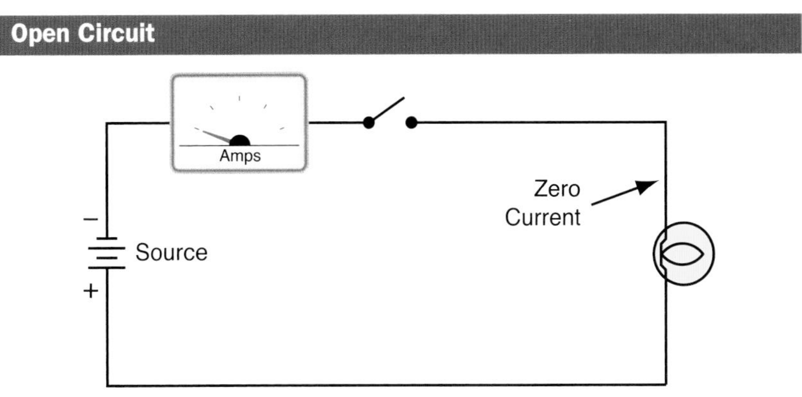

Figure 3–2 An open circuit prevents current flow, and the current is zero.

Another scenario is the short circuit. This is an application in which current takes every available path back to the source. You may hear the phrase "current takes the path of least resistance" back to the source. Although this is true, not all of the current will take that path. Instead, it is more correct to say that current tries to take any path back to the source (Figure 3–3).

In the case of a short circuit, current will find an easier path back to the source if it does not have to go through a load opposition. In this case, there is very little opposition to the current flow, and the current is very high. We might say it is the maximum current the supply can provide. When we actually calculate the value, you will see that there are limits. For now, however, we will say that short circuit current is extremely high.

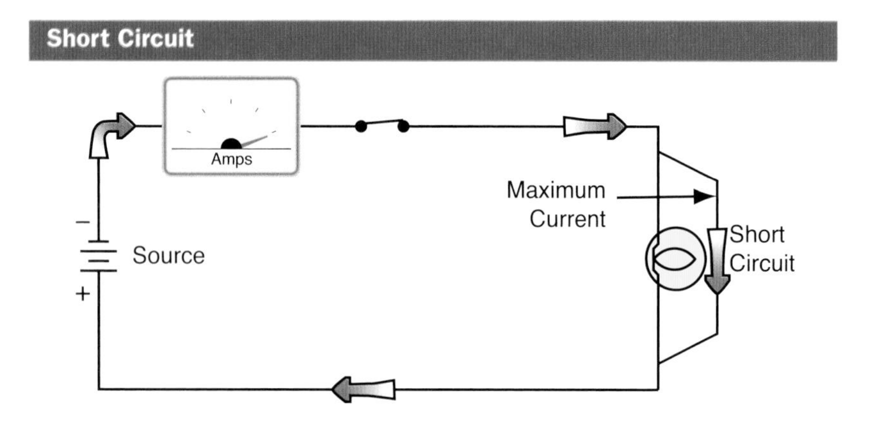

Figure 3–3 A short circuit is a high-current circuit that has a short path back to the source, not necessarily through the load.

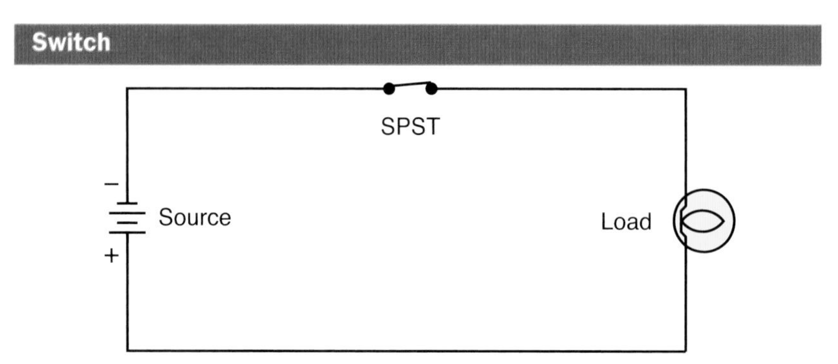

Figure 3–4 A switch is used to make a complete circuit or to break a circuit and create an open circuit.

The term *short circuit* comes from the fact that the current has taken a shorter path back to the source, avoiding the opposition of the load. Back to our map analogy, we avoided the traffic jam and took a shorter route home.

A complete circuit allows current to flow to the intended destination through the load and back to the source. An open circuit has no current flow. A short circuit means maximum current flows but does not reach the intended load before returning to the source at maximum value.

SWITCHES

Now that we have created a circuit that allows current to flow in a closed loop, or stops current from flowing when the (loop) circuit is opened, we need to see how to deliberately create this process with switches. A switch is a mechanical device designed to allow current to flow in a circuit or to interrupt current flow in a circuit. Referring to Figure 3–4, you can see that when we close the switch we would complete the conductive path for the circuit.

At this point of our learning we can place the switch at any point in the path to the load or the path back to the source from the load. It does not matter where we break the circuit, only that we interrupt the current path. Switches are designed in many varieties with the same purpose and intent: to make or break the current path.

SWITCH DESIGNATIONS

Switches are designated by how many paths from the source they make or break and the number of ways they can be moved to compete those circuits. The first notation is the number of poles a switch has in making or breaking circuit paths from the source. For instance, a simple switch that either makes or breaks one circuit path at a time is referred to as a single-pole switch (Figure 3–4).

This switch will complete the circuit path or interrupt the circuit path. This switch also has a number of throws. We can "throw" open the switch or "throw" it closed. The number of actions called "throw" that allows us to close the switch is referred to as the "throws" of the switch. In Figure 3–4, there is one throw. That switch now has the designation of a single-pole/single-throw (SPST) switch.

Another standard type of switch is a single-pole/double-throw (SPDT) switch (Figure 3–5). As you can see in the figure, the number of current paths from the source is one (a single pole). The switch can be closed in two directions to complete the path back to the source. Therefore, this is a double-throw switch. This switch is used extensively and is also known as a three-way switch.

As we add to the switch capabilities—the number of circuit paths from the source we can control—we add to the number of poles. As we add to the number of switch positions we can close, we add to the throws. See Table 3–1 for switch designations and physical characteristics. The style of switch pattern that provides the same function as a DPDT switch is modified to provide the same function with a different operating mechanism. This switch is commonly referred to as a four-way switch. It provides the double-pole/double-throw characteristic. In the toggle wall switch, either position provides the throw function because either position closes a circuit path. As you will see in the section regarding switch patterns, this switch is used to help in multi-location switching. It is typically used in combination with two three-way switches.

Switches are also classified by their physical appearance. The example shown in Figure 3–6 is described as a disconnect or a motor safety switch. This is one type of switch that acts like a knife as it moves into the closed position and seemingly cuts the receiving contact.

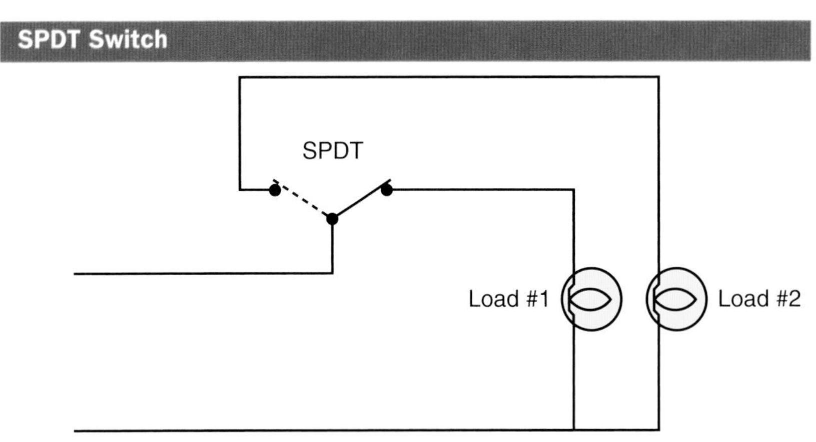

Figure 3–5 A single-pole/double-throw (SPDT) switch has a common center terminal.

Table 3–1

Switch notations with switch diagrams are listed here.

Switch Type	No. Terminals	On/Off Legend
SPST	2	YES
THREE-WAY/SPDT	3	NO
DPST	4	YES
FOUR-WAY	4	NO
DPDT	6	NO

Figure 3–6 A motor safety switch is an example of a knife blade switch.

These switches are available but are used in specific applications and are not seen as such in everyday household wiring. This type of switch is used commercially and in industrial applications as a disconnect switch for motors or large equipment.

A more common style of switch is known as a toggle switch. Toggle switches "toggle" from open to closed and have definite positions where they stop. A switch used in a wall box is known as a flush toggle, meaning that the toggle switch face will be flush with the wall and the switch mechanism behind the cover plate. Only the toggle handle protrudes from the flush mounting. Table 3–2 outlines other types of switches, such as limit switches, safety switches, drum switches, and other configurations.

At this point, we can briefly describe another type of switch that is still a mechanical operation but is not considered a manual switch. A manual switch means that a person provides the mechanical energy to move the switch to a new position. Switches moved by other actions are not considered manual switches. The most common switch moved by electromagnetic means is a relay.

A **relay** is a set of contacts that move when a signal is "relayed" to the operating electromagnet. As seen in Figure 3–7, the contacts can change position when the electromagnet pulls the moveable iron piece. The electromagnet is magnetized only when another circuit relays a voltage to the electromagnet. When the other circuits relay a zero voltage sig-

Table 3–2

Switches are identified by purpose and switch mechanism.

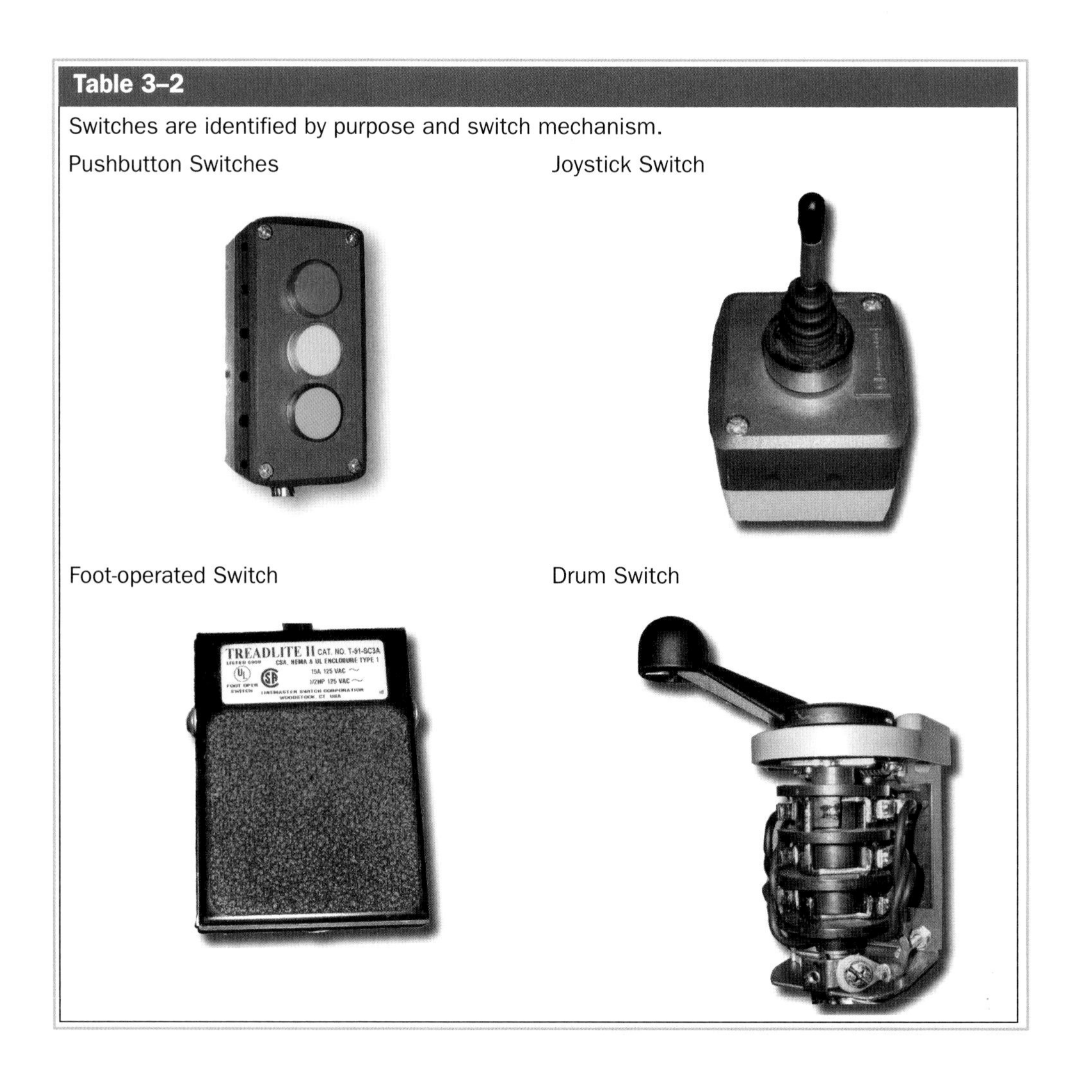

nal to the electromagnetic coil, the electromagnet releases and the relay contacts are opened by spring action. In this situation, one circuit can relay the operation to another circuit. Circuit 1 must be complete to have circuit 2 operate. See Figure 3–8 for a relay type circuit.

SWITCH RATINGS

Switches are designed with ratings to comply with the circuits they must make and break. Standard wall switches used in residential application must meet the voltage and current specifications of the load they are meant to control. A typical switch might be rated for 15 A at 120 VAC. This indicates that the switch will safely conduct 15 A of current when it is closed and can safely break 120 V when it opens. The current rating is used to design the conducting surface and material of the switch contacts. As you know, current through a device can cause heat. There needs to be enough conducting material to allow the current to flow easily without overheating the contacts. The voltage rating tells us that the switch can safely break the circuit without a destructive arc.

Whenever we try to open a circuit while current is flowing, the current tries to continue flowing even if it has to flow through the air in an ionized path or an arc. This arc will continue to allow current to flow until we separate the switch contacts far enough that the arc is extinguished—in other words, the ionized path is too difficult to maintain. If the contacts do not move far enough apart, the arc will continue and the circuit current path will continue to flow. The arc is also very hot and will burn and melt the switch contacts.

Relay

Figure 3–7 One type of relay is assembled into a package commonly referred to as an "ice cube."

Relay Circuit

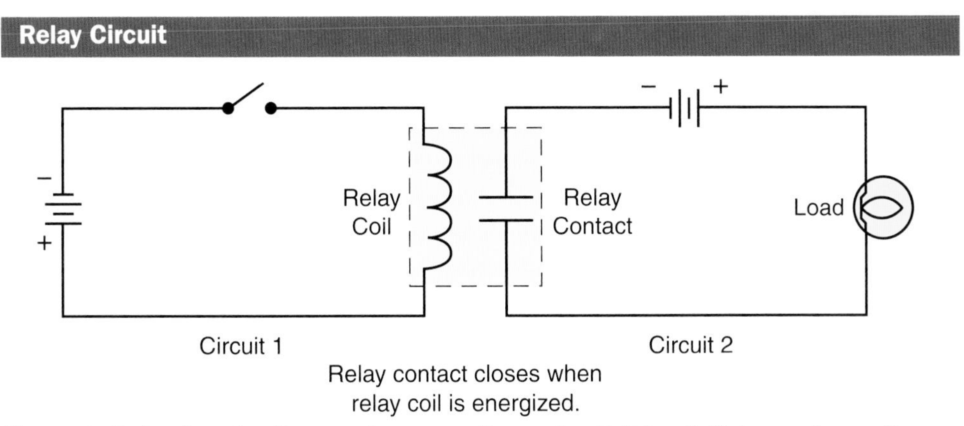

Figure 3–8 A relay circuit can relay one voltage circuit (Circuit 1) to another voltage circuit (Circuit 2).

Relays are very common. One such typical relay is found in a typical heating or cooling system for a home. The fan that moves the heated or cooled air operates at 120 V. The thermostat on the wall has only 24 V connected for power. As the thermostat calls for heat, a contact closes that completes a circuit to a fan relay. The fan relay has a 24 V electromagnet coil that pulls a switch contact closed in the 120 V fan circuit (Figure 3–9). The thermostat has relayed the start signal to the higher-voltage fan motor.

The voltage rating determines the distance we must separate the contact surfaces. The higher the voltage, the farther we must separate the contacts.

Another consideration for switches is the AC/DC rating. Because DC is a continuous current flow until it is interrupted, the arc produced by DC is more destructive than an AC arc of the same voltage and current. The arc formed when breaking a DC flow is more difficult to extinguish. As you will notice on many switch ratings, the AC ratings might be higher than the DC ratings. Some switches are rated for AC only and cannot be used in DC circuits. This rating has to do with the mechanical design of the switch. A DC switch's contacts must move apart faster and farther than an AC switch with the same current and voltage rating. The same switch can handle more AC values than DC values for the same switching action (Figure 3–10).

SWITCH CONNECTIONS IN CIRCUITS

Switches are used throughout the electrical industry to make and break circuits from millivolts and microamps to kilovolts and kiloamps. The methods of making circuits or breaking circuits may vary, and the medium that separates the conducting contacts may vary from air to sulfur hexafluoride to a vacuum. We will concentrate on typical switches you will encounter on a routine basis.

Fan Relay

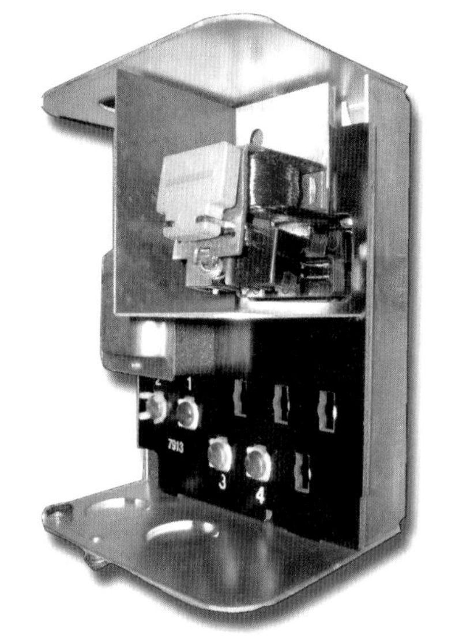

Figure 3–9 A fan relay in a furnace relays a 24 V circuit action to a 120 V circuit to turn on a fan.

Rated Switch

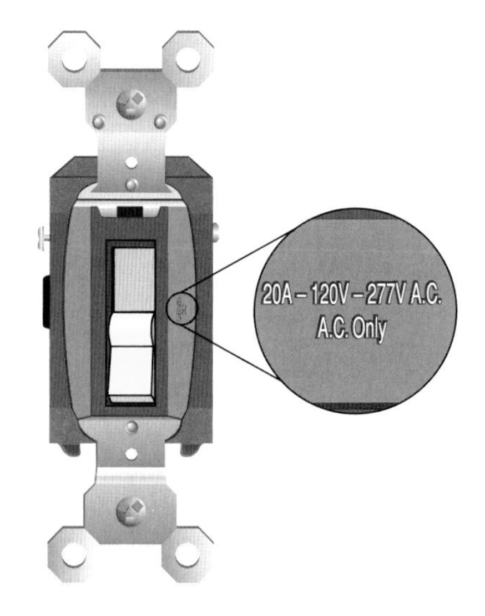

Figure 3–10 A switch is rated for maximum current and maximum voltage.

We will use a simple lamp as a load to illustrate the circuit patterns in the schematic view of the switches. An SPST switch is a simple schematic.

When controlling a light load from two positions, we use two three-way switches. Following the switch positions, we see that either switch can make or break the circuit. The nomenclature of "three-way" seems to come from the fact that there are three terminals on the switch even though it controls from only two locations. This misnomer of three-way has stuck as a switch identity. An identifying feature of this toggle switch is its three terminals, one a more darkly marked terminal. The switch handle has no indication if it is on or off because it cannot be established without knowing the position of the second switch.

Note that in the standard switch pattern the current is brought to the first switch's common terminal (Figure 3–11). It has a common connection to the other two screw terminals. This common terminal is also called the *shunt* terminal. The term comes from the idea that this terminal can shunt, or direct, the current to either of the other two terminals. The other two terminals are referred to as *traveler* terminals. They are connected to the traveler wires that travel between the two switches to carry the current to the next three-way switch. The current is then shunted to the common terminal and continues on the path to the load. See Figure 3–12 for possible switch combinations.

Variations of this "two-switch" (two three-ways) are sometimes known as the "California three-way." In this pattern, an advantage is that it can provide a hot and a neutral conductor to a remote location as well as switch a light circuit. The concepts of hot conductor and neutral conductor are explained in detail later. At this point, we can think of a hot conductor as the supply to the load and the neutral as the return from the load. In typical wiring, we need to keep these designations uniform. See Figure 3–13 for a "California three-way" switch pattern.

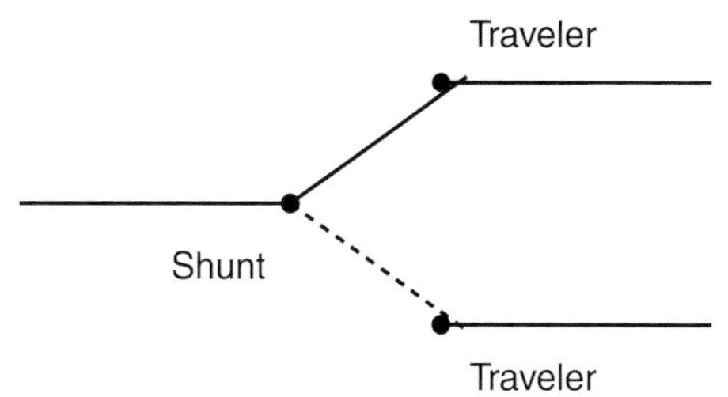

Figure 3–11 A three-way switch has one shunt terminal and two traveler terminals.

Possible Switch Connections

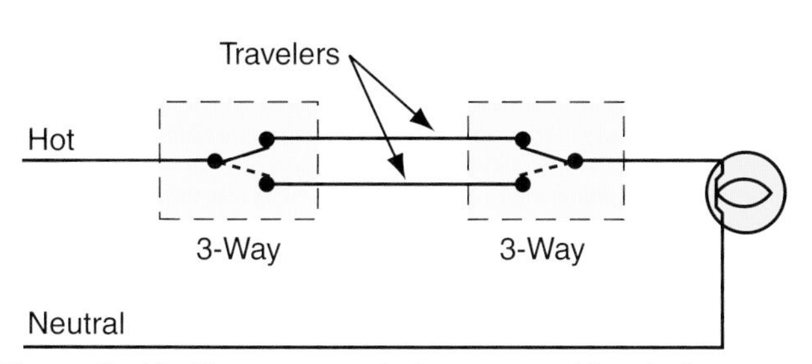

Figure 3–12 Three-way switches control loads from two locations by shunting current to travelers.

"California Three-Way" Switch

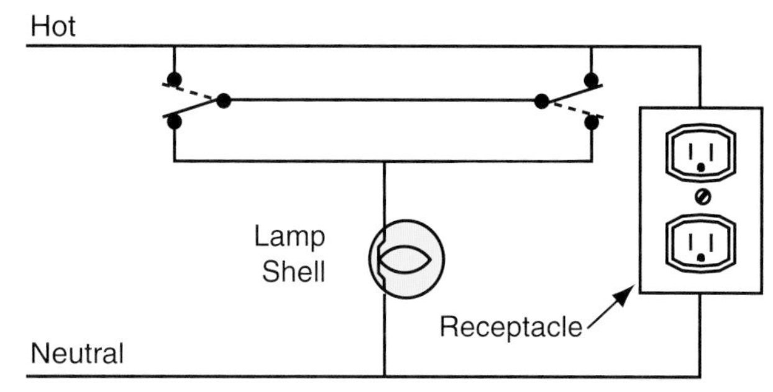

Figure 3–13 A "California three-way" is a unique connection for three-way switches.

Still another variation you might see is sometimes referred to as an "illegal three-way" (Figure 3–14). This pattern was used for years in the wiring of houses. It saved a wire where the light on a stairway was between the switch at the bottom and the switch at the top of the stairs. This switch pattern is now illegal because it changes the polarity of the circuit on the screw shell of the lamp holder. (You will study the code requirements later.) This is not safe and has been rejected by the *NEC*® for many years. We mention it here because it can be confusing if you are troubleshooting or remodeling switch circuits with old wiring systems.

The last standard circuit involves the use of two three-way switches and one or more four-way switches. The misnomer of "four-way" is thought to have started because of the four terminals on the switch. It really has only two positions of connection. This pattern allows you to switch loads from more than two locations. As you can see in Figure 3–15, the current enters the first three-way shunt terminal and is shunted to one of the travelers. The traveler carries the current to the traveler terminals of a four-way. The four-way directs the current to another set of traveler wires and the current continues to the second three-way traveler terminals.

The current is shunted back to the common terminal and continues on its way to the load. Note in Figure 3–15 that there are many combinations that will complete the circuit. The four-way switch does not have a marking on the handle of ON/OFF because it cannot be determined unless the position of the other switches is known. In Figure 3–16, the addition of another four-way into the traveler circuit will create another control point. In fact, we can add as many four-way switches into the traveler circuits as we want and switch the load from any number of points. We always start and end the switch patterns with three-ways.

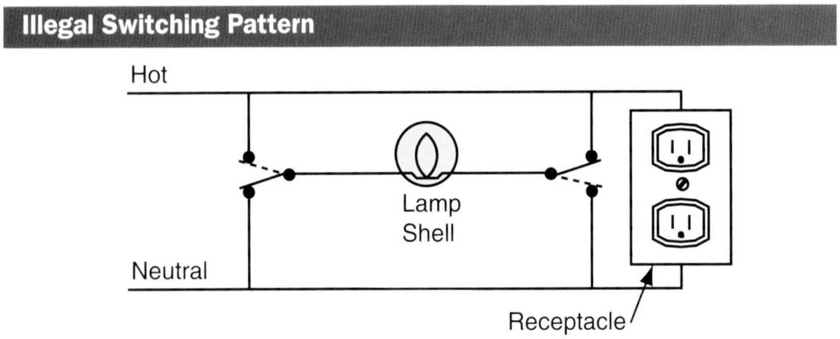

Illegal Switching Pattern

Hot

Lamp Shell

Neutral

Receptacle

Figure 3–14 This now illegal, three-way switching pattern was used in the past.

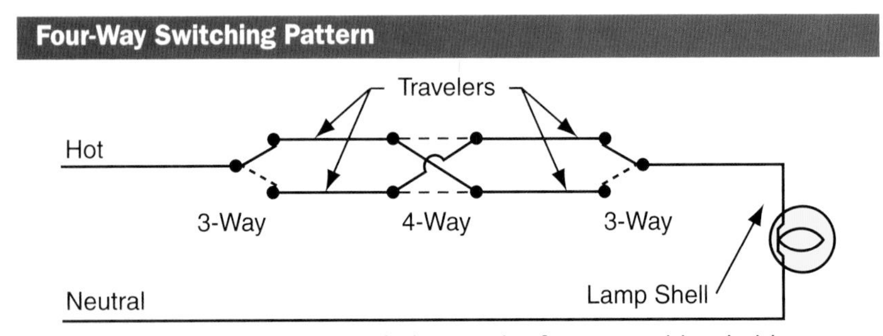

Figure 3–15 Two three-way switches and a four-way add switching locations.

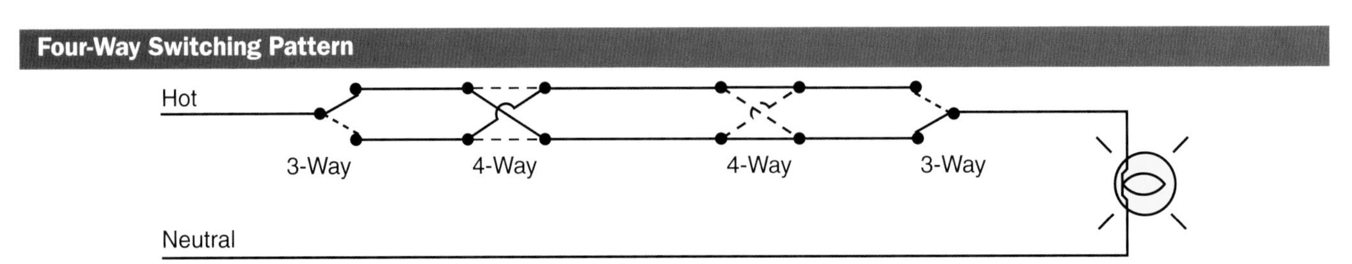

Figure 3–16 As in Figure 3–15, any number of four-way switches can be added to increase switching locations.

TROUBLESHOOTING SWITCH PROBLEMS

When troubleshooting electrical systems, frequent sources of a problem are the switching systems or connection points. When troubleshooting switch problems, familiarity with switch characteristics is helpful. When a switch is closed and the contacts are making solid contact, the opposition of the connection should be near 0 ohms. When measuring across the switch out of the circuit, the resistance measured with an ohmmeter (as described in Chapter 1) will read near 0 ohms with the switch closed (on) and infinite ohms with the switch open (off).

With this knowledge, and using Ohm's law and a voltmeter, you can diagnose switch problems. When the voltage is read across an open switch (Figure 3–17) and the power is on, the voltmeter will read approximately source voltage. The rationale is that there are so many ohms in an open switch that current cannot flow, and there is no difference of potential between the source and the switch terminals. With the switch closed, the resistance of the switch is essentially 0 ohms and no voltage drops across the switch.

Occasionally, the switch will close but not make good electrical connection, creating a high-resistance connection. In this case, the switch will actually drop voltage and reduce the current to the load. The switch will also become warm or hot to the touch. A faulty switch is the problem. Sometimes the mechanism fails to close the switch even when the indicator shows that it is on. Near-line voltage measured across the switch indicates that the switch is open. See if you can diagnose the problem in Figure 3–18 with the voltmeter readings. **Answer: With the voltmeter reading line voltage and the lamp not lit, we can conclude that the three-way switch is faulty and has an open circuit.**

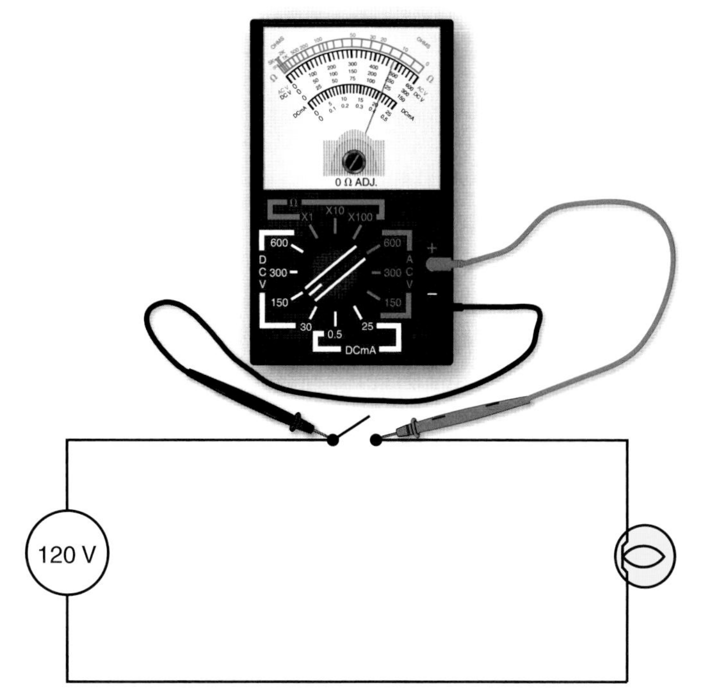

Figure 3–17 In an energized circuit, a voltage reading across a switch indicates that it is open.

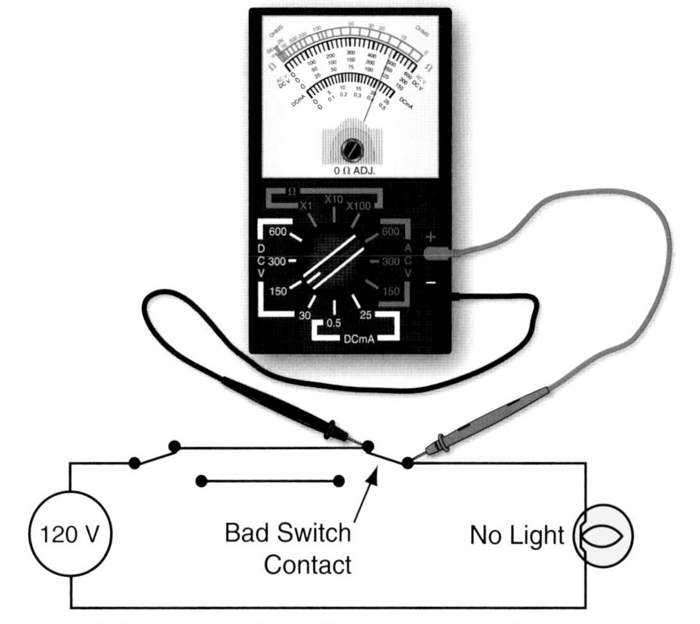

Figure 3–18 Troubleshooting with a voltmeter will indicate where a switch might be open, even though it may appear to be closed.

Conductor size in the trade is expressed in a shortened version. Electricians will just ask for a number 12 or 14 as a standard size. To avoid confusion when asking for a bigger wire, meaning a larger diameter wire, and receiving a larger number wire, just state the wire size needed. Wires *larger than* AWG #6 are normally stranded. This means that #6 *and smaller* can be stranded or solid, and thus when ordering smaller wire, you should specify either stranded or solid.

As the wires increase in size above AWG 1, the numbering system goes to 1/0. This is stated as one-ought. The next size is 2/0 or two-ought, then 3/0 or three-ought, and then 4/0 or four-ought. The wire sizes larger than that are simply referred to by the 1,000 CM. For example, 250 KCM simply means 250,000 CM.

CONDUCTOR SIZE

Between the power supply, the switches, and the load, conductors must be sized to carry the expected current. In the United States, conductors are sized according to the American wire gauge (AWG) standard or by the actual number of circular mil (CM) for the wire. In other countries, the measure is often based on millimeters squared. In the *NEC*®, both measurements are given. CM is a conventional electrical measuring system designed especially for circular (round) wire. A conductor that is 0.001 inches in diameter is considered 1 CM, or 1 mil-inch in diameter.

CROSS-SECTIONAL AREA OF CONDUCTORS

The cross-sectional area of American conductors is given in CM (see Figure 3–19). One mil is equal to 0.001 inch. The area of the conductor is calculated as the square of the diameter. This is the national standard agreed upon by the American National Standards Institute (ANSI). For example, if the diameter of a given conductor is equal to 64.11 mils (0.06411 inch) the cross-sectional area is equal to $(64.11)^2$, which is equal to 4,110 CM.

For larger conductor sizes, it is more convenient to express the cross-sectional area in thousands of CM. This measurement used to be called an MCM, where M stood for 1,000 and CM was circular mils. You will find this designation of MCM, still meaning 1,000 CM, on conductors that were installed some time ago. Modern usage refers to it as a KCM, which is a more conventional denotation for 1,000 CM. Table 3–3 shows that American conductor sizes larger in diameter than 4/0 are actually given in KCM;

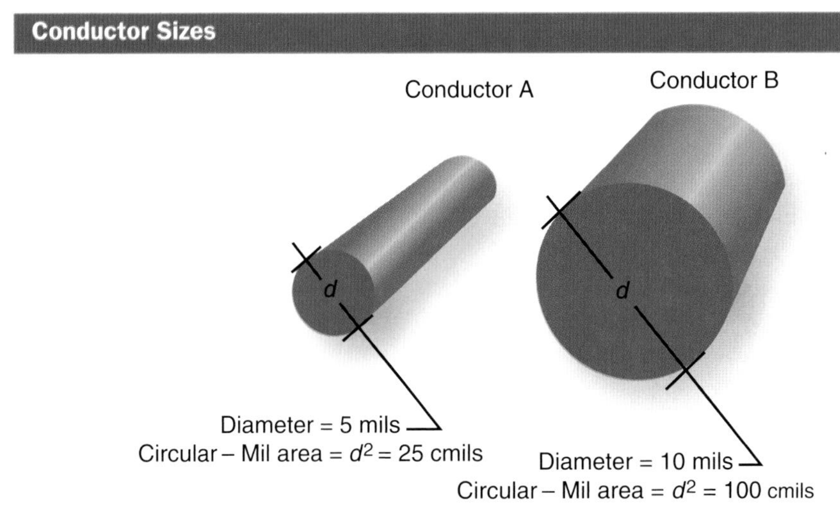

Figure 3–19 A conductor's cross-sectional area is measured in mil-inches squared, or CM.

4/0 and smaller conductors are given in AWG sizes, so cross-sectional areas must be looked up or remembered.

To calculate the cross-sectional area of a conductor in CM, you measure the diameter of the conductor in inches, multiply the result by 1,000, and then square it. For example, a conductor that measures 0.25 inch in diameter of conducting cross-sectional area is 0.25 inch $\times$ 1,000 = 250 mils in diameter $(250 \text{ mils})^2 = 62{,}500$ CM. According to Table 3–3, the wire is about an AWG #2.

Table 3–3

This table compares AWG size numbers with CM areas.

Size (AWG)	Area (CM)	Size (AWG/KCM)	Area (CM)
18	1,620	250	250,000
16	2,580	300	300,000
14	4,110	350	350,000
12	6,530	400	400,000
10	10,380	500	500,000
8	16,510	600	600,000
6	26,244	700	700,000
4	41,740	750	750,000
3	52,620	800	800,000
2	66,360	900	900,000
1	83,690	1,000	1,000,000
1/0	105,600	1,250	1,250,000
2/0	133,100	1,500	1,500,000
3/0	167,800	1,750	1,750,000
4/0	211,600	2,000	2,000,000

Example

You measure a conductor diameter and find that it is 0.5 inch. What is the cross-sectional area of the conductor?

Solution:

This would be calculated as follows:

$$A_{CM} = (d_{(inches)} \times 1{,}000)^2$$

$$A_{CM} = (0.5 \times 1{,}000)^2$$

$$A_{CM} = (500)^2$$

$$A_{CM} = 250{,}000 \text{ CM}$$

Note that 250,000 CM = 250 KCM.

As the world becomes smaller, American electricians will see more international units of measure being used. Whereas ANSI and the National Electrical Manufacturers Association (NEMA) determine many of the standards for U.S. manufacturing, much of the world's electrical manufacturing follows standards from the International Electrotechnical Commission (IEC; headquartered in Europe). Internationally, most conductor areas are expressed in square millimeters (mm^2). In this system, the area is calculated using the familiar formula $A = \pi r^2$, where r is the radius in millimeters and $\pi = 3.14159$.

By way of comparison, a wire that is 150 mm^2 is almost 300,000 cm, or 300 KCM. The cross-sectional areas of IEC conductors are actually given in square millimeters, such as 100, 150, or 200. To calculate the cross-sectional area of a conductor in mm squared, you measure the diameter in mm, divide it by 2, square it, and multiply the result by π.

Example

You measure a conductor using a millimeter scale and find that the conductor diameter is very close to 16 mm. What is the cross-sectional area in mm^2? (Note that if d is the diameter, the radius is $r = d/2$.)

Solution:

This would be calculated as follows:

$$A = \left(\frac{d}{2}\right)^2 \times \pi$$

$$A = \left(\frac{16}{2}\right)^2 \times \pi$$

$$A = 8^2 \times \pi$$

$$A = 64 \times 3.14159$$

$$A = 201 \text{ mm}^2$$

This conductor is probably a 200 mm^2 conductor because that is a standard metric size.

VARIATION OF RESISTANCE WITH AREA

All conductors have some resistance to current flow. The resistance is directly proportional to length of the conductor. In other words, the longer the conductor, the more the internal resistance increases. As current flows through the conducting atoms, some of the energy is given up as the electrons strike each other. This effect of resisting the free flow of electrons is the effect of resistance.

Another variation is that the resistance of a conductor is inversely proportional to its area. This is because a larger cross-sectional area will have more valence electrons available to participate in the current flow. Because the conductor has more conducting atoms, there is a lesser effect of resistance based on the CM of the conductor. Therefore, a 500 KCM conductor that is 1,000 feet long will have half as much resistance as the same 1,000 feet of a 250 KCM conductor that is half the conducting area in cross-sectional area.

Cross-sectional area is just the conducting area of the conductor as viewed at the end of the wire. Examine Table 3–4 and compare the resistance of 1,000 feet of a 1/0 copper conductor at 68°F at 0.0983 Ω to the resistance of 1,000 feet of a 4/0 conductor at the same temperature at 0.049 Ω. A 4/0 wire is approximately twice the CM of a 1/0 conductor. If you check the chart, every fourth wire size difference is approximately twice the CM of the previous size.

For instance, an AWG #14 has 4,107 CM. Another four sizes larger is an AWG #11 at 8,234 CM, and an AWG #8 is yet another four sizes larger at 16,510 CM. Note that the physical cross-sectional area gets larger, the smaller the wire number. This occurs until size 0, then two zeros or 2/0, then three zeros or 3/0, and finally four zeros or 4/0. After that, the size is actually measured in thousands of CM, starting at 250 KCM up to 2,000 CM.

TechTip!

Some wires, or conductors, rather than being one solid piece of copper or aluminum, are actually made up of many separate small-gauge wires, bundled together to form one "stranded" conductor.

Small signals that pass through fine stranded wire cables of this type may exhibit variations in amplitude. Even though the cable appears to be in good condition, it may be defective. Over time, especially where stretching and bending occur, strands of a wire conductor may break. These broken strands "make," and "break" as environmental conditions change. The amplitude fluctuations, or noise, experienced at the cable output results from the variations of series-circuit resistance within the faulty conductor. Conductor corrosion may produce the same outcome.

MATERIAL TYPE EFFECTS ON CONDUCTOR RESISTANCE

In your study of the structure of atoms, you found that each atom has a different atomic composition and that one characteristic of each atom is its unique electron configuration. Because of the varying atomic structures of different materials classified as conductors, each of these materials has a different resistance to current flow. To perform calculations to determine circuit resistance for a conductor, some information must be known about the particular properties of the material being used as the conductor. The necessary information is expressed as specific resistance.

Table 3–4

			Conductors							Direct-Current Resistance at 75°C (167°F)					
			Stranding			Overall				Copper					
Size	Area			Diameter		Diameter		Area		Uncoated		Coated		Aluminum	
(AWG or kcmil)	mm^2	Circular mils	Quantity	mm	in.	mm	in.	mm^2	$in.^2$	ohm/km	ohm/kFT	ohm/km	ohm/kFT	ohm/km	ohm/kFT
18	0.823	1620	1	—	—	1.02	0.040	0.823	0.001	25.5	7.77	26.5	8.08	42.0	12.8
18	0.823	1620	7	0.39	0.015	1.16	0.046	1.06	0.002	26.1	7.95	27.7	8.45	42.8	13.1
16	1.31	2580	1	—	—	1.29	0.051	1.31	0.002	16.0	4.89	16.7	5.08	26.4	8.05
16	1.31	2580	7	0.62	0.024	1.85	0.073	2.68	0.004	10.3	3.14	10.7	3.26	16.9	5.17
12	3.31	6530	1	—	—	2.05	0.081	3.31	0.005	6.34	1.93	6.57	2.01	10.45	3.18
12	3.31	6530	7	0.78	0.030	2.32	0.092	4.25	0.006	6.50	1.98	6.73	2.05	10.69	3.25
10	5.261	10380	1	—	—	2.588	0.102	5.26	0.008	3.984	1.21	4.148	1.26	6.561	2.00
10	5.261	10380	7	0.98	0.038	2.95	0.116	6.76	0.011	4.070	1.24	4.226	1.29	6.679	2.04
8	8.367	16510	1	—	—	3.264	0.128	8.37	0.013	2.506	0.764	2.579	0.786	4.125	1.26
8	8.367	16510	7	1.23	0.049	3.71	0.146	10.76	0.017	2.551	0.778	2.653	0.809	4.204	1.28
6	13.30	26240	7	1.56	0.061	4.67	0.184	17.09	0.027	1.608	0.491	1.671	0.510	2.652	0.808
4	21.15	41740	7	1.96	0.077	5.89	0.232	27.19	0.042	1.010	0.308	1.053	0.321	1.666	0.508
3	26.67	52620	7	2.20	0.087	6.60	0.260	34.28	0.053	0.802	0.833	0.254	0.245	1.320	0.403
2	33.62	66360	7	2.47	0.097	7.42	0.292	43.23	0.067	0.634	0.194	0.661	0.201	1.045	0.319
1	42.41	83690	19	1.69	0.066	8.43	0.332	55.80	0.087	0.505	0.154	0.524	0.160	0.829	0.253
1/0	53.49	105600	19	1.89	0.074	9.45	0.372	70.41	0.109	0.399	0.122	0.415	0.127	0.660	0.201
2/0	67.43	133100	19	2.13	0.084	10.62	0.418	88.74	0.137	0.3170	0.0967	0.329	0.101	0.523	0.159
3/0	85.01	167800	19	2.39	0.094	11.94	0.470	111.9	0.173	0.2512	0.0766	0.2610	0.0797	0.413	0.126
4/0	107.2	211600	19	2.68	0.106	13.41	0.528	141.1	0.219	0.19996	0.0608	0.2050	0.0626	0.328	0.100
250	127	—	37	2.09	0.082	14.61	0.575	168	0.260	0.1687	0.0515	0.1753	0.0535	0.2778	0.0847
300	152	—	37	2.29	0.090	16.00	0.630	201	0.312	0.1409	0.0429	0.1463	0.0446	0.2318	0.0707
350	177	—	37	2.47	0.097	17.30	0.681	235	0.364	0.1205	0.0367	0.1252	0.0382	0.1984	0.0605
400	203	—	37	2.64	0.104	18.49	0.728	268	0.416	0.1053	0.0321	0.1084	0.0331	0.1737	0.0529
500	253	—	37	2.95	0.116	20.65	0.813	336	0.519	0.0845	0.0258	0.0869	0.0265	0.1391	0.0424
600	304	—	61	2.52	0.099	22.68	0.893	404	0.626	0.0704	0.0214	0.0732	0.0223	0.1159	0.0353
700	355	—	61	2.72	0.107	24.49	0.964	471	0.730	0.0603	00184	0.0622	0.0189	0.0994	0.0303
750	380	—	61	2.82	0.111	25.35	0.998	505	0.782	0.0563	0.0171	0.0579	0.0176	0.0927	0.0282
800	405	—	61	2.91	0.114	26.16	1.030	538	0.834	0.528	0.0161	0.0544	0.0166	0.0868	0.0265
900	456	—	61	3.09	0.122	27.79	1.094	606	0.940	0.0470	0.0143	0.0481	0.0147	0.0770	0.0235
1000	507	—	61	3.25	0.128	29.26	1.152	673	1.042	0.0423	0.0129	0.0434	0.0132	0.0695	0.0212
1250	633	—	91	2.98	0.117	32.74	1.289	842	1.305	0.0338	0.0103	0.0347	0.0106	0.0554	0.0169
1500	760	—	91	3.26	0.128	35.86	1.412	1011	1.566	0.02814	0.00858	0.02814	0.00883	0.0464	0.0141
1750	887	—	127	2.98	0.117	38.76	1.526	1180	1.829	0.02410	0.00735	0.02410	0.00756	0.0397	0.0121
2000	1013	—	127	3.19	0.126	41.45	1.632	1349	2.092	0.02109	0.00643	0.02109	0.00662	0.0348	0.0106

Notes:
1. These resistance values are valid only for the parameters as given. Using conductors having coated strands, different stranding type, and, especially, other temperatures changes the resistance.
2. Formula for temperature change: $R_2 = R_1 [1 + \alpha(T_2 - 75)]$ where $\alpha_{cu} = 0.00323$, $\alpha_{AL} = 0.00330$ at 75°C.
3. Conductors with compact and compressed stranding have about 9 percent, respectively, smaller bare conductor diameters than those shown. See Table 5A for actual compact cable dimensions.
4 The IACS conductivities used: bare copper = 100%, aluminum = 61%.
5. Class B stranding is listed as well as solid for some sizes. Its overall diameter and area is that of its circumscribing circle.

FPN: The construction information is per NEMA WC8-1992 or ANSI/UL 1581-20001. The resistance is calculated per National Bureau of Standards Handbook 100, dated 1966, and Handbook 109, darted 1972.

Specific resistance is also called **resistivity** and is usually abbreviated with the Greek letter ρ (rho) or by the capital letter *K*. The *K* of materials is listed for English measurements as ohms per mil-foot. This *K* tells us what the resistance in ohms would be for a conductor one mil in diameter that is 1 foot long. Actual resistance of the same material conductor is always in proportion to this value.

Although you could experimentally measure these values, they have already been determined for materials that may be used to conduct electricity. Table 3–5 lists the resistivity for many common materials in both the ANSI (American or British) and IEC (metric) systems. Once the specific resistance (*K* value) for a conductor material is known, that value can be used in determining the overall resistance of any conductor made from the same material.

The relationship is such that the overall resistance of a conductor is proportional to the specific resistance of that conductor. Because the cross-sectional area of IEC standard conductors is specified in mm^2, Table 3–5 shows metric values of ρ in terms of a conductor that is 1 meter long and 1 millimeter in diameter (called 1 millimeter-meter).

Table 3–5

This table shows the resistivity or ohms per mil-foot for different conducting materials.

K (ρ) at 68°F (20°C)			
Material	**American (English) Ω/Mil-Foot**	**Metric Ω/Millimeter-meter**	**Resistance Temperature Coefficient (Ω/°C)**
Aluminum	17.7	0.0265	.004308
Copper	10.4	0.0168	.004041
Lead	126.0	0.22	.0043
Mercury	590.0	0.98	.00088
Nichrome	600.0	1.00	.00017
Platinum	66.0	0.106	.003729
Silver	9.7	0.0159	.003819
Tungsten	33.8	0.056	.004403

MATERIAL LENGTH EFFECTS ON CONDUCTOR RESISTANCE

For any given conductor, the resistance of that conductor is directly proportional to the length of the conductor. This means that the resistance for that conductor increases as the length increases. If a certain conductor has a resistance (*R*) for a given length of that conductor, a conductor of the same diameter and material that is two times longer than the original length will have a resistance two times the original resistance (or 2*R*). Conductors 5 and 10 times longer will have respective resistances of 5*R* and 10*R*.

Example

If 500 feet of #14 copper conductor has 1.57 ohms, find the resistance of 750 feet of the same conductor. Use a direct proportion ratio.

Solution:

$$\frac{1.57\ \Omega}{500\ \text{ft}} = \frac{x\ \Omega}{750\ \text{feet}}$$

x = 2.36 Ω for 750 feet of the same size and type of wire.

TEMPERATURE EFFECTS ON CONDUCTOR RESISTANCE

The specific resistance of any conductor is related to the temperature of that conductor. That is, the specific resistance changes as the temperature changes. Specifically, the resistance increases if the temperature coefficient is positive, and the resistance decreases if the temperature coefficient is negative. Most conductors have positive temperature coefficients, whereas most insulators have negative temperature coefficients. Carbon is one of the few materials classified as a conductor that has a negative temperature coefficient. Table 3–5 lists the specific resistances for the various materials when their temperature is 68°F (20°C).

Example

Find the resistivity of copper at 90°F.

Solution:

Because Table 3–5 gives the temperature coefficient in $\frac{\Omega}{°C}$, it will be easier if you first convert 90°F to degrees Celsius (°C). This can be done with the following formula:

$$°C = \frac{5}{9} \times (°F - 32)$$

$$°C = \frac{5}{9}(90 - 32) = 32.22°C$$

The resistivity of copper at 20° C is 0.0168, as indicated in Table 3–5. Because the coefficient for copper is 0.004041 ohms for every degree above 20°C, the change in resistivity will be:

$$0.004041 \times (32.22° - 20°) = 0.04938102$$

Because the coefficient is positive, you add it to the coefficient at 20°C. Thus, the resistivity of copper at 90°F = 0.0168 + 0.04938102 = 0.06618.

DETERMINING THE TOTAL RESISTANCE OF A CONDUCTOR

Once all information is gathered on a specific conductor's physical attributes, it becomes a simple task to determine the resistance for any length of that conductor. The resistance (R) in ohms is defined as being equal to the specific resistance (K or ρ) times the length (L) divided by the area (A). When written out, the formula takes the following form:

$$R = K \times \frac{L}{A} \text{ or } R = \frac{K \times L}{A}$$

You will also see it as $R = K \times \frac{L}{CM}$ or $R = \frac{K \times L}{CM}$. Note that you must use the proper units for the type of conductor with which you are working. For example, in the American system you must use the English value for K, feet for L, and CM for A. In the metric (IEC) system, use the metric value of K, meters for L, and square millimeters for A.

When the conductor material, length, and area are known for a conductor, the resistance is found by substituting values into the formula and then solving for the resistance. The AWG information in Table 3–4 shows characteristics of solid copper conductor up to 4/0 in size.

Example

If a 6-gauge conductor is made of copper and is 1,000 feet long, what is the resistance of the conductor at 68°F?

Solution:

$$R_{(Ohms)} = \frac{\Omega/\text{mil foot} \times 1{,}000}{\text{Area in } CM}$$

$$R_{(Ohms)} = \frac{10.4 \times 1{,}000}{26{,}250}$$

$$R_{(Ohms)} = \frac{10{,}400}{26{,}250} = 0.396\ \Omega$$

This same formula is used quite frequently in a different form for determining voltage drops in conductor calculations. Because resistance is an essential part of any electrical circuit, the ability to be able to calculate resistance in conductors is an important part of the knowledge required when working with them. Note in Table 3–4 that the resistance of 1,000 feet of wire is listed at different temperatures.

To confirm the previous calculation, read across from the #6 copper conductor to the 68°F column. Listed is the ohmic value of $\frac{0.395\ \Omega}{1{,}000\ \text{feet}}$.

This table value is very close to the calculated value of 0.396 Ω. They often do not match exactly because we round our figures when we calculate to make it easier. If you want to calculate the ohms of 450 feet of the same wire, you use a simple ratio and proportion calculation: $\frac{450\ \text{feet}}{1{,}000\ \text{feet}} = \frac{X}{0.395}$. Because resistance is directly proportional to length, we use a straight proportion. This means that fewer feet will have proportionally less resistance. In this example, 450 feet of #6 wire has 0.178 ohms.

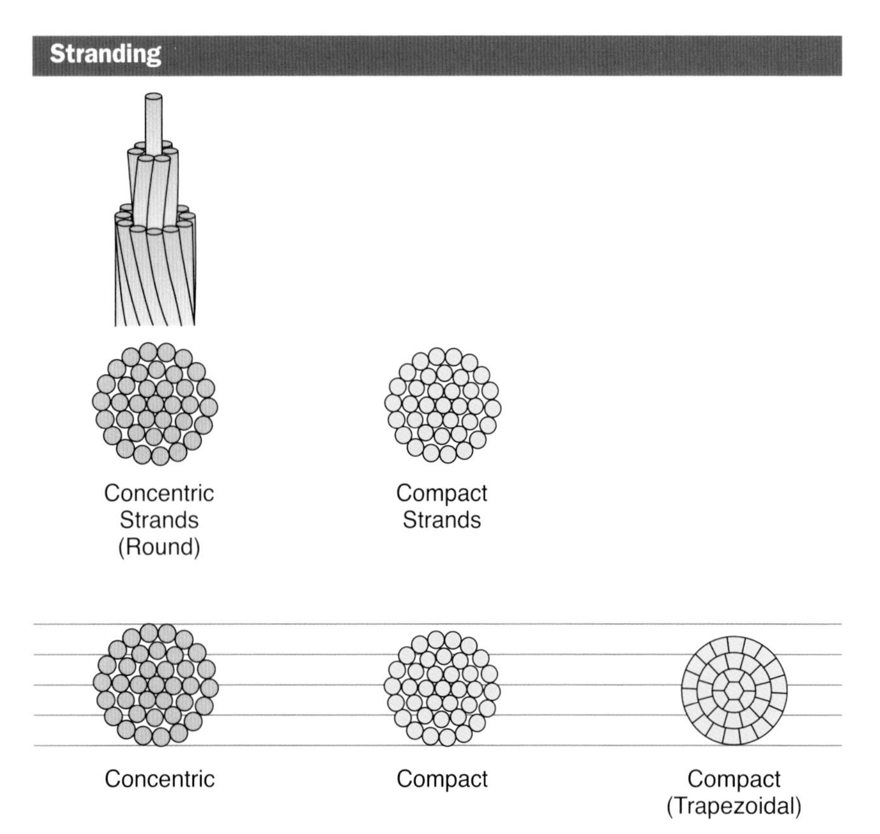

Figure 3–20 Concentric strands take more physical space than compact strands.

CIRCULAR MILS AND SQUARE MILS

Round wire can be either one solid conductor or many strands of smaller wire. In either case, the total cross-sectional area of the conductor must equal the total *CM* area of the wire gauge. The stranding makes the wire easier to bend and pull through conduit. In most cases, the stranded wire is constructed in a process called concentric stranding. This simply means that each layer of the wire has strands added to the original core in concentric circles.

CIRCULAR MILS

The first single core is one wire at the center. The concentric circle has 6 strands, for a total of 7. Then the next circle has 12 additional wires, for 19 strands total. If you refer to Table 8 in Chapter 9 of the *NEC*® you will note that the stranding patterns are consistent. This creates a little larger overall diameter of the wire compared to solid wire because the strands do not fit perfectly in the area of the wire. You will note that the two categories for conductor size listed in the *NEC*® have slightly different overall dimensions.

Sometimes concentric stranded wire may necessitate the use of larger conduit size just to accommodate the larger diameters. In that case, there is another type of stranding called compact stranding (Figure 3–20). The shape of the compact strands is designed to fit exactly in the stranding process and does not provide any air space or wasted unfilled space. These stranded conductors are much like a solid conductor but are still more flexible than a solid conductor.

SQUARE MILS

Conductors are not always round. Often large conductors that carry higher current are rectangular. These conductors are usually identified as bus bars (Figure 3–21).

These conductors are found in large service equipment, switch gears, and motor control centers. Because rectangular bars do not fit the definition of *CM*, we can convert the cross-sectional area measured in square mils to *CM* for use in our normal formulas.

To convert from square mils to *CM*, we use a constant of 0.7854. The area in square mils is first calculated by measuring the length and width of the bar's cross-section in mil-inches. That is, the number of inches × 1,000 equals mil-inches. The length in mil-inches times the width in mil-inches equals square mils (Figure 3–21).

To convert square mils to *CM*:

$$\frac{\text{Square mils}}{0.7854} = CM$$

The *CM* of a circular conductor 1 mil in diameter is 1 *CM*. The area of a square conductor 1 mil on a side is 1 square mil.

$$\text{Area of a circle is } \pi r^2 \text{ or } \pi \left(\frac{d}{2}\right)^2 \text{ or } \frac{\pi d^2}{4}$$

$$\text{Square area of a circle} = \frac{d^2 \pi}{4}$$

The square mil area of a circle 1 mil in diameter is $\frac{1^2 \pi}{4}$ or $1^2 \times \left(\frac{3.1416}{4}\right)$ or $1^2 \times 0.7854$. Therefore, the actual area of a circular shape that is 1 mil in diameter has 0.7854 × the area of a square that is 1 mil across.

Bus Bar

Figure 3–21 A bus bar is a conductor, but not a round conductor made for pulling into place.

CONDUCTOR RESISTANCE, VOLTAGE DROP, AND WATTAGE LOSS

At this point, we will introduce the effects of current through a conductor. As you learned in Chapter 1, with a little exposure to Ohm's law we know that whenever current flows through a resistance, some of the electrical pressure is lost. We use Ohm's law $E = I \times R$ to tell us the voltage *(E)* when we know the current *(I)* and the resistance *(R)*. Now that we have calculated the resistance of the conductor itself, we can determine the amount of voltage dropped in the conductors.

Example

How many *CM* is a rectangular bus bar that measures ½ inch × 3 inches in cross-section?

Solution:

½ inch × 1,000 = 500 mil-inches, 3 inches × 1,000 = 3,000 mil-inches

500 × 3,000 = 1,500,000 sq mils

$$\frac{1{,}500{,}000 \text{ sq mils}}{0.7854} = 1{,}909{,}854\ CM \text{, or nearly } 2{,}000\ KCM$$

This is one of the largest round conductors listed in the conductor tables in the *NEC*®.

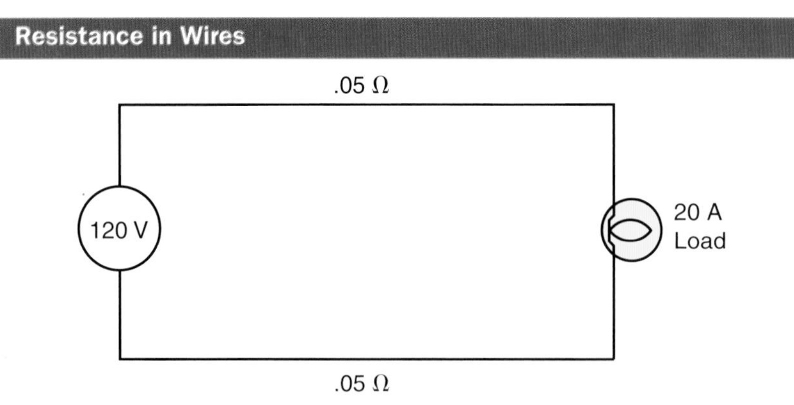

Figure 3–22 A circuit with connecting conductors has resistance in the wires.

This calculation is called line drop (voltage drop), which is due to conductor resistance. We simply need to calculate the resistance of the conductor needed to bring the current from the source to its load and back to the source. Remember that we must have pressure to get current to and from the load. See Figure 3–22 for an example of a simple circuit with resistance in the wires.

The line drop in this simple circuit can be calculated by a formula much the same as the conductor resistance formula. The difference is that we can automatically assume we need two conductors: one to and one from the load. We calculate the resistance of the conductor to the load and double the length to get the total distance. The only other component we need is the actual current in the circuit. The formula now looks as follows:

$$VD = 2 \times \left[\frac{(K \times L)}{CM}\right] \times I$$

VD is the total voltage drop of the line conductors to and from the load.

2 is a constant used to double the resistance one way.

K is the resistivity of the conducting material.

L is the number of feet from the source to the load.

I is the current in amps of circuit current.

CM is the number of CM from Table 3–4.

Example

Find the voltage drop of a circuit with 20 A to a load 500 feet away, using #12 copper wire.

Solution:

$$VD = \frac{2\ KL\ I}{CM}$$

$$VD = \frac{2\ (10.4 \times 500 \text{ feet} \times 20 \text{ A})}{6{,}530\ CM}$$

$$VD = \frac{2\ (104{,}000)}{6{,}530}$$

$$VD = 31.85 \text{ V}$$

The total voltage drop for the entire circuit wire is 31.85 V. This means that 15.925 V is lost to the load and another 15.925 V is dropped on the return conductor.

POWER LOSS IN CONDUCTORS

As discussed in Chapter 1, there is also an energy loss in the form of heat. In Chapter 2, we found that as electrons move through a material they give up energy and it is sensed as heat. Conductors do lose heat and the heat is measured in the form of watts (W). Therefore, the energy lost in moving electricity is in the form of a watt loss.

Conductors do have a watt loss when conducting current. Usually, we do not want the conductor to heat. We deliberately size the conductors large enough so that there is minimum heat loss. However, we can calculate the loss as watts. The term we use for watts lost in the lines is *line loss.* The formula is an adaptation of the watt formula used with the resistance of the lines. Thus, the formula is as follows:

$$P = I^2 \times R_{(\text{line})}$$

P is the power, measured in watts (W), that is lost in the line conductors.

I is the circuit current.

R is the resistance of the conductor involved.

Example

Find the wattage loss in a circuit with 20 A to a load 500 feet away using #12 copper wire.

Solution:

Use the same resistance of the wire in the previous example.

$$R = \frac{KL}{CM}$$

$$R = \frac{10.4 \times 500 \text{ feet}}{6{,}530\ CM}$$

$$R = 0.796\ \Omega \text{ each way}$$

Then:

$$P = I^2 \times R$$

$$P = 20^2 \times 0.796\ \Omega$$

$$P = 318.4$$

The total for the circuit both to and from the load is $2 \times 318.4 = 637$ W.

We can also confirm this figure by using $E \times I$ for the wires involved and using the previous line drop of 31.85 V × 20 A = 637 W for the entire line wires to and from the load that was 500 feet away. This would be considered excessive line drop and line loss. The remedy for this circuit would be to increase the size of the wire. We would use a smaller number on the wire-gauge chart. By increasing the diameter of the conductor, we decrease the resistance of the wire and reduce some of the line drop (voltage) and line loss (wattage).

INCANDESCENT LAMPS

In this chapter you have learned about switches and connecting conductors. An introduction to the actual load on the system and how it interacts with the other parts of the circuit is needed. Incandescent lamps show the relationships among current, voltage, resistance, and power. Lamps have both a voltage rating and a power rating. Common values used in homes are $\frac{120\text{ V}}{100\text{ W}}$, $\frac{120\text{ V}}{75\text{ W}}$, $\frac{120\text{ V}}{60\text{ W}}$, and $\frac{120\text{ V}}{40\text{ W}}$. These lamps are also available in other voltage and wattage ratings. Note that even though incandescent lamps are intended for use in AC circuits, they are rated in such a way that the power and current ratings are the same as in DC circuits.

FieldNote!

There are many different types of power units, and you should be familiar with a few of them. You have covered one already: the watt. However, the watt is a relatively small number when compared to the voltage and current ratings of commercial electrical circuits and components. To make the numbers a little more manageable, the commercial electrical industry has adopted the use of the kilowatt (thousand-watt) unit. One kilowatt is the same as 1,000 watts. An easy relation to this number can be observed in your own home. The rating for incandescent light bulbs is in watts. A standard light might be 100 watts. Thus, you can see that it would not take many lights to use 1 kilowatt (10, to be exact).

RATINGS

Different wattage ratings are achieved for lamps with the same voltage ratings by changing the value of resistance. Higher-wattage bulbs will have a lower resistance and therefore a higher current. Typical resistances include the 100 watt lamp (144 ohms), 75 watt lamp (192 ohms), and 40 watt lamp (360 ohms).

Example

What is the resistance of a 120 V, 60 W lamp?

Solution:

According to the previous information and the resistance quadrant in Figure 1-40, the equation to use is as follows:

$$P = \frac{E^2}{R}$$

or

$$R = \frac{E^2}{P}$$

$$R = \frac{120\text{ V}^2}{60\text{ W}}$$

$$R = 240\ \Omega$$

LOST POWER

In an incandescent lamp, the current flowing through the filament causes the lamp filament to heat to incandescence, which means that it emits visible light by glowing white hot. Generally, the heat generated by the lamp can be classified as a loss because it is unwanted. Because the intended purpose of the lamp is to provide light, not heat, the power loss resulting from this unwanted heat is often identified as an I^2R loss. However, the resistive components in a circuit are the only ones that will generate this heat. Only resistance can use power in an electrical circuit. Even the apparent resistance of an electric motor can be calculated. Consequently, the power used to supply an electric motor can be calculated with these same formulas.

EFFICIENCY

Efficiency has become a very important concept in our modern world. Efficiency is a measure of how much energy (or power) is being used for useful work, such as light, and how much is being wasted. If a 75 W incandescent lamp uses 55 W to generate light and the other 20 W are dissipated as heat, the percentage of efficiency for the lamp would be:

$$\% \text{ Efficiency} = \left[\frac{(75 \text{ W} - 20 \text{ W})}{75 \text{ W}}\right] \times 100$$

$$= 73.3\%$$

Thus, 73.3% of the energy is converted to light and 26.7% of the energy is wasted as heat. From this you can see that quite a large portion of the energy you pay for in your home goes into heating rooms, not lighting them.

The kilowatt is a common unit of electrical power, but it is more helpful for commercial electric companies to know how much energy has been used over a period of time (such as every month). Because the kilowatt is the amount of work being performed per unit time (second), we simply multiply by a unit of time (hour) and derive a new unit: the *kilowatt-hour*. This new unit is really just a different way of saying how much total energy has been used. Each month, the electric company in your area will read the kilowatt-hour meter on your home to let you know (and bill you for) how much energy your home has used.

SUMMARY

This chapter reinforced the concept of complete circuits that are required to connect the power source to the intended load. The circuit is a closed (complete) circuit or an open (incomplete) circuit. A short circuit is a circuit that has maximum current flow but the current does not flow to the right destination; instead, it takes a shorter path back to the supply. To control the circuit current with predictable results, you studied switches that are created for different applications and with different switch patterns. Switches either open or close the circuit for current to flow and must do it safely. Thus, switch ratings must be adhered to.

The switches, power source, and intended load must be connected in order to move the energy from the source to the load. Conductors, usually in the form of round wire and typically made of copper, are the standard conductor. You learned that there are many factors that influence the ability of a wire to conduct current. These include wire diameter, length, material, temperature influence, and the opposition to current flow within a conductor. All conductors have some opposition to current flow.

The resistance of a specific conductor can be calculated by observing the variables. We also found that the resistance did create a difference in potential (voltage) along the length of the conductor. We calculated the line drop as a voltage drop due to conductor resistance and the load current it was carrying. The loss in the conductor is not only in voltage but in wattage loss, called line loss. Our objective is to keep the line drop and line loss to a minimum. Finally, we added a load to the circuit. In this case, the incandescent lighting load is the purpose of all the generation, transmission, distribution, and controls we encounter with electrical systems. This is just an introduction to a complete circuit concept.

CHAPTER GLOSSARY

Relay An electromagnetic coil mechanism that mechanically opens or closes a contact to control a separate electrical circuit.

Resistivity Specific resistance is defined for both American and IEC units. Under American (ANSI) standards, the resistance of a conductor is based on a conductor 1 foot long and 1 millimeter in diameter. Under IEC standards, the resistance of a conductor is based on a conductor 1 meter long and 1 millimeter in diameter.

REVIEW QUESTIONS

1. Define the concept of a short circuit compared to an open circuit.
2. Explain how a schematic diagram can be used when installing a building's wiring.
3. Draw a DPDT switch and explain how many circuits can be completed.
4. Draw the schematic diagram of two three-way switches controlling a lighting load.
5. Explain how you could check a switch using an ohmmeter on a switch out of the circuit.
6. Explain how to calculate the *CM* of a circular wire if you know the diameter in inches.
7. Find the *CM* equivalent of a bus bar that is 0.5×1.5 inches in cross-sectional area.
8. What is the next largest conductor after four-ought?
9. Write the formula for wire resistance. Identify each of the variables.
10. What is a standard resistance in ohms per mil-foot for copper conductors?
11. Find the line drop of an AWG #14 copper wire that is 800 feet long. The current in the wire is 10 A.
12. Find the line loss for the same conditions in question 11.
13. Explain what is meant when we say that a light bulb is an incandescent lamp.
14. What is the lost power in a 60 W lamp at 120 V if it generates 48 W of light energy?
15. A 60 W incandescent lamp operating at 120 V has a current of 0.5 A. What switch ratings would you observe to control this light?

PRACTICE PROBLEMS

Refer to Figure PP3–1 to answer the following questions on troubleshooting a relay circuit.

1. If the relay is operational and the circuit of the relay coil has an open circuit, will the load be on or off?
2. Explain what would happen to the load if the relay contacts were shorted.
3. What would be the result for circuit 2 in the diagram if the load were shorted?
4. Would the contacts move if there were an open in the load circuit?
5. Explain what ratings you might need to know about the relay coil.
6. What ratings would you need to know about the relay contact ratings when it is performing the function of a switch?
7. What might be a problem if the relay seems to operate but the load does not respond?

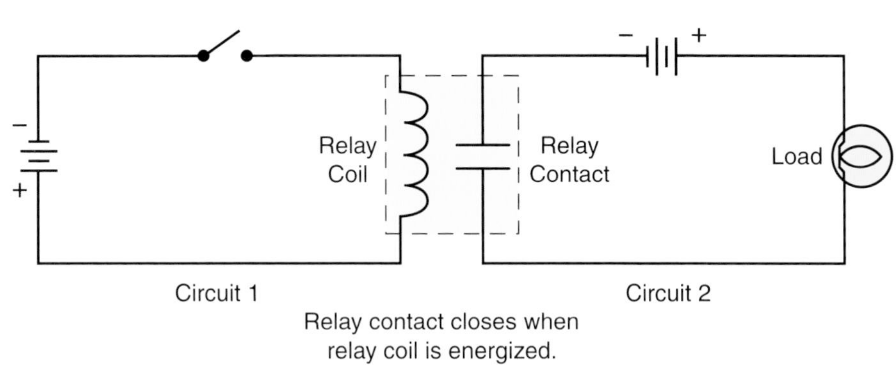

Figure PP3–1

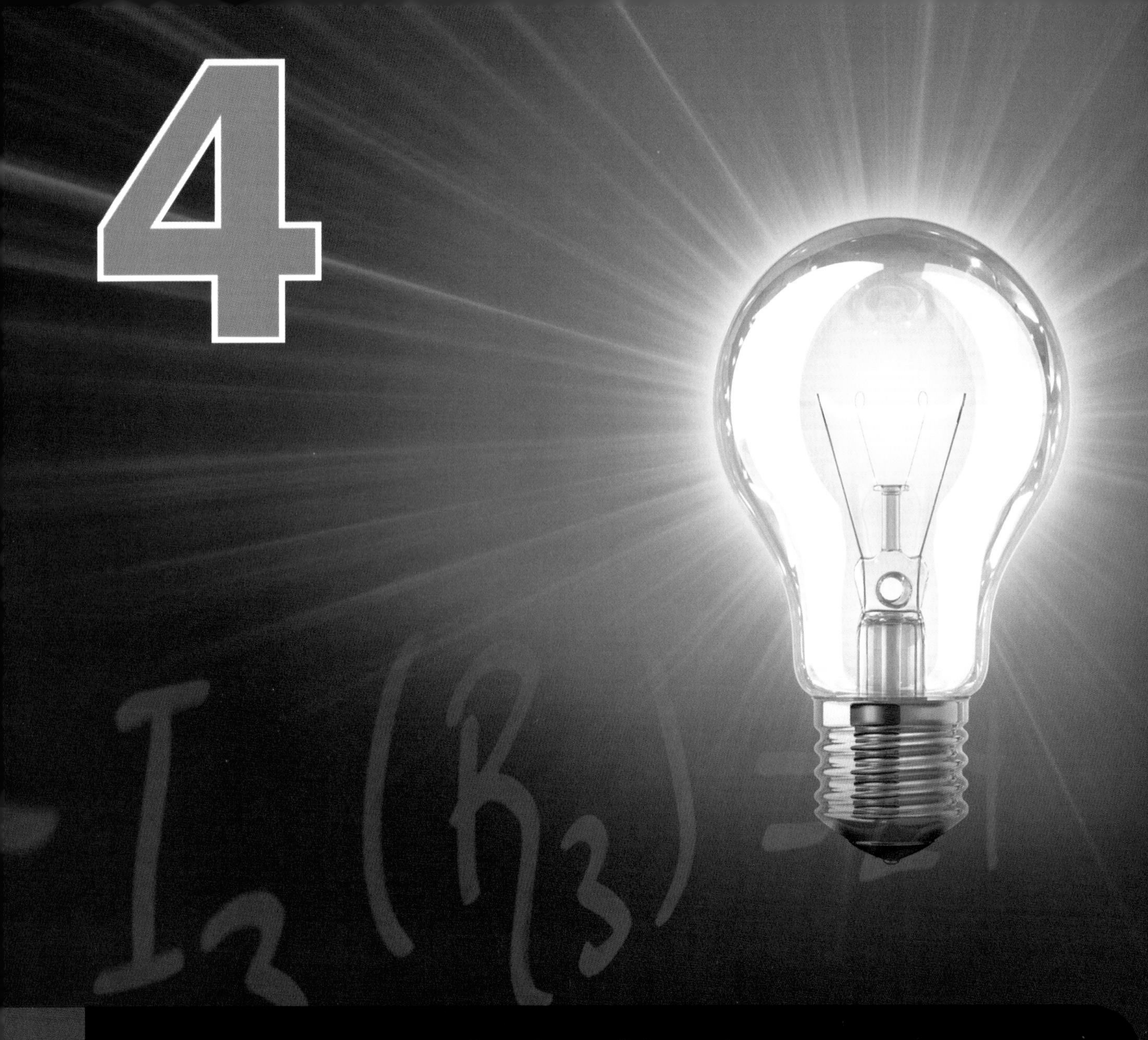

4 Series Circuits

TOPICS

- INTRODUCTION
- DIAGRAMS
- COMPONENT DIAGRAMS
- SCHEMATIC SYMBOLS
- TYPES OF RESISTORS
- FIXED RESISTORS
- VARIABLE RESISTORS
- THE SERIES CIRCUIT
- CALCULATING RESISTANCE IN A SERIES CIRCUIT
- DETERMINING PROPORTIONAL RELATIONS IN A SERIES DC CIRCUIT
- PROPORTIONAL VOLTAGE DROP
- WATTAGE CALCULATIONS IN SERIES DC CIRCUITS
- KIRCHHOFF'S LAWS
- VOLTAGE SOURCES

OVERVIEW

Electrical circuits can be classified into three basic formats: series, parallel, and combination. In this chapter you will learn about the first type, series circuits. A series circuit is one in which all of the various components are connected to form only one path for current to flow. As current leaves the power source, it moves through all components of the circuit and returns to the source. That is, a series circuit is one in which the same current flows through all devices in a serial (series) fashion. The rate of flow is the same at every point of a series circuit.

Before you can analyze circuits, you must understand something about the symbols used to represent them in drawings. The first part of this chapter discusses the symbols used to represent various types of components and how to interpret schematic diagrams in terms of the component drawing. In series circuits there are rules for total resistance, total voltage, current, and total wattage. This chapter explains each of these quantities as related to series circuits.

OBJECTIVES

After completing this chapter, you will be able to:

- Describe a series DC circuit
- Identify which specific component of a circuit is being represented by a specific symbol on a schematic drawing
- Draw the correct symbol for electrical or electronic components when making schematic drawings
- Determine a resistor value and tolerance by using its color bands
- Calculate total resistance in a series circuit
- Explain what the proportional rules are for a series circuit in relation to resistance and voltage drop
- Describe the type and construction of various standard resistors, including the resistance value, wattage rating, and tolerance
- Determine wattage loads and total watts of a series circuit
- Explain how to use Kirchhoff's laws
- Describe how to determine circuit voltages with multiple voltage sources

INTRODUCTION

The most basic circuits we construct as electrical professionals is the series circuit. This circuit has all components needed to create a complete circuit. In this chapter we identify the various requirements and methods for calculating what is expected to happen. This chapter introduces the various necessary components of any circuit and allows you to become familiar with the basic concepts and to build the essential skills of analysis.

The circuit diagrams are explained, along with the standard component symbols. Then the circuits are analyzed according to the current, resistance, voltage, and power characteristics associated with series circuits. Along with circuit characteristics, various resistor styles and methods used to mark ohmic values based on a standard color code are introduced. This will allow you to work with experimental circuits in a lab setting to experience the effects of variables introduced into the circuit.

DIAGRAMS

As you learned in the last chapter, the schematic diagram for a circuit is a representation of how electricity would flow, drawn in a logical sequence. The schematic is a graphic representation of the actual circuit and as such uses standard graphic symbols to represent the actual components. The actual physical circuit may appear quite different. If you have ever looked at any DC circuits, you have probably noticed a lot of components that look the same from circuit to circuit. There are also a large number of components that are different. Some are very different in size and rating, whereas others are only slightly different. It is learning these differences that will aid you in becoming a professional electrical worker.

Circuit Component Diagram

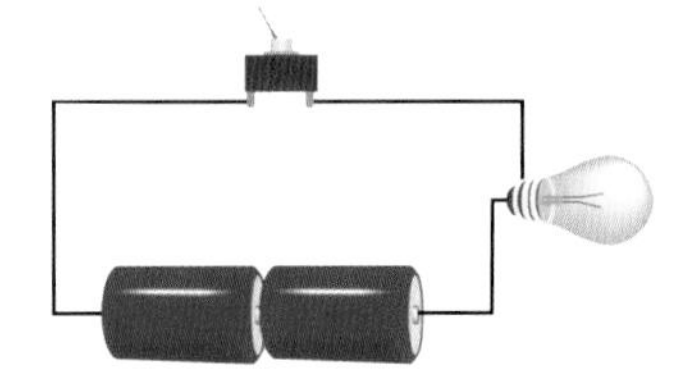

Figure 4–1 A series circuit component diagram shows realistic components.

COMPONENT DIAGRAMS

One of the first steps is learning how the components are represented in circuit drawings. There are general representations electricians use to represent the **schematic diagrams** of circuits. One method of drawing the circuit is to actually draw the components themselves, as shown in Figure 4–1. This drawing is a representation of the components in a flashlight and shows four distinct components:

1. The switch used to turn the lamp on and off
2. The batteries used to supply power to the circuit
3. The lamp, which emits visible light when current flows through it
4. The wires or conductors, which carry the current throughout the circuit

Component diagrams are sometimes used to show the physical relationship between components or to show components of a circuit when the reader might not be familiar with those components. Component diagrams are also called **pictorial diagrams**.

Figure 4–2 shows a schematic diagram of the same flashlight. A schematic diagram shows a circuit in such a way that makes it easy to follow the flow of current and to determine how the circuit is logically connected. You will note that the schematic uses symbols instead of a drawing or picture of the actual component. The only component that is the same as in Figure 4–1 is the wire or conductors. All others are represented by their schematic symbols.

Schematic Diagram

Switch
Batteries
− + − +
Lamp

Figure 4–2 A schematic diagram of circuit components uses schematic symbols.

The schematic symbols represent the components in a standard way, using common figures. Using these symbols allows you to draw, interpret, and communicate with other electrical personnel worldwide using circuit drawings. As you use schematics more and more, it will become easier for you to trace the current paths through the circuit. Tracing the main and alternate current paths is a skill you will need over and over throughout your career as an electrical worker.

International symbols make it easier to interpret circuits created in various countries. Even though you may not be able to read the word descriptions, you will be able to ascertain the electrical intent and operation of the electrical circuits. In Figure 4–3, you can see that the diagrams can be very readable in the language of electrical circuits but not necessarily in the English language.

SCHEMATIC SYMBOLS

Before you can easily interpret a schematic, you must be able to identify each component. Many components have more than one symbol to represent them, depending on the type of schematic used. The symbols you will study in this lesson will be used throughout the DC theory course. Table 4–1 provides a short description with each symbol to help you understand where that component may be used in a circuit.

TYPES OF RESISTORS

There are two general types of resistors: fixed and variable. Their names describe the main difference between the two. A fixed resistor has a value that is set and is not adjustable.

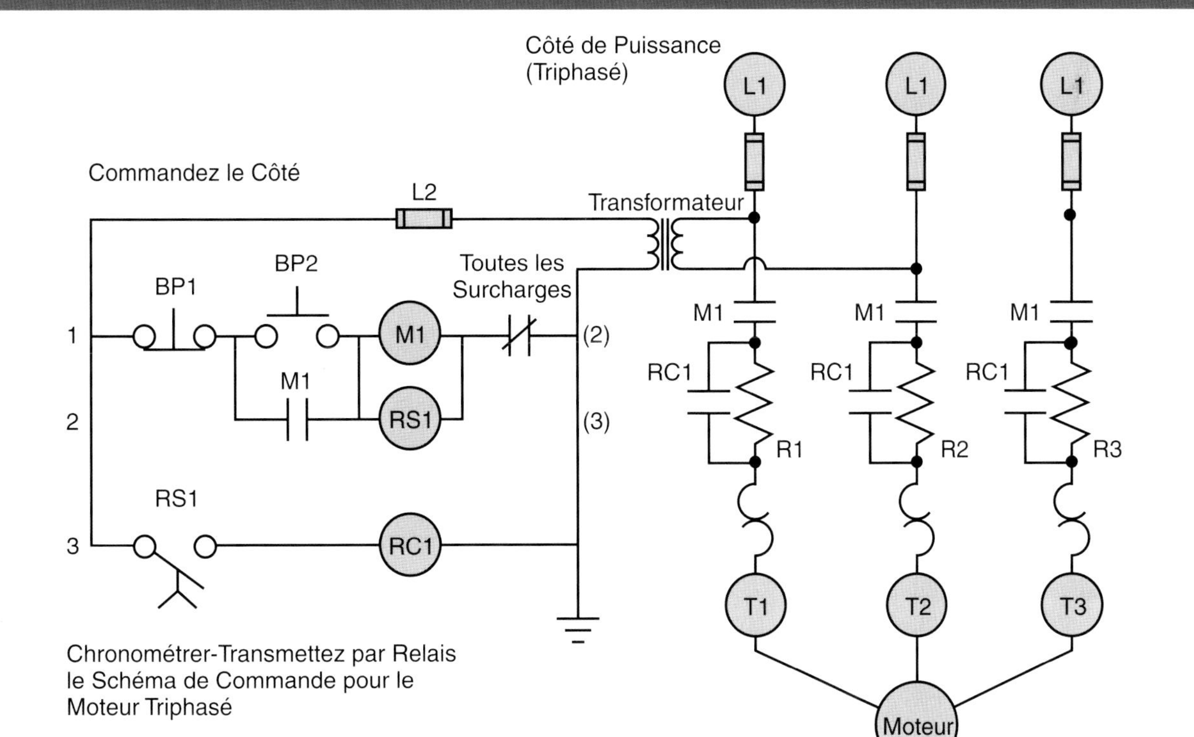

Figure 4–3 Schematics may look the same worldwide, even though the descriptions are not always written in English.

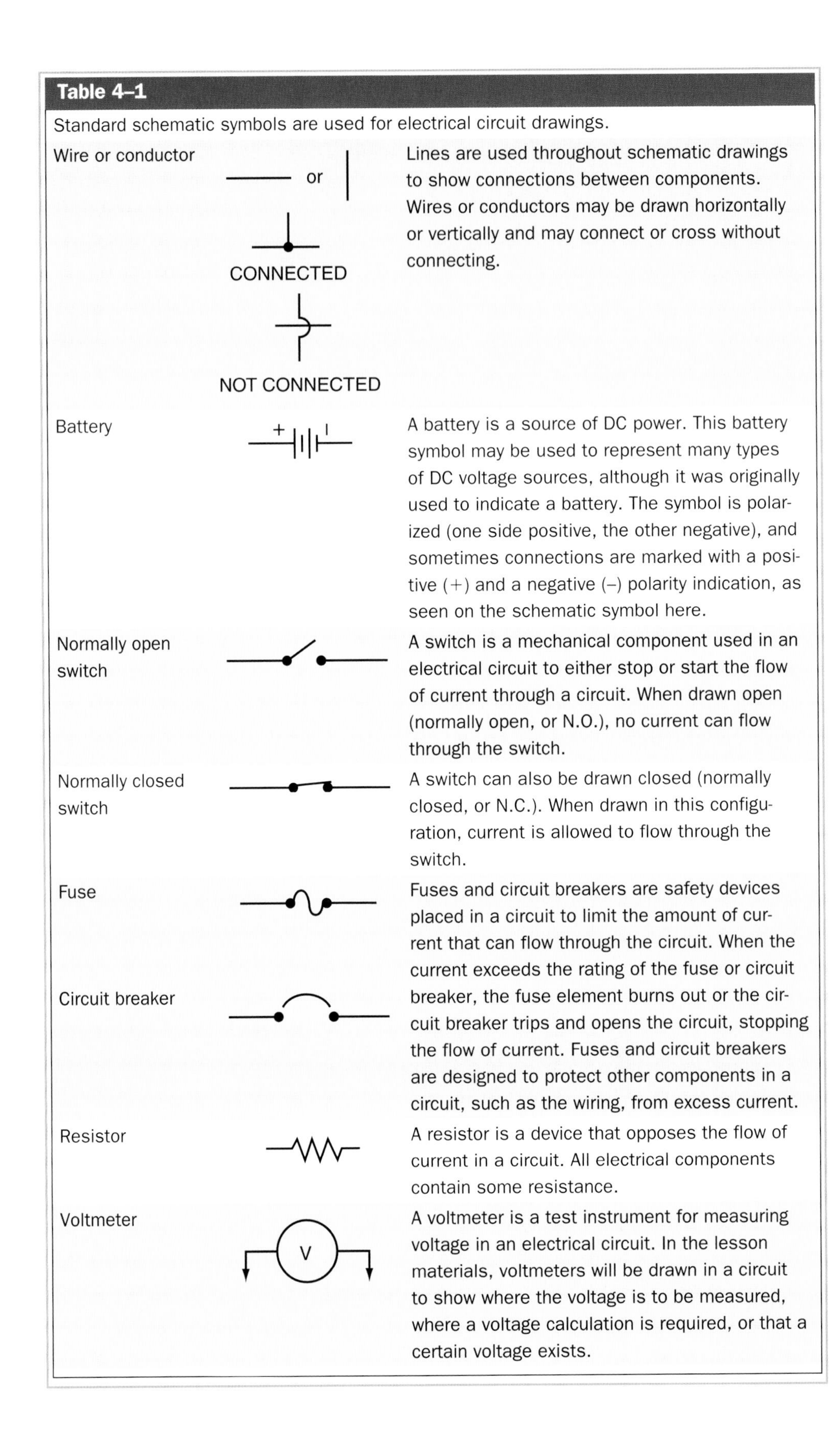

Table 4–1

Standard schematic symbols are used for electrical circuit drawings.

Component	Symbol	Description
Wire or conductor	or CONNECTED NOT CONNECTED	Lines are used throughout schematic drawings to show connections between components. Wires or conductors may be drawn horizontally or vertically and may connect or cross without connecting.
Battery	+	A battery is a source of DC power. This battery symbol may be used to represent many types of DC voltage sources, although it was originally used to indicate a battery. The symbol is polarized (one side positive, the other negative), and sometimes connections are marked with a positive (+) and a negative (–) polarity indication, as seen on the schematic symbol here.
Normally open switch		A switch is a mechanical component used in an electrical circuit to either stop or start the flow of current through a circuit. When drawn open (normally open, or N.O.), no current can flow through the switch.
Normally closed switch		A switch can also be drawn closed (normally closed, or N.C.). When drawn in this configuration, current is allowed to flow through the switch.
Fuse		Fuses and circuit breakers are safety devices placed in a circuit to limit the amount of current that can flow through the circuit. When the current exceeds the rating of the fuse or circuit breaker, the fuse element burns out or the circuit breaker trips and opens the circuit, stopping the flow of current. Fuses and circuit breakers are designed to protect other components in a circuit, such as the wiring, from excess current.
Circuit breaker		
Resistor		A resistor is a device that opposes the flow of current in a circuit. All electrical components contain some resistance.
Voltmeter	V	A voltmeter is a test instrument for measuring voltage in an electrical circuit. In the lesson materials, voltmeters will be drawn in a circuit to show where the voltage is to be measured, where a voltage calculation is required, or that a certain voltage exists.

Table 4–1 *(continued)*

Ammeter	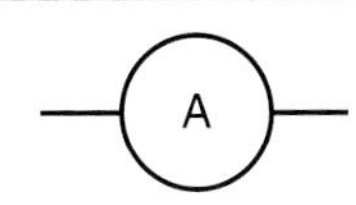	An ammeter is a test instrument for measuring the current flowing through a branch of an electric circuit. Ammeters are sometimes drawn in a circuit to show where the current is to be measured, where a current calculation is required, or that a certain current exists in the circuit.
Lamp		*Lamp* is the common electrical term for a lightbulb. Often, fluorescent tubes or high-pressure sodium or metal halide bulbs are also referred to as lamps. Normally, this symbol is used only in reference to various sizes of incandescent bulbs. A lamp could also be accurately represented as a resistance in a circuit because the filament in the bulb gives off light as a result of the heat generated by the current flow through the resistance provided by the filament.
Rheostat		A rheostat is a special type of two-terminal resistor that can be adjusted to have different resistance values. Normally, rheostats are used to adjust the current in a branch of a circuit. The symbol shown here is sometimes used to represent a variable resistor.
Potentiometer	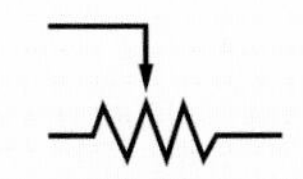	A potentiometer is a special type of three-terminal resistor. The third terminal on the potentiometer is connected in such a way that it can be adjusted to make contact with the resistor at any point, allowing adjustment in the resistance value. Potentiometers are often used to adjust the voltage to a load. The term *potentiometer* comes from the concept of potential (voltage) difference.
AC voltage source	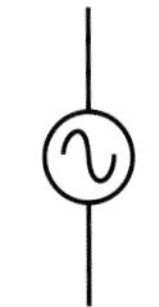	The AC voltage source is a supply of AC power in a circuit. AC is the second type of current you will learn about later in your study of electricity. AC is different from DC in that the current alternates polarity (positive to negative). The changing from positive to negative and back again takes place at a fixed rate, measured in cycles per second (hertz). DC does not have this change in polarity.
Coil		A coil, or inductor, is an electronic component used to store electrical energy as a magnetic field. The magnetic field alternates polarity in AC. The coil simply represents a coil of wire. This coil can be wrapped around a core of iron or other material that enhances the coil's ability to store the energy. A coil is sometimes called a choke.

A variable resistor can have its value changed. In addition to the two types of resistors, there are three important numbers associated with any type of resistor: the resistance value, the power rating (wattage rating), and the tolerance. A fourth value for reliability can sometimes be used.

RESISTANCE VALUE

The resistance value is the measurement of the amount of opposition a resistor offers to current flow. This value is measured in ohms and is represented by the symbol Ω (omega).

WATTAGE VALUE

The second number that is important when working with resistors is the wattage value or power rating of the resistor. The power rating of the resistor is the amount of power that can be dissipated by the resistor without damaging the resistor or affecting its operation. The power rating is measured in watts (W). Using a resistor with too low a power rating can result in destruction of the device or variations in the resistor value due to the effects of heat generated by the device itself. The power rating is determined by the I^2R calculation for the resistor, using the current flows through the resistor and its resistance. The power (in watts) dissipated by the resistor can also be determined by the voltage drop and the current flow and the formula $E \times I = P$.

TOLERANCE

Variations in resistance values may show up among resistors. Even resistors of exactly the same type and specified resistance will be somewhat different because of variations in material, quality control, and other such issues. This means that if you measure the resistance of 100 resistors of the same type and specified resistance, you will find that they are all slightly different from one another. The allowed maximum variation is called the tolerance and is equal to the allowed percentage variation in the resistor's specified resistance. For example, if a family of resistors is specified as 100 Ω ± 2%, the measured value of any one of those resistors could be as low as 98 Ω or as high as 102 Ω. Any value in this range for that type of resistor is acceptable. Common resistance tolerances include 1%, 2%, 5%, 10%, and 20%. Higher precisions are available if needed.

A 100 Ω resistor that has been manufactured for a 2% tolerance and measures 104 Ω would have been acceptable under the 20% tolerance requirement but not under the 2% tolerance. Resistor tolerances are marked on many resistor cases by colored bands, numerical value, or some other identifiable code. As today's sophisticated electrical and electronic equipment continues to evolve, it is important to initially choose and maintain consistency in the replacement of components of the design tolerance for which they were manufactured.

FIXED RESISTORS

CONSTRUCTION

One of the most common types of fixed resistors is the **composition carbon resistor**. It is made of a compound of carbon graphite and a bonding resin. The value of the resistor is determined by the ratio of the resin to the carbon graphite. An insulating material surrounds the mixture (Figure 4–4). They are the most popular resistors because they are durable, cheap, and readily available. However, they can change value with age or overheating. They are made in a wide range of values and ratings.

The wattage (power) rating is determined by their ability to release heat. More surface area means they can release more watts of heat energy, and therefore the size of the resistor is typically larger if it is a higher wattage. In terms of the dimensions of various wattage ratings, 0.5 W resistors are $\frac{1}{8}$ inch in diameter and $\frac{3}{8}$ inch long, 1 W resistors are $\frac{1}{4}$ inch in diameter and $\frac{7}{16}$ inch long, and 2 W resistors are $\frac{5}{16}$ inch in diameter and $\frac{11}{16}$ inch long.

Composition Carbon Resistor

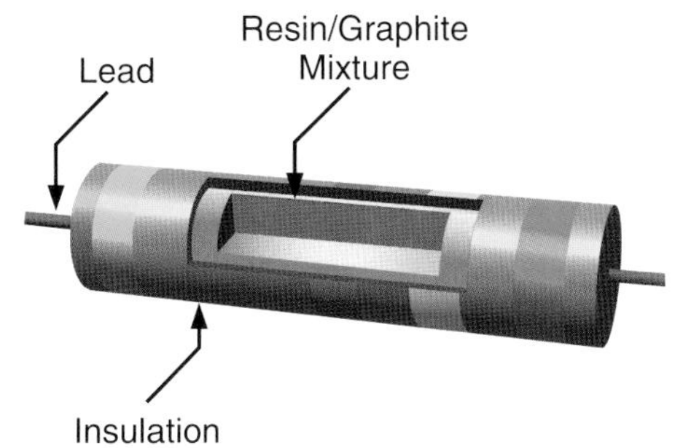

Figure 4–4 Composition carbon resistors are surrounded by insulating material.

Fixed Resistor

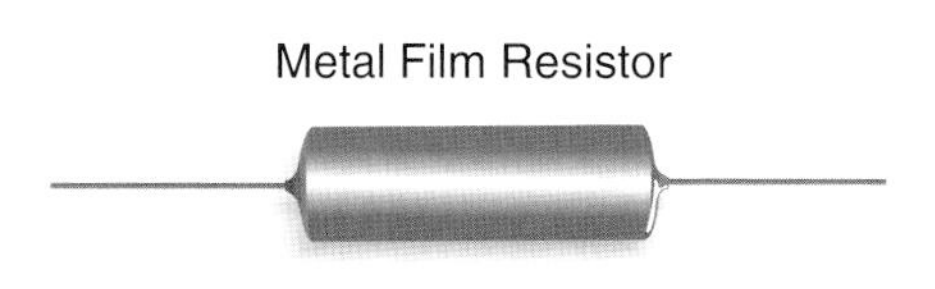

Figure 4–5 This fixed resistor is constructed with metal film.

Another type of fixed resistor is the **metal film resistor** (Figure 4–5). These are made by applying a metal film to a ceramic rod in a vacuum. The resistance value is determined by the thickness of the metal film applied and the type of metal used. The thickness of the metal film can range from 0.00001 to 0.00000001 inch. Metal film resistors are better than the composition carbon resistor because their values change less over time and they have a closer tolerance. Whereas composition carbon resistors can deviate as much as 20%, metal film resistors vary from 0.1% to 2%. The manufacturing costs of metal film resistors are much higher than those of composition carbon types.

The **metal glaze resistor** is similar to the metal film resistor. It is constructed by combining metal with glass and then applying this to a ceramic core as a thick film. The ratio of the metal to the glass determines the rating value of the resistor. The metal glaze resistor also has good tolerances of 1% to 2%.

Wire-wound resistors (Figure 4–6) are formed by a resistive wire-wound around an open core and may or may not be covered with a protective outer layer or insulating material. The wire-wound resistors are best used in applications requiring high temperature and high power. They can withstand more heat than any other type of resistor. They are expensive to construct and require a large amount of space for mounting. In addition, because it is a coil, a wire-wound resistor will add a certain amount of inductance in higher-frequency circuits. Inductance is covered in the AC portion of your training.

Wire-Wound Resistor

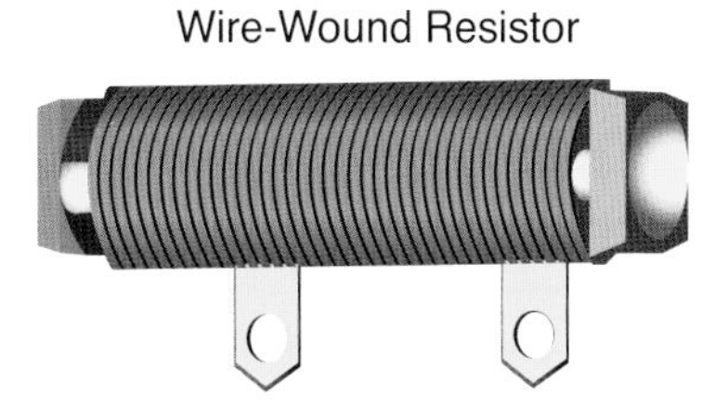

Figure 4–6 A wire-wound resistor is often used for higher wattage ratings.

CAUTION: Wire-wound resistors, when operated under normal conditions, can run hot enough to inflict serious injuries or burns from the heat generated by the device. One of the uses of the wire-wound resistor is as a heating element whose sole purpose is to generate heat to surrounding equipment or areas.

COLOR CODING

Fixed resistors have a color code to indicate the value of the resistor using the colored rings around the resistor. There are two common color band configurations that an electrical worker should become familiar with: a four-band configuration and a five-band configuration. The most common configuration is a four-band configuration.

Figure 4–7 indicates the coding of a four-band and standard five-band resistor. By looking at the bands on the resistor and using the chart shown in Figure 4–7, the value of the resistor can be determined. The first band, closest to the end of the resistor, represents the first digit of the resistance value. The second colored band represents the second digit. The third band represents the multiplier in 10^{Nth} power. The fourth band represents the tolerance rating as a percentage. For example, if the color coding were red-black-orange-silver, the resistor value would be red = 2, black = 0, orange = 10^3 (or × 1,000), and silver = 10% tolerance. The value of the resistor is 20 × 1,000 = 20,000, or 20 kΩ ± 10%. The resistor will have a value anywhere between 18,000 and 22,000 Ω. If the resistor is red-red-red and a gold fourth band, the resistor value is red = 2, red = 2, red = 10^2 (or × 100) for a base value of 2,200 or 2.2 kΩ ± 5%, or 110 Ω tolerance. The resistor, then, has a value anywhere between 2,090 and 2,310 Ω.

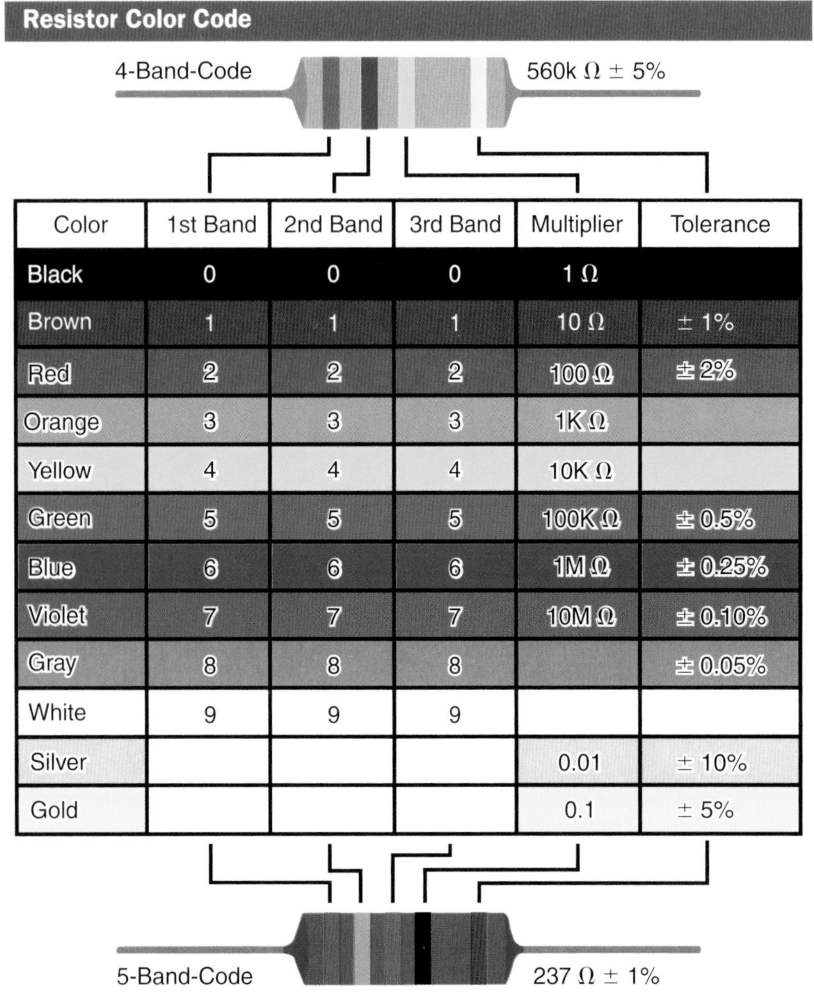

Color	1st Band	2nd Band	3rd Band	Multiplier	Tolerance
Black	0	0	0	1 Ω	
Brown	1	1	1	10 Ω	± 1%
Red	2	2	2	100 Ω	± 2%
Orange	3	3	3	1K Ω	
Yellow	4	4	4	10K Ω	
Green	5	5	5	100K Ω	± 0.5%
Blue	6	6	6	1M Ω	± 0.25%
Violet	7	7	7	10M Ω	± 0.10%
Gray	8	8	8		± 0.05%
White	9	9	9		
Silver				0.01	± 10%
Gold				0.1	± 5%

Figure 4–7 A standard color code is used to represent numerical values of resistance.

The fourth band is brown, gold, silver, or none. This represents a 1%, 5%, 10%, or 20% tolerance, respectively, of the resistor's marked value. Of course, the lower the percentage tolerance the more precise the resistor value and the higher the cost. If the third band is gold or silver, the multiplier times the first two digits is 0.1 and 0.01, respectively. This is used for resistors of very small ohmic values.

On some specialty resistors, you may see an additional (fifth) band that represents a reliability factor. This should not be confused with the standard five-band configuration discussed next, where the fifth band represents a tolerance factor. The last (fifth) band, when used with a four-band configuration, represents the reliability of the resistance value. This indicates the failure rate of the resistors as a percentage of failures per 1,000 hours of use. The fifth band on these specialty resistors will be brown = 1%, red = 0.1%, orange = 0.01%, yellow = 0.001%, respectively. For example, if the fifth band were orange the failure rate after 1,000 hours of use would yield 0.01% of all resistors. If you use 10,000 resistors of that size in your circuits, an average of one would fail.

A standard five-band configuration is very similar to a four-band configuration. With this configuration the third color band also represents a numeric value and the fourth band represents the multiplier. The fifth band represents the tolerance. (See Figure 4-7.)

A mnemonic system is a memory word association used to remember a sequence of items and is often used to help remember the color of the resistor bands that represent digits. Beginning electricians learn various versions of mnemonics. Two such versions are detailed in Table 4–2. You may make your own version to help remember the sequence.

Table 4–2

A mnemonic memory aid helps keep track of color progression with numbers.

Color of Stripe	Represented Digit	Mnemonic	Mnemonic
Black	0	Black	Big
Brown	1	Brown	Boys
Red	2	Rover	Race
Orange	3	Often	Our
Yellow	4	Yelps	Young
Green	5	Growls	Girls
Blue	6	But	But
Violet	7	Very rarely	Violet
Gray	8	Gets	Generally
White	9	Wild	Wins
Gold	Multiplier of 0.1 if third band; 5% tolerance if fourth band	—	Gold
Silver	Multiplier of 0.01 if third band; 10% tolerance if fourth band	—	Silver
None	Tolerance of 20% if no fourth band	—	No metals

Variable Resistor

Figure 4–8 A three-terminal variable resistor can be used as a resistor or a potentiometer.

Variable Resistor

Resistor

Wiper Arm

A X B

Figure 4–9 This diagram depicts the construction of a three-terminal variable resistor, with fixed resistance available between terminals A-B and variable resistance available between terminals A-X and B-X.

VARIABLE RESISTORS

CONSTRUCTION

Variable resistors are resistors whose resistance value can be changed by some method. Figure 4–8 shows a standard variable resistor. Figure 4–9 shows a diagram of its electrical components. A stem connected to a wiper arm controls the variance in resistance values.

As the wiper arm is turned, the resistance measured between the wiper arm terminal and the terminal at the end of the resistor changes. Such resistors are sometimes called potentiometers.

Variable resistors may be either composition or wire-wound. Like fixed wire-wound resistors, variable wire-wound resistors can handle more power than composition resistors. Such wire-wound resistors could be used to control large currents in devices such as battery chargers or electroplating processes.

The resistance between points A and B in the variable resistor shown in Figure 4–9 is fixed. The value of the resistance between points A and X (AX) or B and X (BX) varies, depending on where the wiper arm is positioned. Figure 4–9 indicates that points A to X have a higher resistance than points X to B. This is because there is more resistive material between the contacts of AX than between the contacts of BX.

Figure 4–10 shows an example of a multi-turn variable resistor. These are operated with a knob or screw. For example, a five-turn multi-turn resistor would require five full rotations to go from the least to the highest resistance value. These types of resistors provide much finer control and are used in precision applications, such as the alignment of electronic devices.

Variable Resistor

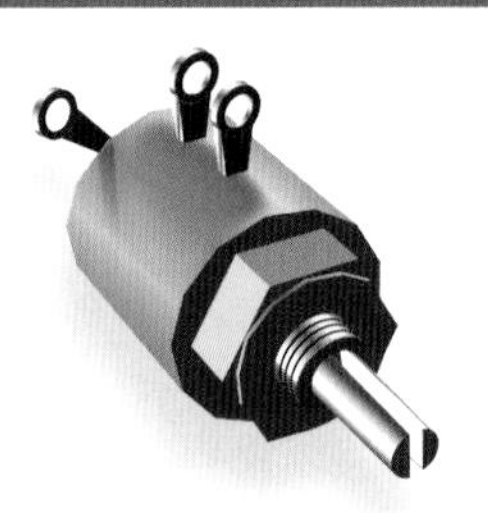

Figure 4–10 Multi-turn variable resistors provide precise control.

POTENTIOMETERS AND RHEOSTATS

Two of the more common terms you will encounter in your career as an electrician are *potentiometer* and *rheostat*. A potentiometer is a variable resistor that has three terminals. Potentiometers are generally used to adjust the voltage applied to a circuit, hence the *potentio*-(potential) prefix combined with *meter* (i.e., potential metering), referring to adjustment of potential through a voltage divider circuit. Figure 4–11 shows a schematic representation of a variable resistor being used as a potentiometer. Voltage divider circuits are discussed in later chapters.

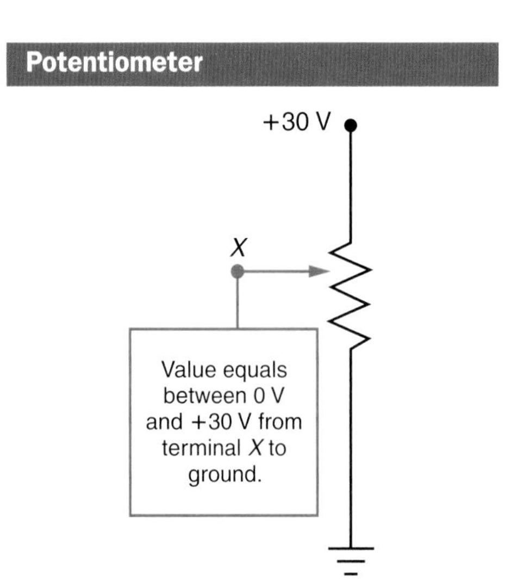

Figure 4–11 A potentiometer (pot) can be used for a voltage divider to adjust voltage.

A rheostat is a variable resistor that has two terminals. A potentiometer may be used as a rheostat if only two of the three terminals are used because it does actually adjust the resistance of the circuits. A rheostat is strictly used as a variable resistor to vary the amount of current that flows to a circuit. With all of the resistance inserted, the ohmic value is high and the current to the circuit is low. As the resistance is bypassed in the rheostat, the resistance values decrease and the current in the circuit increases.

TAPPED RESISTORS

A tapped resistor is a fixed resistor with a tap (or taps) permanently connected between the terminals of a fixed resistor. The resistance value between each tap (terminal) is fixed. Therefore, depending on which tap you use, you can vary the resistance value. However, the resistance between each possible connection point remains fixed. Figure 4–12 shows some of the common schematic representations for popular types of resistors.

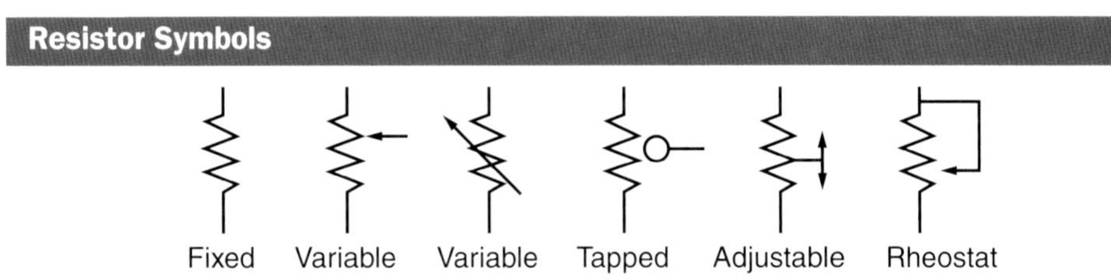

Figure 4–12 Schematic symbols represent various resistor configurations used in circuits.

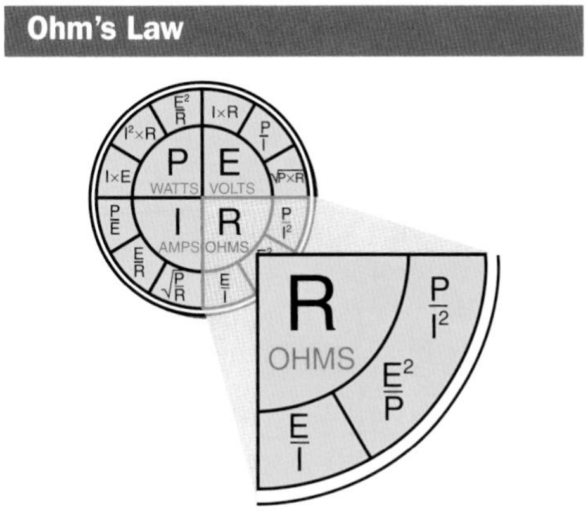

Ohm's Law

Figure 4–13 The Ohm's law pie chart involves finding ohms by means of other factors.

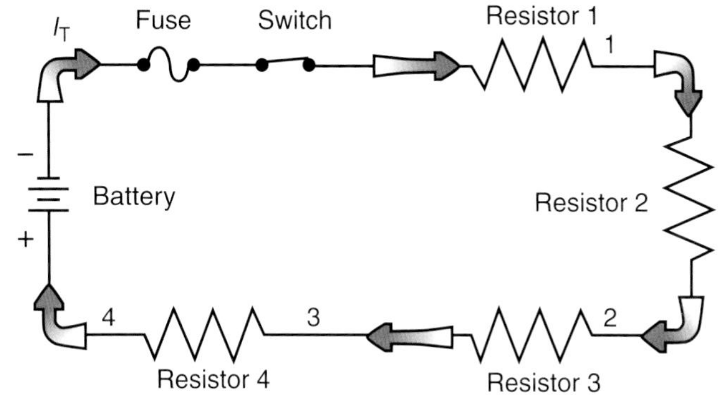

Figure 4–14 This simple series circuit has four resistors and one current path.

THE SERIES CIRCUIT

The DC series circuit is a building block for many other concepts in the electrical sciences. There are four basic rules that every series circuit is subject to as we determine what values to expect. The rules are:

1. Current is the same wherever it is measured in the circuit ($I_T = I_1 = I_2 = I_3 = ... I_n$).
2. The total voltage of the circuit is the sum of all the component voltage values in the circuit ($E_T = E_1 + E_2 + E_3 + ... E_n$).
3. The total resistance of the circuit is the sum of all of the individual resistor values ($R_T = R_1 + R_2 + R_3 + ... R_n$).
4. The total power is the sum of the individual circuit watt dissipations ($P_T = P_1 + P_2 + P_3 + ... P_n$).

In Chapter 1, you learned about Ohm's law and used a pie chart (Figure 4–13) to examine the relationships among resistance, current, voltage, and power. In this section, you will concentrate on the resistance slice of the pie chart. Applications of these resistive relationships are presented in a series circuit.

A series circuit is one in which there is only one path for current to flow. In Figure 4–14, all four resistors are connected end to end so that all electrons leaving one resistor are entering the next. This flow continues through all resistors, the switch, fuse, and battery. Even though each resistor in a series circuit offers some opposition to the flow of current in the series circuit, the current flow level is the same throughout the circuit. If you were to take an ammeter (a meter designed to measure current flow) and place it in series with the components of the series circuit, usually by opening the circuit and connecting the leads of the ammeter, the amount of current measured would be the same regardless of where you opened the circuit to connect the meter.

As you use the formulas for calculating various circuit values, it is easiest to use full units of measure. The formulas are written for ohms, amperes, volts, and watts. When using prefixes such as milli, micro, kilo, and mega, it can be confusing—at least when starting. If you convert milliamps (mA) to amps and kilovolts (kV) back to volts first before using them in the formula, you will be able to keep track of the true values better. As you become more accustomed to the prefixes and how to manipulate the powers of 10 in the formulas, you will find it easier to use the formulas with the prefixes to the unit values.

You could also measure the current using a DC clamp-on meter, as shown in Figure 4–15. This is a very important concept to remember. The current is the same throughout the series circuit. It is represented in circuit equations by the following equation:

$$I_T = I_1 = I_2 = I_3 = \dots I_N$$

This means that wherever you measure the series circuit current, up to (N) locations, the current will be equal. This is a "current law" about a series DC circuit. All current measurements throughout the circuit will be the same. Note also the subscripts used with the letter designations in the circuit. The subscripts allow circuit designers to accurately label each circuit component with its own unique alphanumeric label in a circuit. This is very important in more complex circuits.

Each resistor in the series circuit adds to the total resistance and restricts the current flow through the entire circuit. The connecting wires, the fuse, and the battery provide some resistance. However, compared to the resistors in the circuit, they account for a very small amount of the total resistance and are neglected in most calculations.

Clamp-On Meter

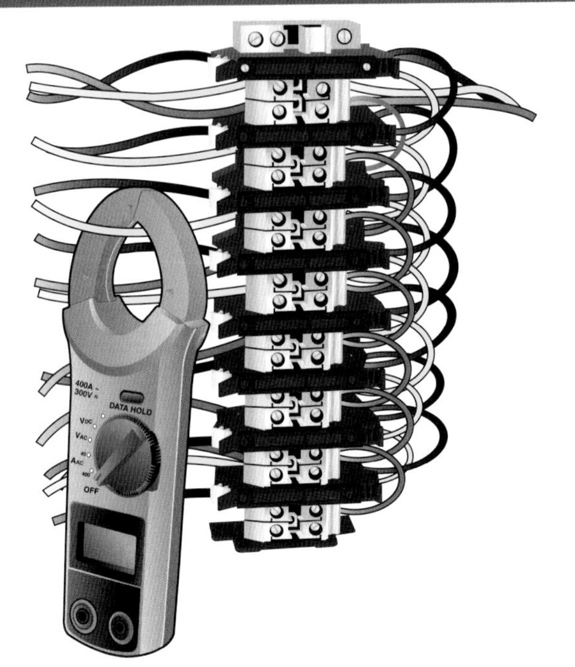

Figure 4–15 A DC-style clamp-on meter may be used to measure the same current at any point of a series circuit.

CALCULATING RESISTANCE IN A SERIES CIRCUIT

All components in a circuit provide some value of resistance to the current flow. The resistors shown in Figure 4–14 provide almost all of the total circuit resistance. To keep the concepts simple, we will assume that the total resistance in Figure 4–14 is due only to the resistors. In any series circuit, the total circuit resistance in ohms is the sum of the individual resistors' in ohms.

$$R_T = R_1 + R_2 + R_3 + R_4 + \dots R_n$$

Here, n represents any additional resistors in the circuit and R_T = total resistance. The total resistance can also be calculated by using the other circuit total values and Ohm's law. This equation is:

$$R_T = \frac{E_T}{I_T}$$

R_T is the total circuit resistance.

E_T is the circuit supply voltage.

I_T is the circuit total current.

Example

What is the total resistance of the series DC circuit shown in Figure 4–16?

Solution:

Using the equation for series resistance:

$$R_T = R_1 + R_2 + R_3 + R_4$$

$$R_T = 25\ \Omega + 30\ \Omega + 50\ \Omega + 35\ \Omega$$

$$R_T = 140\ \Omega$$

Series Circuit

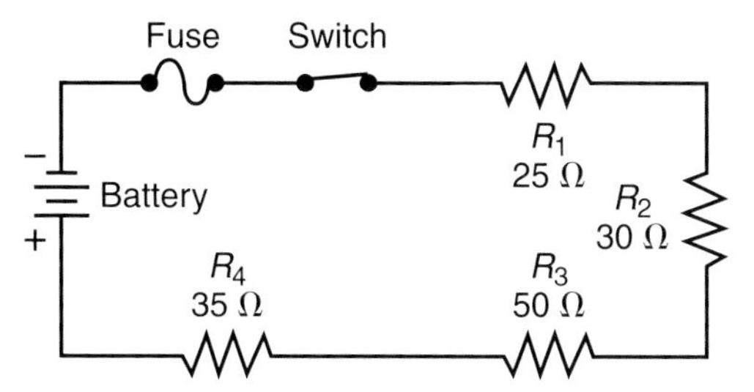

Figure 4–16 A series circuit with four resistors also includes a single-pole/single-throw switch.

Example

Calculate the total resistance of the circuit shown in Figure 4–17.

Solution:

Using the series circuit total resistance equation:

$$R_T = R_1 + R_2 + R_3$$

$$R_T = 5\ \text{k}\Omega + 150\ \Omega + 3.5\ \text{k}\Omega$$

$$R_T = 5{,}000\ \Omega + 150\ \Omega + 3{,}500\ \Omega$$

$$R_T = 8{,}650\ \Omega$$

or

$$R_T = 8.65\ \text{k}\Omega$$

Series DC Circuit

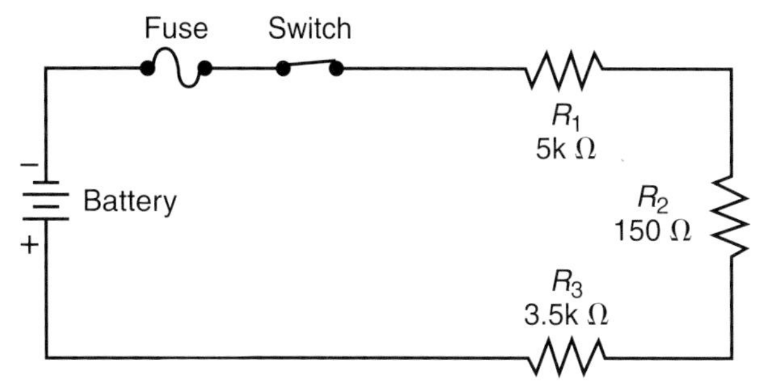

Figure 4–17 This series DC circuit has three different resistors.

As you make the calculations for circuit values, make sure you go back to verify the values you have found. This is an important step, especially as the circuits become longer and more calculations are used to determine the electrical characteristics and parameters of the circuits. There is often more than one way to calculate the circuit values, and the more methods you learn, the easier it is to verify the correct values that may be used for further calculations.

DETERMINING PROPORTIONAL RELATIONS IN A SERIES DC CIRCUIT

You learned that the total resistance in a series circuit is determined by the addition of all resistive components in that circuit. To keep things simple, we defined those components as just resistors. In keeping with this definition, then, resistors cause opposition to current flow, and the total opposition to current flow is the total resistance in the circuit.

Thus, to find the total current in a circuit we must know two things: the supply voltage (E_T) and the total resistance (R_T). This means that the total current is affected when the total resistance or the supply voltage changes. If any component increases by 10 Ω, the total resistance is increased by 10 Ω, which affects the total current.

What would happen to the total current if the source voltage shown in Figure 4–18 increased to double its original value?

Solution: Because the voltage has changed, the total current must be determined.

$$I_T = \frac{E_T}{R_T}$$

Original circuit:

$$E_T = 30 \text{ V}$$

$$R_T = R_1 + R_2 + R_3$$

$$R_T = 6{,}000\ \Omega + 150\ \Omega + 3{,}500\ \Omega$$

$$R_T = 9{,}650\ \Omega$$

$$I_T = \frac{30 \text{ V}}{9{,}650}$$

$$I_T = 0.003108 \text{ A}$$

If the circuit voltage doubles:

$$I_T = \frac{60 \text{ V}}{9{,}605\ \Omega}$$

$$I_T = 6.22 \text{ mA or } 0.00622 \text{ A}$$

If the circuit voltage increases by twice the original voltage, the circuit current will also double. The increased current through each component will double the voltage drop across each component. Let's look at this using Ohm's law. The total current in a circuit is directly proportional to the total voltage (E_T) and inversely proportional to the total resistance (R_T). The formula is:

$$I_T = \frac{E_T}{R_T}$$

If E_T *increases*, I_T *increases* by the same factor or rate. For example, if voltage doubles and resistance stays the same, the current will double. The voltage and current are directly proportional. If R_T *increases* and E_T stays the same, I_T *decreases* by the same factor or with an inverse ratio. An increase in resistance by a factor of three will cause the current to decrease to one third its original value. Use the inverse of the ratio of resistance increase to determine the factor that determines the new lower current. This is the inverse ratio (inversely proportional). Let's look at some examples using Figures 4–19 and 4–20.

Voltage in a Series Circuit

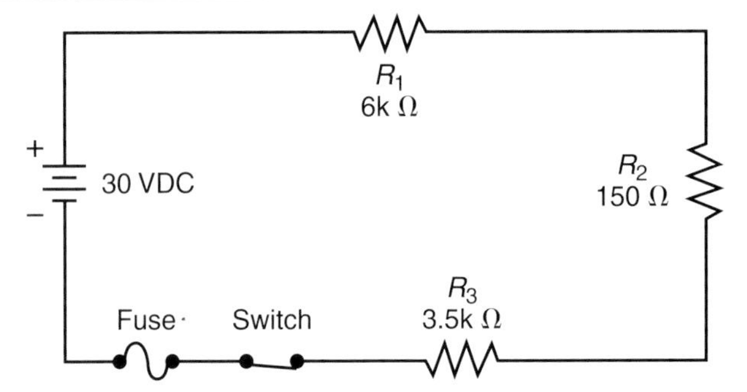

Figure 4–18 Changing the voltage in a series circuit changes the total current and the individual voltage drops.

Determining Total Current

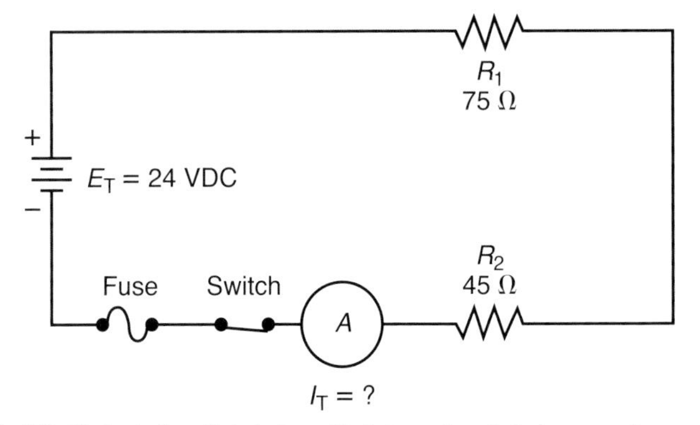

Figure 4–19 Using circuit totals will determine total current.

Example

Referring to Figure 4–19, calculate the total current in the circuit. Find total resistance:

$$R_T = R_1 + R_2$$

$$R_T = 75\ \Omega + 45\ \Omega$$

$$R_T = 120\ \Omega$$

Using Ohm's law:

$$I_T = \frac{E_T}{R_T}$$

$$I_T = \frac{24 \text{ V}}{120\ \Omega}$$

$$I_T = 0.2 \text{ A}$$

Example

Referring to Figure 4–19, determine the value of the current (I_T) if R_2 doubles in resistance. In this example, we have to recalculate R_T to find the new value of I_T.

$$R_T = 75\ \Omega + (45\ \Omega + 45\ \Omega) \text{ if } (R_2 \text{ doubles in value})$$

$$R_T = 75\ \Omega + 90\ \Omega$$

$$R_T = 165\ \Omega$$

$$I_T = \frac{E_T}{R_T}$$

$$I_T = \frac{24\ \text{V}}{165\ \Omega}$$

$$I_T = 0.145\ \text{A}$$

Example

Apply what you have learned about series circuit current and determine the total current in the circuit shown in Figure 4–20. Because the current through a series circuit is the same no matter where it is measured, the current through R_2 will be equal to the total current. Thus:

$$I_T = I_{R2} = \frac{V_{R2}}{R_2}$$

$$I_T = I_{R2} = \frac{12\ \text{V}}{45\ \Omega}$$

$$I_T = I_{R2} = 0.266\ \text{A}$$

Remember, total resistance and the total voltage affects I_T. If either changes, recalculate the changed value and reapply Ohm's law for current.

Determining Total Current

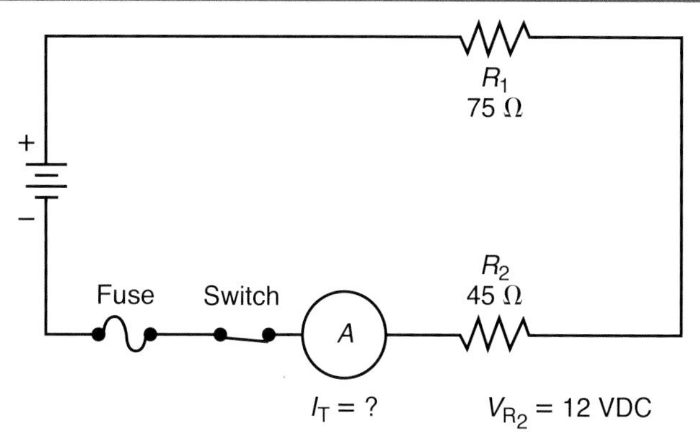

Figure 4–20 Using component values of *E* and *R* will determine total current.

Two Equal Resistors

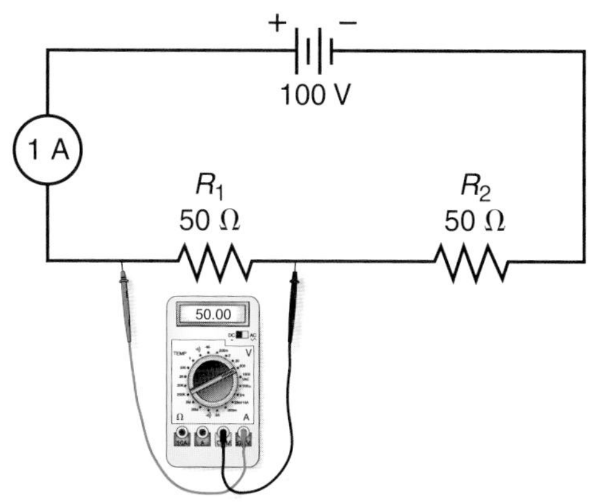

Figure 4–21 The ratio of each resistor to the total resistance is the same ratio of the individual voltage drop to the total voltage.

PROPORTIONAL VOLTAGE DROP

The amount of voltage dropped (the difference in electrical pressure) across a circuit resistance is proportional to its proportion to the total resistance of the circuit. Consider the circuit of Figure 4–21 with two equal resistors.

Because the current flow in the series circuit is the same throughout, we can easily calculate the voltage drop at each resistor.

The circuit has a total voltage or electromotive force (EMF) of 100 V at the source. The two resistors are 50 ohms each, creating a total resistance of 100 Ω, each resistor representing 50% of the total resistance. The circuit current of 1 ampere flows through each resistor. Ohm's law determines that each resistor drops 50 V, using $E = I \times R$. Whenever we have a series circuit, we must account for all voltage applied at the source. In this case, we can account for all 100 V. One-half of the total voltage drops on R_1 (50%), and one-half (50%) drops across R_2.

We find that if the resistor is 50% of the total resistance, it will drop 50% of the total voltage.

The same proportional rule applies to the circuit shown in Figure 4–22. With three resistors you can determine what percentage of the total resistance each resistor represents. For example, the three resistors equal 100 ohms. R_1 is 15 Ω or 15/100 or 15% of the total resistance. R_2 is 30 Ω or 30/100 or 30% of the total resistance. R_3 is 55 Ω or 55/100 or 55% of the total resistance. The proportional voltage drop across each resistance will be the same proportion of the total voltage as the individual resistance is of the total resistance. If the total voltage is 100 V, 15% of 100 V (15 V) drops across R_1. Thirty percent of 100 V (30 V) drops across R_2, and 55% (55 V) of the total drops across R_3.

To verify this rule of proportions, we would actually calculate the current in the circuit and the respective voltage drops. With 100 V total and 15 Ω + 30 Ω + 55 Ω = 100 Ω total, the current would be 1 A. One amp flowing through R_1 yields 15 V (15% of the total). R_2 with 1 A and 30 Ω drops 30 V (30% of the total voltage). R_3 drops 55 V (55% of the total). Whatever voltage we use, the percentage stays the same. You will note that the 30 Ω resistor drops twice as much as the 15 Ω resistor. The larger the resistor, the larger the voltage drop in a series circuit.

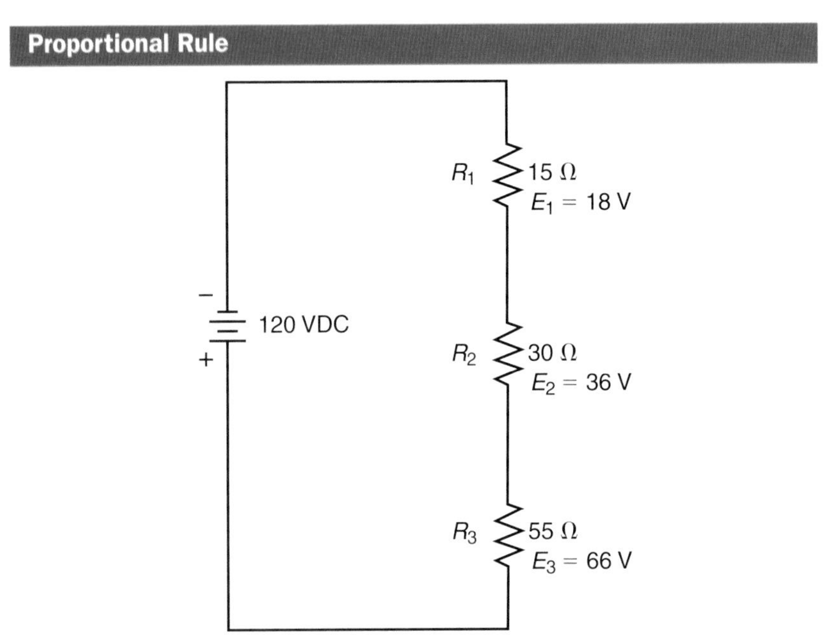

Figure 4–22 The proportional value of resistance is the same as the proportional value of voltage.

WATTAGE CALCULATIONS IN SERIES DC CIRCUITS

The number of watts dissipated by loads in a series circuit are all added together to obtain the total watts dissipated by the circuit.

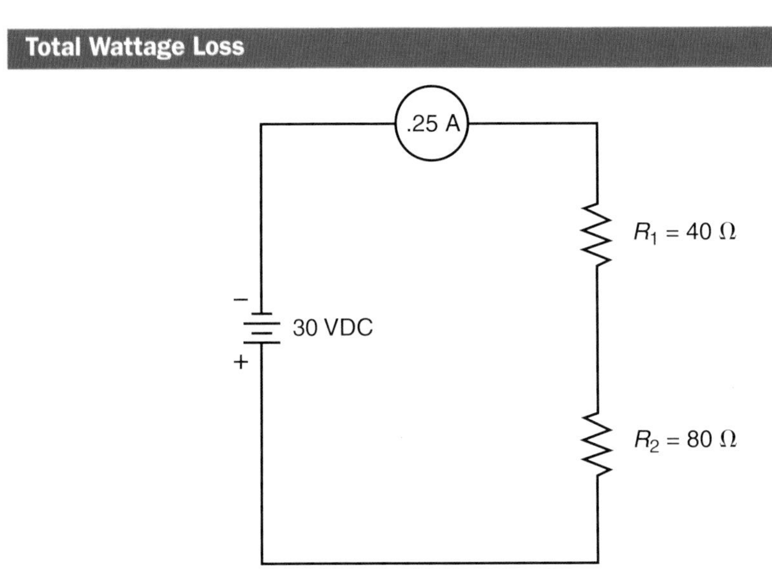

Figure 4–23 The total wattage loss is the sum of the individual wattage losses.

The easiest method is to determine the total current and the total voltage and use the wattage formula $E_T \times I_T = P_T$. Refer to Figure 4–23. Using the total voltage of 30 VDC and the current of 0.25 A, we can determine the total watts of all loads as:

$$P_T = E_T \times I_T$$

$$P_T = 30\text{ V} \times 0.25\text{ A}$$

$$P_T = 7.5\text{ W}$$

To this point, you have been introduced to resistance, current, voltage, and power in a series circuit. You have also been introduced to the four simple rules that apply to all series circuits. When trying to determine unknown values in a series circuit, these concepts and rules can be used to aid in finding the unknown quantities. To find unknown values in a circuit, you must know two of the following values:

- Total current of the circuit
- Voltage drop across the individual components
- Power dissipated by the individual component
- Resistance of the component

In Figure 4–23, you can determine the watts expended by each resistor via the following:

$$P = I^2 \times R$$

$$P_{R_1} = I_T^2 \times R_1$$

$$P_{R_1} = 0.25\text{ A}^2 \times 40\ \Omega$$

$$P_{R_1} = 2.5\text{ W}$$

and

$$P_{R_2} = I_T^2 \times R_2$$

$$P_{R_2} = 0.25\text{ A}^2 \times 80\ \Omega$$

$$P_{R_1} = 5\text{ W}.$$

Then add the individual watt loads together to yield the total watts of the circuit.

$$P_T = P_{R_1} + P_{R_2}$$

$$P_T = 2.5\text{ W} + 5\text{ W}$$

$$P_T = 7.5\text{ W}$$

You should verify your answer by calculating values using different methods to check if each method yields the same results. In this case, the answers coincide. Another way to calculate the total resistance is to use Ohm's law and the circuit total values. To do this, you must know two of the following:

- Total current
- Total voltage of the circuit
- Total power

When you have two of these three, it is simply an application of the equations used in Ohm's law pie chart.

Example

Calculate the total resistance of the circuit in Figure 4–24, using Ohm's law.

Solution:

Because the total current is determined by the total voltage and the total resistance, you can also determine the total resistance if you know the current and the voltage. Using Ohm's law equation for resistance:

$$R_T = \frac{E_T}{I_T}$$

$$R_T = \frac{30\text{ V}}{0.25\text{ A}}$$

$$R_T = 120\ \Omega$$

Determining Total Resistance

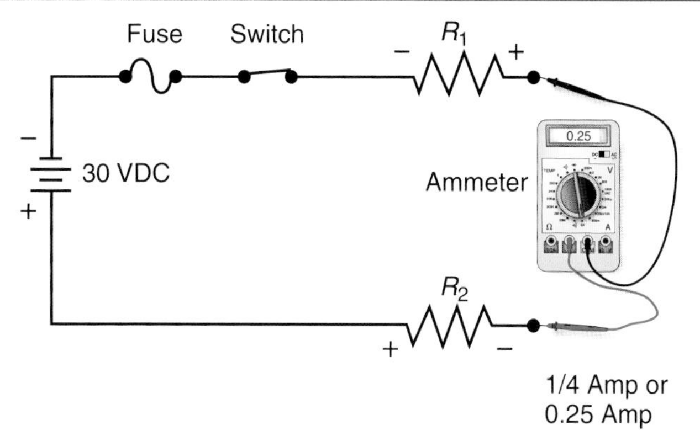

Figure 4–24 If you know the current as shown on an ammeter and the total voltage, you can determine total resistance.

Example

Likewise, if you use the watt formulas as related to voltage and resistance, you can calculate the total resistance for the circuit in Figure 4–25 using Ohm's law.

Solution:

Using Ohm's law and the equation for power, we obtain the following:

$$R_T = \frac{E^2}{P}$$

$$R_T = \frac{28\ V^2}{7\ W}$$

$$R_T = \frac{784}{7}$$

$$R_T = 112\ \Omega$$

Determining Total Resistance

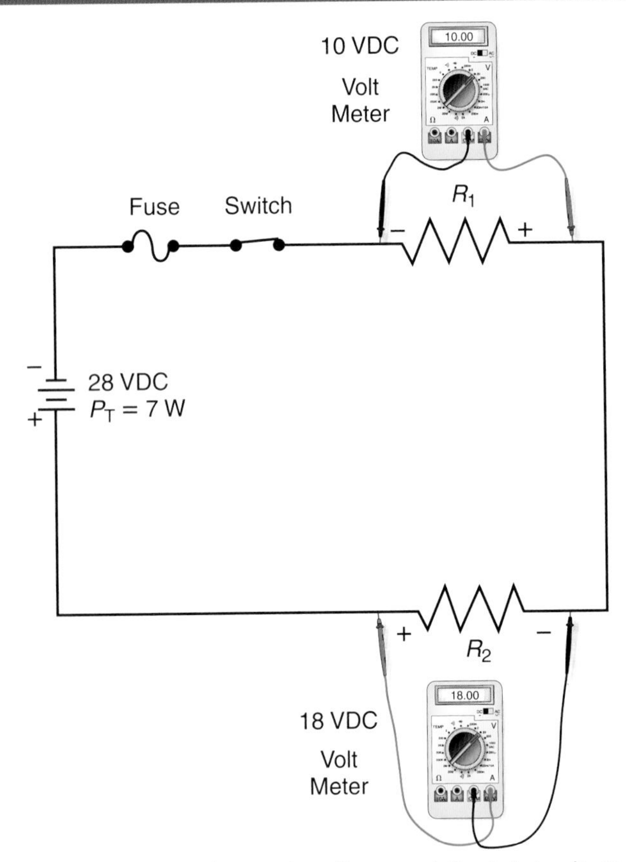

Figure 4–25 If you know the total wattage and the total voltage, you can determine the total resistance.

You may also calculate the individual values of resistance for resistors R_1 and R_2 by using Ohm's law for the individual components. If you know the current through a component and the voltage drop at the component (difference across the component), you can find the resistance of that component by using Ohm's law: $R_{comp} = \frac{E_{comp}}{I_{comp}}$

Example

In Figure 4–25, first find the series current flow. Using Watt's law for DC circuits, you can calculate the current flow using $I = \frac{P}{E}$ as a transposition of the formula.

$$P = E \times I$$

Transposed:

$$I = \frac{P}{E}$$

$$I = \frac{7\ W}{28\ V}$$

$I = 0.25$ A (also expressed as 250 mA)

Then, with 0.25 A in the circuit and the voltage drop across R_1 as 10 VDC, we can determine that the resistance is:

$$R_1 = \frac{E_1}{I_T}$$

$$R_1 = \frac{10\ V}{0.25\ A}$$

$$R_1 = 40\ \Omega$$

In addition:

$$R_2 = \frac{E_2}{I_T}$$

$$R_2 = \frac{18\ V}{0.25\ A}$$

$$R_2 = 72\ \Omega$$

You should go back to verify your calculations. Using the series resistance formula, you can now calculate the total resistance due to the resistors for the circuit shown in Figure 4–25. Using the total current you calculated, verify that $E_T = I_T \times R_T$ for the circuit. First calculate R_T.

$$R_T = R_1 + R_2$$

$$R_T = 40\ \Omega + 72\ \Omega$$

$$R_T = 112\ \Omega$$

Then:

$$E_T = I_T \times R_T$$

$$E_T = 0.25\ \text{A} \times 112\ \Omega$$

$$E_T = 28\ \text{V}$$

(which matches the starting value of the circuit as given).

As another verification check, you may use the total watts and the total resistance to verify the given voltage, using $W_T = \frac{E_T^2}{R_T}$ as the formula, transposed to:

$$E_T^2 = P_T \times R_T$$

or

$$E_T = \sqrt{P_T \times R_T}$$

$$E_T = \sqrt{7\ \text{W} \times 112\ \Omega}$$

$$E_T = \sqrt{784}$$

$E_T = 28$ V (which matches the value of voltage as given in the circuit)

KIRCHHOFF'S LAWS

Kirchhoff's laws provide you with the tools necessary to solve more complex circuits than is possible using Ohm's law. Applying these laws allows you to gain a better working knowledge of circuit operations necessary for solving multiple-source circuits. Kirchhoff's laws are not replacing Ohm's law but are going beyond the basics and giving you new skills in circuit analysis. Kirchhoff's two laws are as follows:

- The algebraic sum (Σ) of the currents entering and leaving any node (junction point) is zero.
- The algebraic sum (Σ) of the voltages around any closed path is zero.

KIRCHHOFF'S CURRENT LAW

Kirchhoff's current law states that the sum of any currents entering and leaving any given point in a circuit must be equal to zero. When summing the values of current entering and leaving a specific point (node) within the circuit, the currents entering the point are positive and the currents leaving the point are negative. With reference to Figure 4–26, Kirchhoff's current law can be written algebraically as:

$$I_{R1} + I_{R2} = 0$$

As you examine Figure 4–26, you might realize that the current flow through R_2 is actually flowing away from node A. This is the beauty of both of Kirchhoff's laws: As long as you are consistent with your assumptions, the direction of current flow will work out correctly when you do the algebraic solution. As you can determine from this diagram, as current flows toward a point, you can assign it a positive value (meaning that it is gaining electrons).

As it flows away from the point it is a negative value, meaning that it is losing electrons. As you create a sum of these currents, the mathematical sum must be zero. This seems obvious with this simple circuit, but as the circuits become more complex, Kirchhoff's laws can be valuable in determining circuit behavior and can aid in troubleshooting. Of the two laws, you will probably use Kirchhoff's voltage law more often than Kirchhoff's current law. The same principle applies in the voltage law, as you will see in the next section.

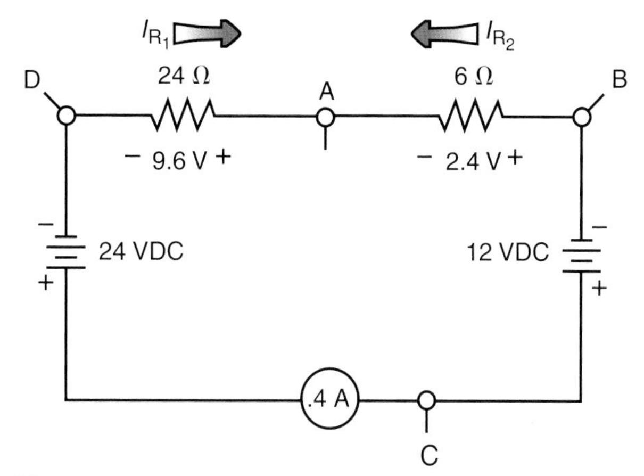

Figure 4–26 A circuit with nodes is used to illustrate Kirchhoff's laws.

KIRCHHOFF'S VOLTAGE LAW

Refer to Figure 4–26 again. The circuit has only one loop for current to flow, identified by the node points of ABCD. Note that as current flows around its loop it will create a voltage drop across the resistors. In resistor R_1, the voltage drop will be caused by the circuit current through the resistor. Although the choice of polarities for the voltage drops is not critical, it makes sense to select them logically. In Figure 4–26, each voltage drop is labeled and assigned a polarity. The polarities are chosen as follows:

1. E_{R_1} is chosen with its negative end located where the current flows in. This is a reasonable guess and is the way most of the polarities should be chosen.
2. E_{R_2} is chosen with its negative end located where the current flows in.
3. Start at the positive side of the 24 V source on the left of the circuit in Figure 4–26. Continue clockwise around the circuit, adding the voltage sources and the voltage drops across the resistors with the polarity as shown. As you add the voltages with the polarities shown, the total voltage will be zero: $(+\ 24\ V) + (-9.6\ V) + (-2.4\ V) + (-12\ V) = 0\ V$.

As long as you are consistent with the voltage polarities, with the node where current enters represented by a negative voltage and the node where current leaves the source represented as a negative (opposite the voltage drops at the loads), the net sum of the voltages must be zero. This means that all of the source voltage must be accounted for in the circuit. With this simple circuit, that seems obvious, but with circuits that have multiple voltage sources or circuitry that is more convoluted, the amount of voltage and the polarity might not be as obvious.

Pictorial of Two Batteries

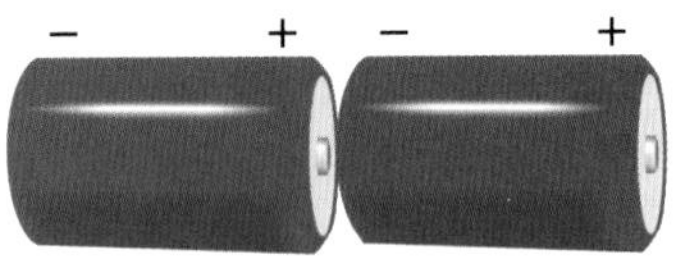

Figure 4–27 Two batteries in series have an additive effect.

Schematic of Two Batteries

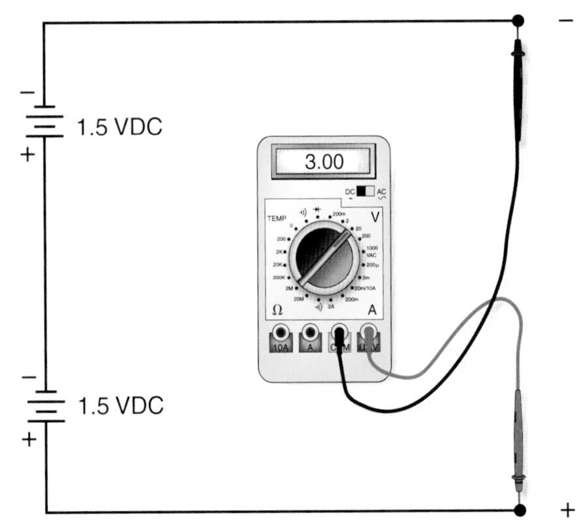

Figure 4–28 Two batteries in series have additive voltages.

VOLTAGE SOURCES

So far, you have studied circuits containing one simple battery. What if the circuit requires a larger voltage source than is available from a single battery? One circuit that almost always requires a larger voltage than is available in an individual battery is a household flashlight. A common flashlight requires two D cell batteries (Figure 4–27).

Each battery is 1.5 VDC. This means that the lamp in the flashlight is a 3 V light. Figure 4–28 shows the schematic equivalent of two 1.5 VDC batteries connected in series. Note that the negative terminal of one battery is connected to the positive terminal of the next battery. This causes the voltage of one battery to be added to the next. This is called a series-additive power source. As the voltmeter shows, the total voltage is 3.0 VDC across both batteries. In other words, the total voltage is the sum of the voltages added in series, regardless of their values.

$$E_T = E_1 + E_2 + E_3 + \ldots. E_n$$

If batteries are connected in series but are not connected positive terminal to negative terminal, we call them series-opposing power sources. Figure 4–29 shows an example of series-opposing batteries. In this case, the total voltage is equal to the larger voltage minus the smaller voltage.

$$E_T = E_{Larger} - E_{Smaller}$$

Figure 4–30 shows a combination of series-additive and series-opposing power sources. Note that in this circuit two of the voltages are connected in one direction or polarity, and the other two voltages (although matching each other) are connected in the opposite direction relative to the first group. To solve for the total resulting voltage from all sources, the voltage sources must be grouped by polarity and then each group added separately. In the circuit shown, the 3.0 V and 6.0 V are both the same polarity, whereas the 1.5 V and the other 1.5 V both have the same polarity but are connected in the opposite direction relative to the first group. These two groups must be added, and then the smaller value must be subtracted from the larger. The following are polarity group 1:

$$E_{Group}\ 1 = E_1 + E_4$$

$$E_{Group}\ 1 = 1.5\ \text{V} + 1.5\ \text{V}$$

$$E_{Group}\ 1 = 3.0\ \text{V}$$

The following are polarity group 2:

$$E_{Group}\ 2 = E_2 + E_3$$

$$E_{Group}\ 2 = 3.0\ \text{V} + 6.0\ \text{V}$$

$$E_{Group}\ 2 = 9.0\ \text{V}$$

$$E_{(Total)} = \text{Larger Group} - \text{Smaller Group}$$

$$E_{(Total)} = E_{Group}\ 2 - E_{Group}\ 1$$

$$E_{(Total)} = 9.0\ \text{V} - 3.0\ \text{V}$$

$$E_{(Total)} = 6.0$$

This shows that by adding the various voltages in groups and then subtracting the smaller group total from the larger group total, the total amount of resulting voltage supplied to the circuit from all the various voltage sources can be calculated. We can solve the same problem by using Kirchhoff's voltage laws. Start at any point in the circuit and add the voltages as you see them, with the polarities as indicated for the voltage sources.

For instance, starting at point A and going clockwise, we encounter a negative 1.5 V. Add to that a positive 3.0 V, a positive 6.0 V, and a negative 1.5 V.

Schematic of Two Batteries

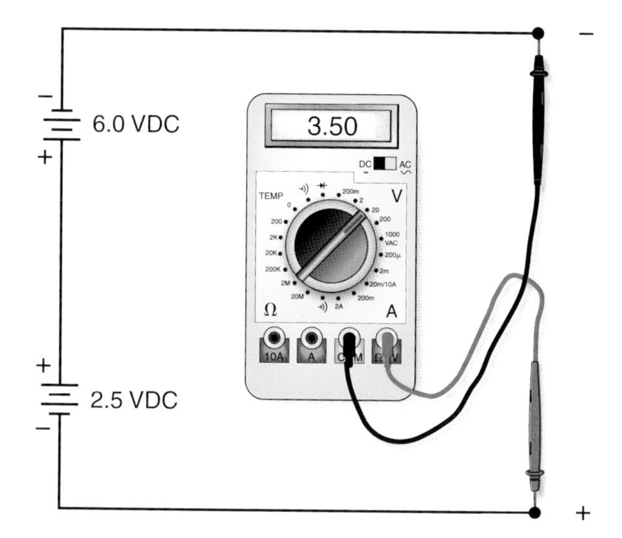

Figure 4–29 Voltage is subtracted when two batteries have opposing polarities.

Combination of Power Sources

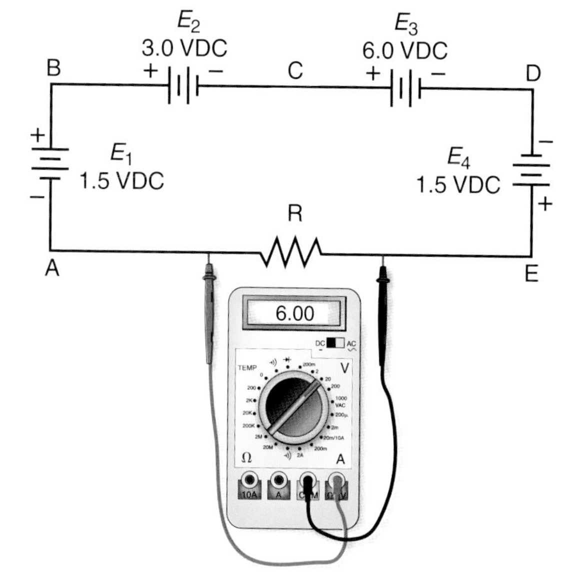

Figure 4–30 Circuits with multiple voltages can add or subtract from the total.

The voltage from A to E is:

$$(-1.5) + (+3.0) + (+6.0) + (-1.5) = 6$$

A is + 6 V compared to E. Kirchhoff's law tells us that all of the voltages have to equal 0 V by the time we get back to the starting point. Therefore, the voltage drop across the resistor must be 6 V positive at point A and negative at point E. The sum is as follows:

point A (–1.5) + (+ 3.0) + (+ 6.0) + (–1.5) point E + (resistor) = 0

Because we keep going in the same clockwise direction, the resistor must be (–6.0) in the series circuit, with the negative on the right.

SUMMARY

In this chapter you have been introduced to the standard method of drawing circuit diagrams as schematic diagrams. The symbols used are almost universal, with just a few variations. This allows electrical personnel worldwide to understand each other's work. As the series DC circuit is the most basic, we started with the rules for series circuits. These rules stay constant for series DC circuits and are only slightly modified for AC circuits.

We know that there is only one path for current to flow in a series circuit. The total resistance of the circuit is the sum of the individual resistors. The voltage drop on the individual loads is proportional to the resistance, and all voltage must be accounted for (as in Kirchhoff's law). Power consumed or dissipated by the loads is added to determine the total wattage load. Series circuits can include multiple voltage sources, as in batteries connected in series. The electrical polarity of the DC sources dictates whether the voltages add or subtract from the total.

Resistor styles and types were introduced to allow you to conveniently work with resistors in testing circuits. Resistors have several parameters of which to be aware. Often the values of these parameters are coded by a color stripe system and the standard color code system.

CHAPTER GLOSSARY

Component diagram (pictorial diagram) A component diagram is a drawing that shows the interconnection of system components by using photographs or drawings of the actual components. This is also known as a pictorial diagram.

Composition carbon resistor A resistor that derives its resistance from a combination of carbon graphite and a resin bonding material.

Metal film resistor A resistor that derives its resistance from a thin metal film applied to a ceramic rod.

Metal glaze resistor Similar to a metal film resistor except that the film is much thicker and is made of metal and glass.

Schematic diagram A schematic is a structural or procedural diagram, especially of an electrical or mechanical system, using special symbols to represent the actual physical components. The schematic shows the "scheme" of the current flow in a systematic representation.

Wire-wound resistor A resistor made by winding resistive wire around an insulating form.

REVIEW QUESTIONS

1. Review the circuit symbols shown in Table 4–1. Draw a circuit that shows a battery in series with an ammeter, in series with three resistors, and in series with a fuse.
2. Describe the concept of directly proportional voltage and current versus inversely proportional resistance and current.
3. Compare and contrast carbon-composition resistors, metal film resistors, and wire-wound resistors. The elements to include in your discussion are resistor values, tolerances, and power capabilities.
4. Carefully review Figure RQ4–1. What would be the color code for a standard 5-band, 2,200 ohm resistor with ± 1% tolerance? A standard 5-band, 5 megohm resistor with ± 20% tolerance?
5. Discuss the various uses for variable resistors. What advantage does a 10-turn variable resistor have over a one-turn variable resistor?

6. How is a rheostat physically different from a potentiometer?
7. Briefly explain how Kirchhoff's voltage laws could help you in a series DC circuit.
8. List two methods for finding the total resistance of a series circuit.
9. List two methods of finding the total voltage of a series circuit.
10. Examine Figure RQ4–2. All elements in this circuit exhibit some resistance. Why can you generally ignore the resistance of the wire, battery, fuse, and switch?

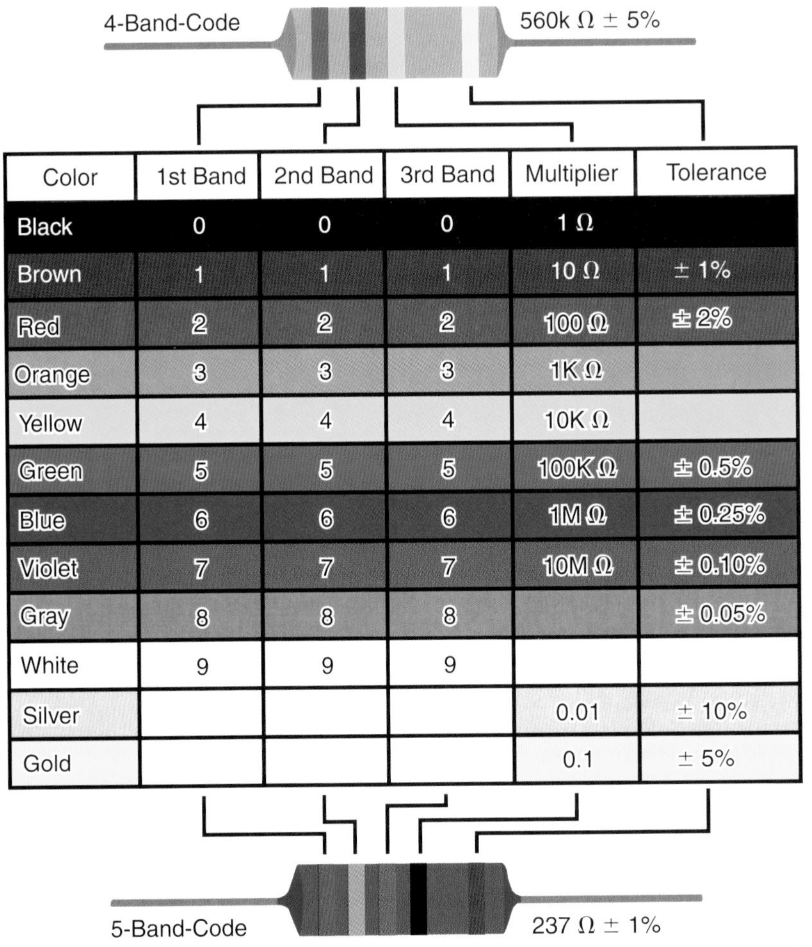

Color	1st Band	2nd Band	3rd Band	Multiplier	Tolerance
Black	0	0	0	1 Ω	
Brown	1	1	1	10 Ω	± 1%
Red	2	2	2	100 Ω	± 2%
Orange	3	3	3	1K Ω	
Yellow	4	4	4	10K Ω	
Green	5	5	5	100K Ω	± 0.5%
Blue	6	6	6	1M Ω	± 0.25%
Violet	7	7	7	10M Ω	± 0.10%
Gray	8	8	8		± 0.05%
White	9	9	9		
Silver				0.01	± 10%
Gold				0.1	± 5%

Figure RQ4–1 A standard color code is used to represent numerical values of resistance.

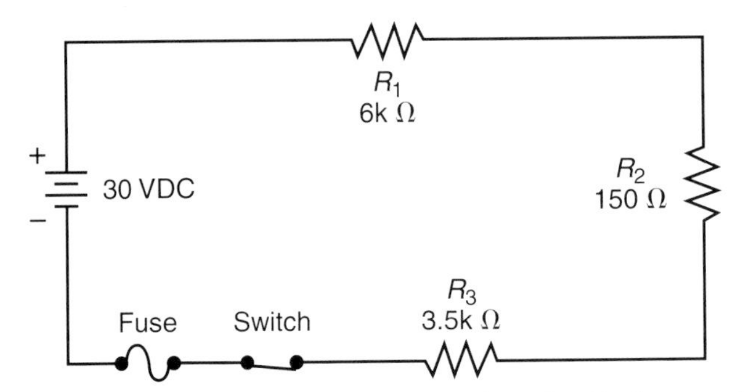

Figure RQ4–2 Changing the voltage in a series circuit changes the total current and the individual voltage drops.

PRACTICE PROBLEMS

1. In Figure PP4–1, what is the resistance and tolerance of each resistor?

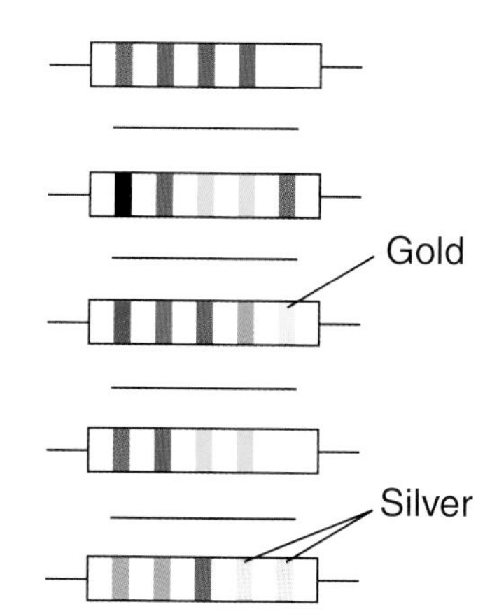

Figure PP4–1

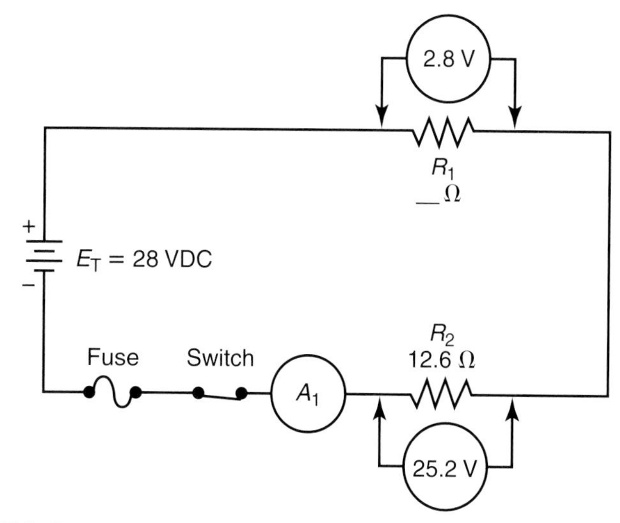

Figure PP4–2

2. Ignoring the individual values shown in Figure PP4–2, assume that the battery voltage is 125 V and that the total power dissipation is 25 W. What is the total resistance of the circuit?
   a. 5 ohms
   b. 20 ohms
   c. 550 ohms
   d. 625 ohms
3. Ignoring the individual values shown in Figure PP4–2, assume that the battery voltage is 125 V and the total current is 10 amperes, what is the total resistance in ohms?
   a. 1.25 ohms
   b. 12.5 ohms
   c. 125 ohms
   d. 10 ohms
4. In Figure PP4–3, what are the missing values for the resistors in three different circuits (Ckt) with the values shown in Table PP4–1? All resistances are given in ohms.

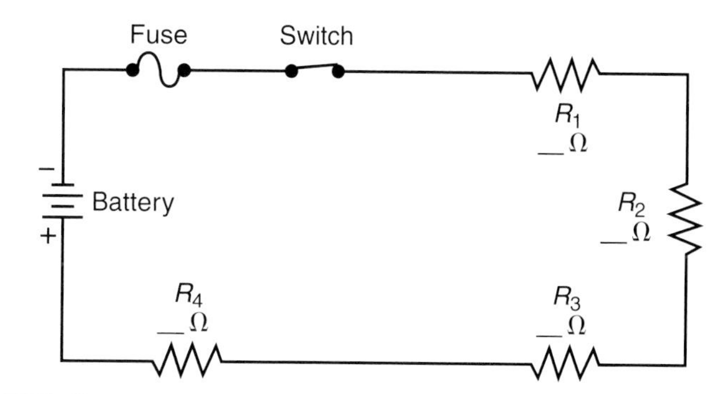

Figure PP4–3

Table PP4–1

	R_1	R_2	R_3	R_4	Total R_T
Ckt 1	5	10	15	20	?
Ckt 2	2,200	?	14,000	600	20,100
Ckt 3	1.5	2	6	?	23.5

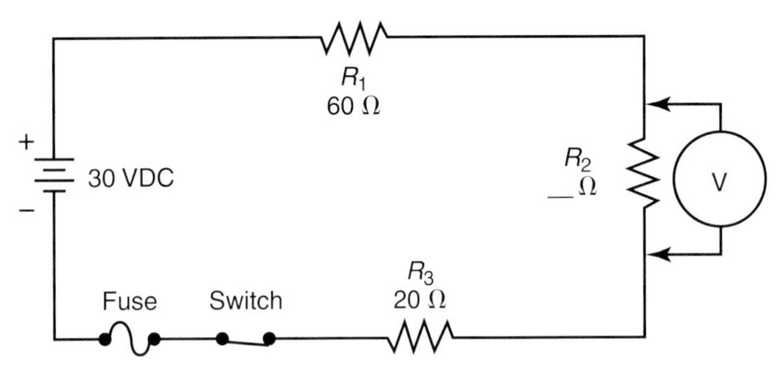

Figure PP4–4

5. Using Ohm's law and the values shown, calculate the resistance of resistor R_1 in Figure PP4–2.
   a. 0.7 ohms
   b. 1.4 ohms
   c. 2.5 ohms
   d. 2.8 ohms
6. In Figure PP4–4, if the voltage across resistor R_2 is measured as 10 V, what is the resistance of R_2?
   a. 20 ohms
   b. 40 ohms
   c. 60 ohms
   d. 80 ohms
7. A resistor is 33,000 ohms and has a ±10% tolerance. What is the lowest resistance that you should read if you measure the resistor with a very accurate ohmmeter?
   a. 29,700 ohms
   b. 33,000 ohms
   c. 36,300 ohms
   d. 25,000 ohms
8. The voltage in a circuit is 100 V and the current is 20 amperes. What is the resistance of the circuit?
   a. 2,000 ohms
   b. 120 ohms
   c. 5 ohms
   d. 1 ohm
9. A certain composition carbon resistor is 5/16 inch in diameter and 11/16 inch long. What is its wattage rating?
   a. 1/4
   b. 1/2
   c. 1
   d. 2
10. Refer to Figure PP4–5 to answer questions a through c.
   a. What will the bottom meter read? ____
   b. What is the value of R_1? ____
   c. R_1 is a 5 W resistor. Will it exceed its wattage rating? ____

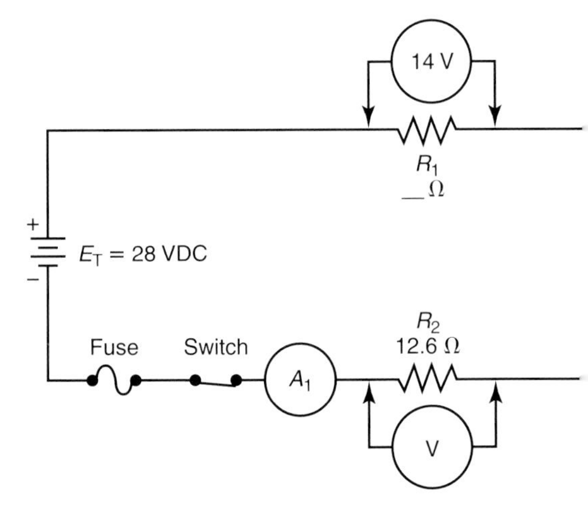

Figure PP4–5

Refer to Table 4-1 on pages 88 and 89 to answer questions 11 through 18.

11. Draw a series circuit schematic diagram with a battery and a lamp controlled by a normally open switch.

12. Construct a complete series circuit with a battery as a DC power source, an ammeter, a rheostat, and a lamp.

13. Calculate the current if the voltage source in the circuit of question 12 is 60 V, the rheostat has 20 Ω of resistance, and the lamp has 20 Ω.

14. Add another identical lamp to the circuit in question 12, with the rheostat adjusted to 5 Ω. Using the other values given in question 13 for the circuit, calculate the new current.

15. Create a circuit with a DC power source, a fuse, a switch, and a coil, all in series. The coil will operate a valve when the coil is energized. Explain how to use a voltmeter to troubleshoot the circuit if the valve does not operate when the switch is closed.

16. Draw a series circuit with a power source, a fuse, an ammeter, a normally open switch, and a lamp. If the lamp "burns out" and the switch is closed, what does the ammeter read?

17. If the circuit in question 16 is operating normally, a short occurs across the lamp, and what would a voltmeter read across the fuse?

18. In an operating circuit with two resistors and a rheostat in series with a DC power supply, what happens to the voltage reading across the rheostat if the rheostat's resistance is decreased?

5

DC Parallel Circuits

TOPICS

- INTRODUCTION
- PARALLEL CIRCUITS: DEFINITION AND PRINCIPLES
- THEORY OF CURRENT IN PARALLEL
- CALCULATING TOTAL RESISTANCE IN PARALLEL CIRCUITS
- VOLTAGE SUPPLY
- MULTIPLE VOLTAGE SOURCES
- POWER CALCULATION IN PARALLEL CIRCUITS

OVERVIEW

This chapter introduces you to DC parallel circuits by showing you how current flows through multiple branches. You already have the tools required to analyze current flow in a series circuit using Ohm's law. As you will see, Ohm's law still holds true but its application to parallel circuits changes. The way current flows through a parallel circuit will dictate how to use which rules to solve for the total circuit values.

A parallel circuit means that there is more than one path for current to flow from the negative source, through the loads, and back to the source. The multiple paths are thought of as parallel paths, although they may not seem parallel (as in geometry) in an actual circuit. Current rules for parallel circuits are explained and tested. Total current is the sum of each branch current.

In this chapter, you will also learn different methods of calculating total circuit resistance in parallel circuits. You will recall that to calculate the total resistance in series you simply added the values of all resistors. It seems sensible to think that whenever resistance is added to a circuit, total resistance will increase and total current will decrease. This is true only for a series circuit. As we add resistors in parallel, the total resistance decreases.

You will learn that in a parallel circuit each component has the same voltage drop (applied voltage) across it. Each branch does not have to have the same resistance. This means that the current flowing through each branch depends on the resistance of that branch. The total current in the circuit is simply the sum of the various currents flowing through each of the branches, but each branch current, voltage, and resistance can be calculated with Ohm's law. Finally, Watt's laws as applied to parallel circuits are explored. As you will see, the rules are very similar to series circuit rules.

OBJECTIVES

After completing this chapter, you will be able to:

- Draw parallel circuits showing alternate current paths in those circuits
- Calculate the currents in individual branches of parallel circuits
- Determine the total current in parallel circuits by determining the equivalent resistance of the circuit and solving for current
- Determine total current by finding the current in each branch and then solving for total current
- Calculate total resistance of a variety of resistance values in parallel
- Determine circuit voltage and branch voltages in parallel circuits
- Use Watt's laws to determine branch wattage and total circuit power in parallel circuits

INTRODUCTION

DC parallel circuits differ from DC series circuits in many ways. The rules for calculating individual circuit values are the same, but the effect each component has on the total circuit is much different. Therefore, new rules for parallel circuits are needed. Methods of calculating total current require us to add each component current to create a sum, rather than the same current (as in series circuits). Resistance is not just added together to create a larger total resistance, and voltage does not drop proportionally at each load (as in series circuits). Power totals are similar to series circuits.

PARALLEL CIRCUITS: DEFINITION AND PRINCIPLES

Parallel circuits have more than one path for current. These paths are called branches. All of the parallel branches have the same voltage applied, and each branch contributes to the total circuit current. The total current is made up of the sum of the current flows from each branch of the circuit (Figure 5–1).

Parallel Circuit

E_T R_1 R_2

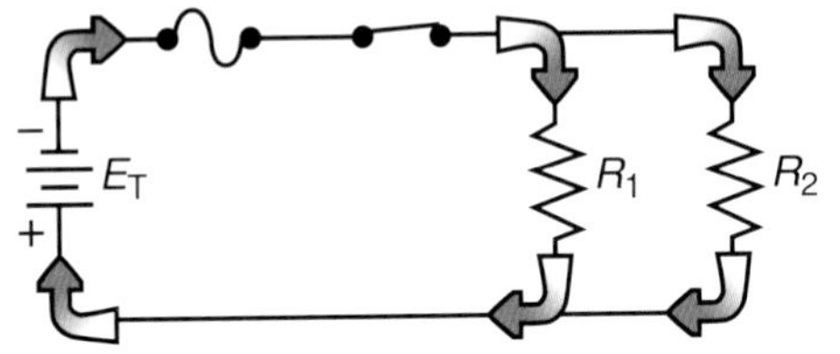

Figure 5–1 Two resistors in a parallel circuit are depicted.

The colored arrows indicate the direction of current flow. You can see that the current takes two paths to move from the negative side of the battery back to the positive side. The circuit is called parallel because the branches are connected side by side (parallel with each other). Both ends of each branch are connected to each side of the battery.

Figure 5–2 shows the same circuit without the current arrows. Note that the two resistors have voltmeters connected across them. The voltmeters are measuring the voltage drop across each branch. Each voltmeter relates to the battery in the same way. The top probe of each voltmeter is connected through the switch and fuse to the negative side of the battery. The bottom probe is connected to the positive side of the battery.

With both voltmeters connected to the same place on the battery, each reads the same voltage. In other words, both voltmeters are reading the battery voltage. You can see that the voltmeters are also reading the voltage drop across each resistor. This means that the parallel branches have the same voltage applied to them: battery voltage.

Parallel Circuit

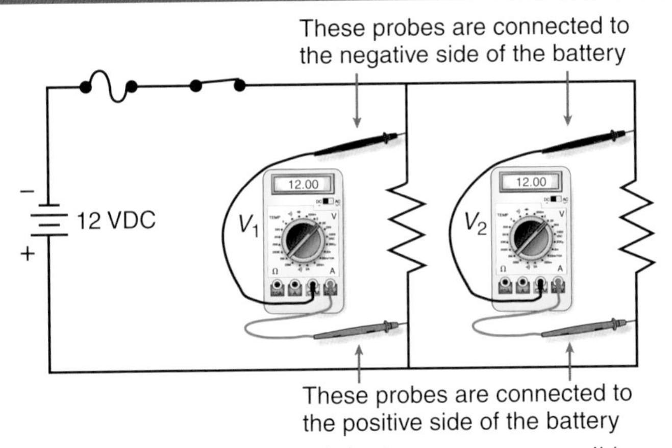

Figure 5–2 The voltage in a parallel circuit is the same across all branches.

The basic fact about parallel circuit branches is that they all have the same applied voltage. This is true regardless of the number of branches connected in parallel. As long as they are all connected to the same voltage source, they will have the same applied voltage:

$$E_T = E_1 = E_2 = E_3 = \ldots.\ E_n$$

Here, n is the number of branches in the parallel circuit. Figure 5–3 shows a simple parallel house circuit as an example of how parallel circuits work. Note that the toaster and the light are connected in parallel with the house voltage supply via the wall receptacle. The defining characteristic of **parallel circuits** is: *There is more than one current path for current to flow, and the total current is the sum of the individual component currents.*

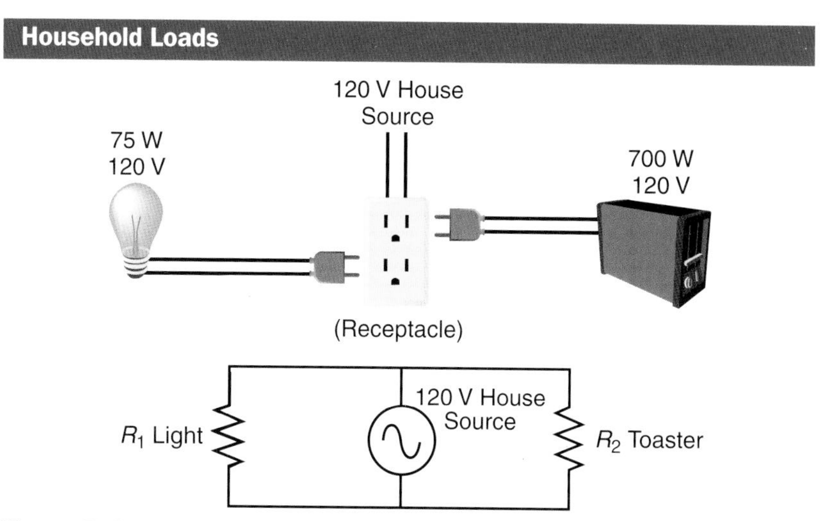

Figure 5–3 Typical electrical household loads are connected in parallel, as shown in the schematic view.

THEORY OF CURRENT IN PARALLEL

As described in previous chapters, current flow in circuits can be thought of as resembling flow through a piping system. Figure 5–4 represents a parallel piping system. The arrows show the relative amount of flow through the pipes. As you move from right to left (A to E) in the upper horizontal main pipe, the amount of flow in that pipe decreases as it "distributes" some of the flow into each parallel branch.

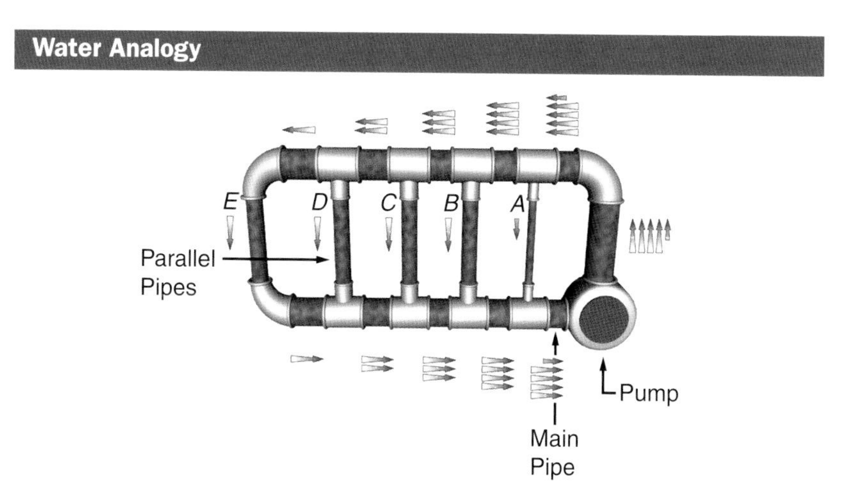

Figure 5–4 A water analogy helps visualize a parallel electrical circuit.

Moving left to right (E to A) in the lower horizontal main pipe shows the flow increasing as it "collects" current flowing back to the source. Everything that leaves the pump goes through the system of pipes and reenters the pump. This means that both the upper and the lower main pipes have the same amount of flow at equivalent points in the circuits.

The path between the two main pipes is the combination of the five parallel pipes (labeled A through E). Think of the main pipe as spreading out into five smaller pipes and then recombining back into one pipe before entering the pump. The *combined* flow through the parallel pipes is equal to the total flow to and from the source through the main pipes. In other words, the sum of the flow of the parallel pipes equals the total flow at the source. Flow in Pipe A + Pipe B + Pipe C + Pipe D + Pipe E = Total Flow.

The parallel pipes of the water analogy are called branches from the main pipes. The same is true about parallel DC circuits. The multiple paths for current to flow are called branches. The total current flow leaving the voltage source is equal to the sum of the currents in all branches.

The **parallel current rule** is: *The sum of the individual branch currents equals the total source current. The formula is:*

$$I_T = I_1 + I_2 + I_3 + \ldots. I_n$$

Here, n is the number of branches in the parallel circuit.

In Figure 5–5, the total current (I_T) is shown leaving the battery and returning to the battery. Between the two points where current is labeled I_T, the current divides into three parallel circuit branches, each containing a resistor (R_1, R_2, and R_3). After traveling through each branch, the current recombines into I_T before reentering the battery.

CALCULATING CURRENT IN PARALLEL

Total current will have to be calculated in parallel circuits in which current through the branches is unknown. Figure 5–6 is the first example.

Example

What is the current through each of the three branches shown in Figure 5–6?

Solution:

To calculate the current, determine what the known values are.

$$R_1 = 15\ \Omega,\ R_2 = 30\ \Omega,\ R_3 = 45\ \Omega$$

$$E_T = 25\ \text{VDC}$$

Voltage across each branch is the same 25 V because the negative and positive terminals of the battery are directly connected to the top and bottom of each resistor. From Ohm's law:

$$I_1 = \frac{E_T}{R_1}$$

$$I_1 = \frac{25\ \text{V}}{15\ \Omega} = 1.667\ \text{A}$$

$$I_2 = \frac{E_T}{R_2}$$

$$I_2 = \frac{25\ \text{V}}{30\ \Omega} = 0.833\ \text{A}$$

$$I_3 = \frac{E_T}{R_3}$$

$$I_3 = \frac{25\ \text{V}}{45\ \Omega} = 0.556\ \text{A}$$

Parallel Circuit

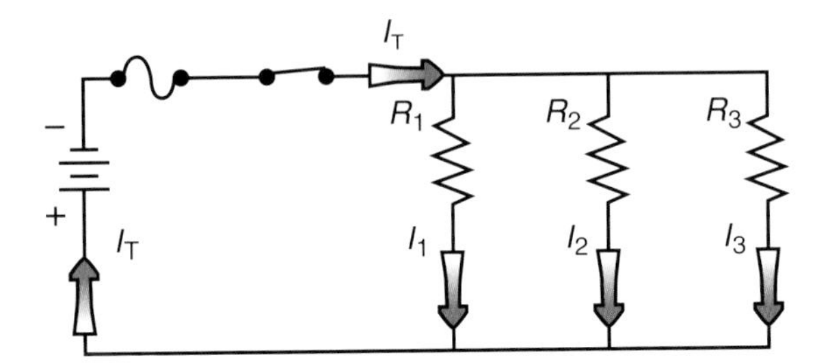

Figure 5–5 Current has multiple paths in a parallel circuit, as shown by the arrows.

Calculating Total Current

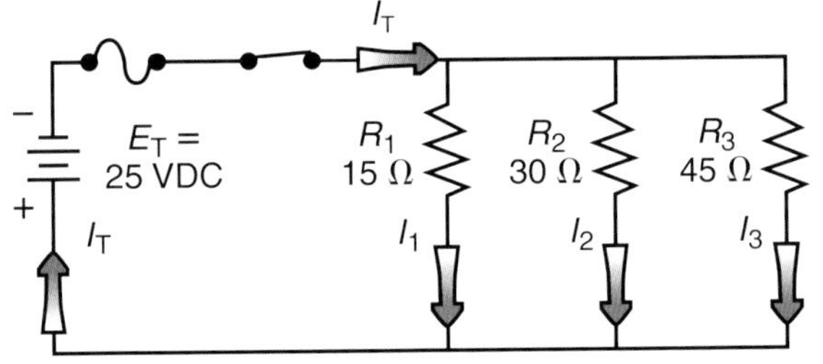

Figure 5–6 The current in each branch is calculated to find the total.

As we add more branches to the circuit, the amount of current that flows increases as long as the power supply can yield enough current at the rated voltage. As we add more paths in parallel (more resistors in parallel), the total current from the source increases ($I_T = I_1 + I_2 + I_3 \ldots$). In parallel circuits, the more resistors we add in parallel, the more total current flows.

Thus, using Ohm's law for circuit totals: $R_T = \frac{E_T}{I_T}$. Note that as I_T increases, the R_T must decrease, as they are inversely proportional. This equates to the idea that the total resistance of the entire circuit decreases as the number of resistors increases. This is the opposite effect of what happens in a series circuit.

Example

What is the total current in Figure 5–6?

Solution:

There are two methods for determining total current. If we know the total resistance of the entire circuit, we can use the totals for resistance and voltage to find the total current. We will provide the total resistance at this point as 8.18 ohms. (We will calculate total resistance using several methods in the following sections.) Then, using the equation for current from Ohm's law and placing total resistance in for R, the total current is:

$$I_T = \frac{E_T}{R_T}$$

$$I_T = \frac{25\ \text{V}}{8.18\ \Omega} = 3.056\ \text{A}$$

Another method for determining total current in Figure 5–6 is to add all branch currents. Each branch current is calculated individually by using the component voltage and the component resistance to provide the component branch current. The parallel current rule (that the total current is the sum of the branches) yields:

$$I_T = I_1 + I_2 + I_3$$

$$I_T = 1.667 + 0.833 + 0.556 = 3.056\ \text{A}$$

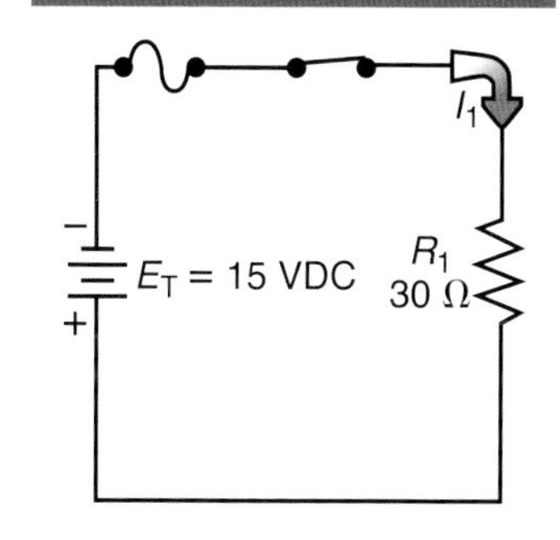

Figure 5–7 A simple circuit with one load resistor is depicted.

CALCULATING TOTAL RESISTANCE IN PARALLEL CIRCUITS

The fact that total circuit resistance decreases as resistance is added to a parallel circuit is easily misunderstood. Figures 5–7 through 5–9 show a circuit with a progression of parallel resistors being added to the circuit. Note that each branch has the same resistance and thus the same current through each resistor.

Also note the effect that adding each resistor has on the total current of the circuit.

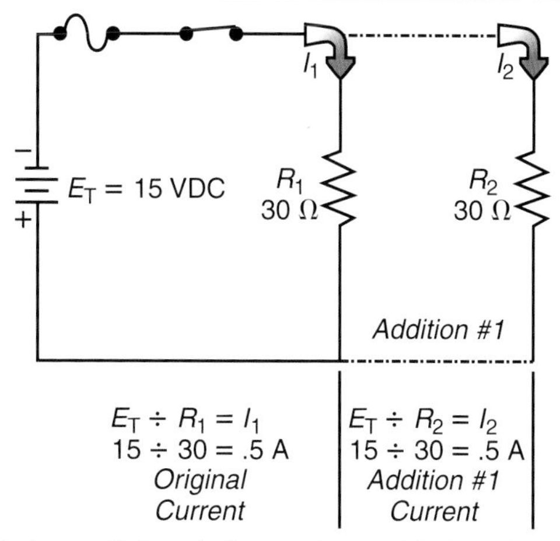

Figure 5–8 A parallel resistive path is added to the circuit of Figure 5–7.

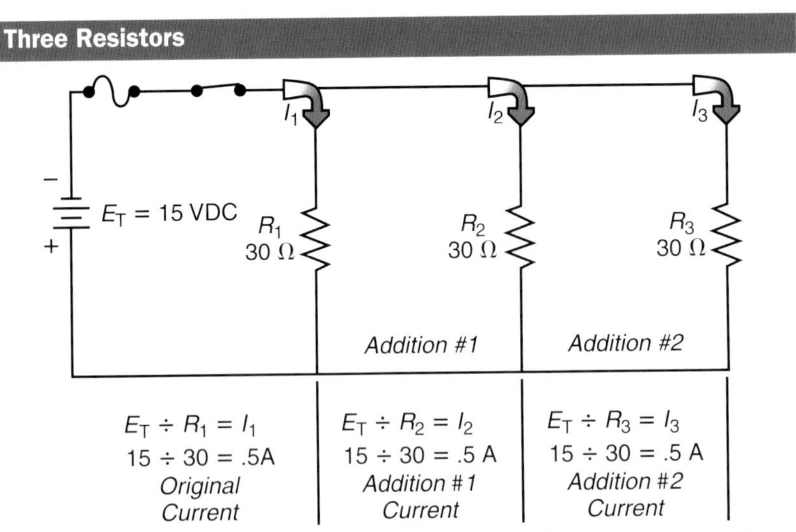

Figure 5–9 Adding a third resistor to Figure 5–8 does not change the voltage but does affect the total current.

In Figure 5–7, you see a relatively simple circuit comprising a resistor in series with a fuse and switch. You can see that the source voltage is the same voltage seen by resistor R_1. Therefore, $E_T = E_1$. Use Ohm's law to calculate the current through the resistor R_1.

$$I_1 = \frac{E_1}{R_1}$$

$$I_1 = \frac{15\text{ V}}{30\ \Omega}$$

$$I_1 = 0.5\text{ A}$$

When you look at the total circuit current in Figure 5–7, you see that the only current in the circuit is passing through resistor R_1. This means that $I_1 = I_T$. Therefore:

$$I_T = I_1 = 0.5\text{ A}$$

Now examine Figure 5–8. A second resistor has been added in parallel with the resistor in the circuit from Figure 5–7. The second resistor has the same voltage across it as the resistor in Figure 5–7. In parallel circuits, the voltage is the same across all branches of the circuit. In Figure 5–8, we have added a second branch that also has the supply voltage E_T across the resistive load. Therefore, $E_2 = E_T = E_1$. To calculate the current through the resistor R_2, you again use Ohm's law.

$$I_2 = \frac{E_2}{R_2}$$

$$I_2 = \frac{15\text{ V}}{30\ \Omega}$$

$$I_2 = 0.5\text{ A}$$

By adding the second resistor, the total circuit current changed. Remember:

$$I_T = I_1 + I_2 + \ldots.\ I_n$$

Therefore:

$$I_T = I_1 + I_2 = 0.5\text{ A} + 0.5\text{ A} = 1\text{ A}$$

By adding a second resistor in parallel with the first resistor, you have also added to the total current of the circuit. Now add a third resistor. Figure 5–9 shows the new circuit with three resistors in parallel. Just as both the first and the second resistors saw the supply voltage E_T across them, the third resistor will see the supply voltage across it. This is one of the basic rules of parallel circuits: All branches in a parallel circuit will see the same voltage. The voltage they will see is the voltage supplying that branch. You now calculate the current in the third branch, again using Ohm's law.

$$I_3 = \frac{E_3}{R_3}$$

$$I_3 = \frac{15\text{ V}}{30\ \Omega}$$

$$I_3 = 0.5\text{ A}$$

With the third resistor, the total circuit current has changed. Remember:

$$I_T = I_1 + I_2 + I_3 + \ldots I_n$$

Therefore:

$$I_T = I_1 + I_2 + I_3$$

$$= 0.5\text{ A} + 0.5\text{ A} + 0.5\text{ A} = 1.5\text{ A}$$

By adding a third resistor in parallel with the first and second resistors, the total current of the circuit has increased. In Figure 5–9, it just so happens that all resistances were equal. However, not all circuits are as convenient.

As resistors are added in parallel, regardless of their value, the total circuit current increases. The current from each branch will make up part of the total of all branches.

Using Ohm's law, the total resistance of all branches of a parallel circuit can also be calculated once the total circuit current is known. The following calculations show how to find the total circuit resistance. Note that the total circuit resistance value is smaller than any of the individual branch resistance values. A **rule for parallel circuit resistance** is: *The total resistance is less than the smallest branch resistor.*

To solve for total resistance when we know current values, we use Ohm's law, which states that $R_T = \frac{E_T}{I_T}$. The parallel current rule determines that $I_1 + I_2 + I_3 = I_T$. Therefore:

$$E_T = 15 \text{ V}$$

$$R_T = \frac{15 \text{ V}}{(0.5 \text{ A} + 0.5 \text{ A} + 0.5 \text{ A})}$$

$$R_T = \frac{15 \text{ V}}{1.5 \text{ A}}$$

$$R_T = 10 \ \Omega$$

PARALLEL RESISTANCE RECIPROCAL EQUATION

To calculate the total resistance in a parallel circuit, a general formula is needed. This formula is known as the *reciprocal equation for parallel resistance*. This equation may be used for any parallel circuit. The following describes how the reciprocal equation is derived. In Figure 5–10, the sum of all branch currents is equal to the total current.

$$I_T = I_1 + I_2 + I_3$$

Using Ohm's law and substituting the equation for current in the previous equation:

$$I_T = (I_1 + I_2 + I_3)$$

Because this circuit is a parallel circuit, all branch voltages are equal to each other and to the total voltage. Thus, using the total current rules, we can use the voltage and the resistance to find the current as follows:

$$\frac{E_T}{R_T} = \left(\frac{E_T}{R_1} + \frac{E_T}{R_2} + \frac{E_T}{R_3}\right)$$

If you divide both sides of this equation by E_T, the equation becomes:

$$\frac{1}{R_T} = \left(\frac{1}{R_1} + \frac{1}{R_2} + \frac{1}{R_3}\right)$$

This is the reciprocal equation, but it results in the reciprocal of total resistance, which is difficult to work with. To get total resistance as a useable number, take the reciprocal of both sides. The equation then becomes:

$$R_T = \frac{1}{\left(\frac{1}{R_1} + \frac{1}{R_2} + \frac{1}{R_3}\right)}$$

The true form of the equation is:

$$\frac{1}{R_T} = \left(\frac{1}{R_1} + \frac{1}{R_2} + \frac{1}{R_3} + \ldots. \frac{1}{R_n}\right)$$

Here, n is the number of branches in the circuit. The more useable form of the reciprocal formula is:

$$R_T = \frac{1}{\left(\frac{1}{R_1} + \frac{1}{R_2} + \frac{1}{R_3} + \ldots. \frac{1}{R_n}\right)}$$

To determine the total resistance of the circuit when you only know the resistance values, take the reciprocal of each branch resistance and add them together to get the sum of the reciprocals. Then take the reciprocal of the sum to yield the total R_T. This is sometimes referred to as taking the reciprocal of the reciprocals.

Different Currents Flow

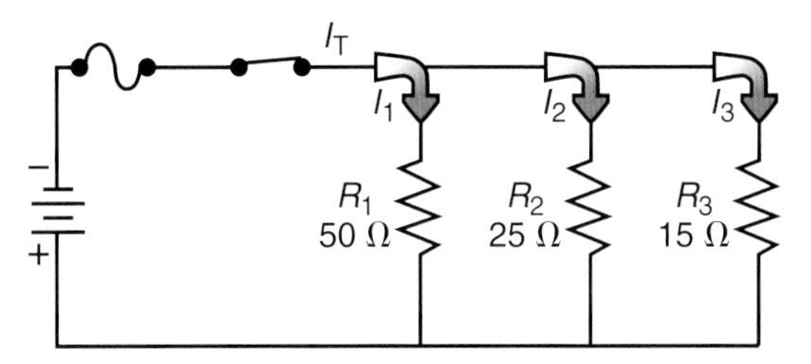

Figure 5–10 Different currents flow in each branch if the branch resistances are different.

Example

What is the total resistance in Figure 5–10?

Solution:

To calculate the total resistance, use the reciprocal equation.

$$R_T = \frac{1}{\left(\frac{1}{R_1} + \frac{1}{R_2} + \frac{1}{R_3}\right)}$$

$$R_T = \frac{1}{\left(\frac{1}{50\ \Omega} + \frac{1}{25\ \Omega} + \frac{1}{15\ \Omega}\right)}$$

$$R_T = \frac{1}{(0.02 + 0.04 + .0667)}$$

$$R_T = \frac{1}{(0.1267)}$$

$$R_T = 7.89\ \Omega$$

Note that the total resistance (R_T) is less than the smallest branch resistance (15 Ω). This is always the case with parallel resistance values. The total resistance of the circuit formed by parallel resistors is always smaller than the resistance of the smallest branch alone.

Example

Find the total resistance of four resistors in parallel. They are 10 Ω, 20 Ω, 40 Ω, and 60 Ω. Remember that the R_T must be smaller than the smallest resistor of 10 Ω.

Solution:

$$R_T = \frac{1}{\left(\frac{1}{R_1} + \frac{1}{R_2} + \frac{1}{R_3} + \frac{1}{R_4}\right)}$$

$$R_T = \frac{1}{\left(\frac{1}{10} + \frac{1}{20} + \frac{1}{40} + \frac{1}{60}\right)}$$

$$R_T = \frac{1}{(0.1 + 0.05 + 0.025 + 0.01667)}$$

$$R_T = \frac{1}{0.19167}$$

$$R_T = 5.2\ \Omega$$

PRODUCT OVER SUM EQUATION

A special formula can be used if the parallel circuit has only two branches. If the parallel circuit has only two branches, a simpler equation can be used. This formula is referred to as the **product over sum method** to determine total resistance.

Look at the reciprocal equation again. If the circuit has only two branches, the equation becomes:

$$R_T = \frac{1}{\left(\frac{1}{R_1} + \frac{1}{R_2}\right)}$$

At this point, it is useful to find a common denominator for the right-hand side of the equation. This becomes:

$$R_T = \frac{1}{\left(\frac{R_2}{R_1R_2} + \frac{R_1}{R_1R_2}\right)}$$

$$= \frac{1}{\left(\frac{R_1 + R_2}{R_1R_2}\right)} = \frac{R_1R_2}{R_1 + R_2}$$

This equation is called the product over sum equation and can be used for any circuit, two branches at a time.

Example

What is the total resistance in Figure 5–11?

$$R_T = \frac{(R_1 \times R_2)}{(R_1 + R_2)}$$

$$R_T = \frac{(40\ \Omega \times 30\ \Omega)}{(40\ \Omega + 30\ \Omega)}$$

$$R_T = \frac{1{,}200}{70}$$

$$R_T = 17.14\ \Omega$$

A way to verify this answer is to take any voltage you choose and determine what the individual current in the branches would be. What would the total current be with that assumed voltage? For example, if we choose 120 VDC as the source voltage, the 40 Ω resistor would allow 3 A and the 30 Ω resistor would allow 4 A of branch current. The total current would be the sum, or 7 A, using the totals of:

$$\frac{E_T}{I_T} = R_T = \frac{120\ \text{V}}{7\ \text{A}} = 17.14\ \Omega$$

This is the same answer determined by the product over sum method.

Another way to verify these answers is to use the reciprocal method.

$$R_T = \frac{1}{\left(\frac{1}{R_1} + \frac{1}{R_2}\right)}$$

$$R_T = \frac{1}{\left(\frac{1}{40\ \Omega} + \frac{1}{30\ \Omega}\right)}$$

$$R_T = \frac{1}{(0.025 + 0.0333)}$$

$$R_T = \frac{1}{(0.0583)}$$

$$R_T = 17.15\ \Omega$$

Note that because of rounding of repeating decimals, the resultant total numbers are not exactly the same.

Example

Find the total resistance of a circuit with a 55 ohm resistor in parallel with an 83 ohm resistor. Without a calculator to find reciprocals conveniently, these odd values may seem more difficult to work with, but the product over sum method can easily accomplish the same task.

Solution:

$$R_T = \frac{(R_1 \times R_2)}{(R_1 + R_2)}$$

$$R_T = \frac{(55 \times 83)}{(55 + 83)}$$

$$R_T = \frac{4{,}565}{138}$$

$$R_T = 33.08$$

FieldNote!

When there are many different sizes of resistors in parallel, it is most convenient to use a calculator with a reciprocal key to find the total value. Add all reciprocals together and then take the reciprocal of the sum of the reciprocals. When you are adding reciprocals of odd values, round the numbers to three significant digits to avoid long, cumbersome additions. The answers are usually close enough to the actual values to provide needed circuit parameters.

Product Over Sum Rule

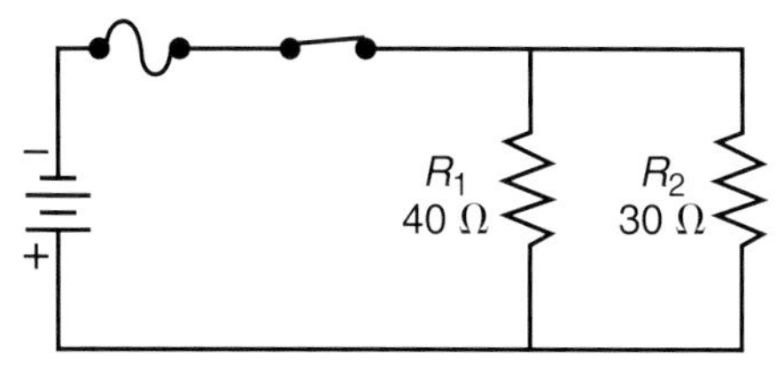

Figure 5–11 By using the product over sum rule, you can find the resistance total for two parallel resistors.

Equal Resistors in Parallel

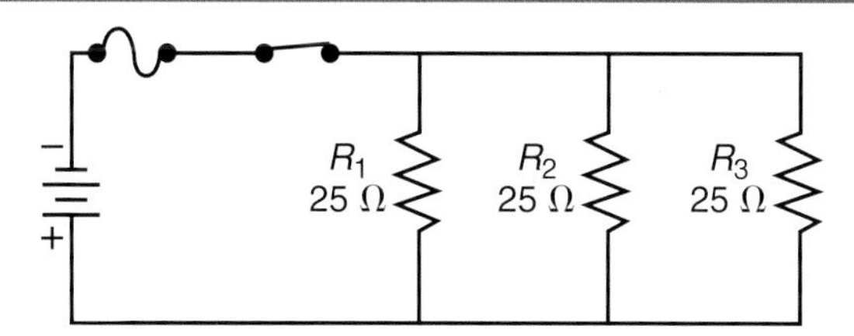

Figure 5–12 By using the resistors of equal value equation, you can find the resistance total for any number of equal resistors in parallel.

RESISTORS WITH EQUAL VALUE EQUATION

Parallel circuits with resistors of equal value are the easiest to calculate. The total resistance is simply the value of one of the resistors divided by the number of branches in the circuit. Refer to Figure 5–12. Use the reciprocal equation to determine how to calculate the total resistance of a parallel circuit whose branches are all of equal resistance.

$$R_T = \frac{1}{\left(\frac{1}{R_1} + \frac{1}{R_2} + \frac{1}{R_3}\right)}$$

Because R_1, R_2, and R_3 are equal, the equation becomes the following after substituting R_1 for the other resistance values:

$$R_T = \frac{1}{\left(\frac{1}{R_1} + \frac{1}{R_1} + \frac{1}{R_1}\right)}$$

$$R_T = \frac{1}{\left(\frac{3}{R_1}\right)}$$

$$R_T = \frac{R_1}{3}$$

The general form of this equation is:

$$R_T = \frac{R_X}{n}$$

Here, R_X is the resistance of any branch and n is the number of branches.

FieldNote!

Assume a chandelier has six lamps on the fixture (luminaire) and each is the same wattage and all are in parallel. If the lamps are 120 V 60 W, the resistance of each lamp can be calculated by the wattage formula $R = \frac{E^2}{P}$ or $R = \frac{120\text{ V}^2}{60\text{ W}}$. The resistance of each lamp is then 240 Ω. If there are six lamps (each 240 Ω), the total resistance of that circuit is $\frac{240\ \Omega}{6}$ or 40 Ω of R_T. We can determine the total current for the six 60 W lamps in parallel with $\frac{E_T}{R_T} = I_T$ or $\frac{120\text{ V}}{40\ \Omega} = 3$ A. This is a quick method of calculating the total resistance of a number of parallel loads if all are the same ohmic value.

Example

What is the total resistance of the circuit in Figure 5–12?

$$R_T = \frac{R_X}{n}$$

Here, $R_X = 25$ and $n = 3$.

$$R_T = \frac{25}{3}$$

$$R_T = 8.33\ \Omega$$

Prove the calculation using the reciprocal method.

$$R_T = \frac{1}{\left(\frac{1}{R_1} + \frac{1}{R_1} + \frac{1}{R_1}\right)}$$

$$R_T = \frac{1}{\left(\frac{1}{25} + \frac{1}{25} + \frac{1}{25}\right)}$$

$$R_T = \frac{1}{\left(\frac{3}{25}\right)}$$

$$R_T = \frac{25}{3}$$

$$R_T = 8.33\ \Omega$$

EQUAL VALUE VARIATION

As you have studied, the equal resistance rule applies when all resistance values in a parallel circuit are the same. A variation of the equal resistance rule can also be used when there are different resistance values in a parallel circuit and when these resistance values are multiples of one another or can be divided evenly into a common multiple.

For example, any single resistor can be considered as two or more equal-value resistors connected in parallel to form an equivalent resistor. A single 12 Ω resistor can be thought of as two 24 Ω resistors. Three 36 Ω resistors connected in parallel can be thought of as one equivalent 12 Ω resistor. An 18 Ω resistor could be thought of as two 36 Ω resistors or three 54 Ω resistors. When we have parallel circuits, a convenient way to perform calculations is to combine like resistors into equivalent resistors to calculate the total effect on the circuit. Always remember to use the actual component values to calculate the actual branch values.

For example: If you have a 12 Ω resistor and an 18 Ω resistor connected in parallel, you can express the circuit as five 36 Ω resistors in parallel (12 Ω = 36 Ω ÷ 3) with two more 36 Ω resistors (18 Ω = 36 Ω ÷ 2) in parallel. Refer to Figure 5–13. This circuit can also be shown as equal-value resistors in parallel. Figure 5–14 shows the equivalent circuit using a common dividend (36) as the value of resistors in parallel.

Equivalent Resistance Circuit

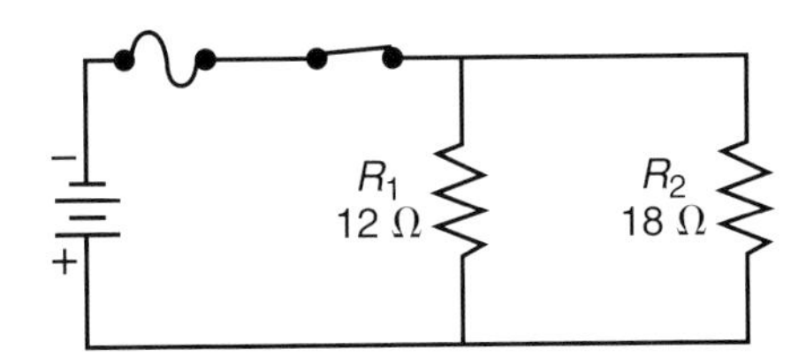

Figure 5–13 Two resistors can be divided to create an equivalent resistance circuit.

Equivalent Resistance Circuit

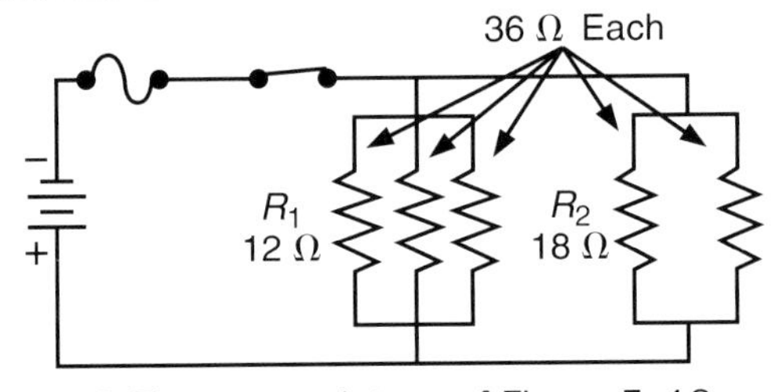

Figure 5–14 The two resistors of Figure 5–13 can be represented as equivalent resistors.

The total resistance can now be calculated using the equal resistance equation.

$$R_T = \frac{36}{5} = 7.2\ \Omega$$

Here, $R_X = 36$ and $n = 5$.

ASSUMED VOLTAGE

Another method of calculating total resistance when you know the branch resistances is to assume a voltage for the circuit. This **assumed voltage method** works with any number of resistor branches of any value.

For any circuit of parallel loads, we can determine what the assumed current would be for each parallel branch with the same assumed voltage applied. We then add all assumed currents together, as we do in any parallel circuit, to obtain an assumed total current. Then this assumed current total and the assumed voltage we chose will provide the actual total circuit resistance. The formula is $\frac{E_{T\ assumed}}{I_{T\ assumed}} = R_{T\ actual}$. For an example, see Figure 5–15 (a 40 Ω, a 60 Ω, and a 120 Ω resistor in parallel).

Select any voltage you would like to use. If you choose a voltage that easily works with your resistor values yielding whole numbers, the process is easier. If we choose 240 V, the current that would flow in branch 1 is 6 A. In branch 2 it would be 4 A, and in branch 3 it would be 2 A. The total current of this particular circuit is 12 A. We then use the assumed current of 12 A and the assumed voltage of 240 V to determine the actual R_T of 20 Ω. No matter which voltage we assume, the total resistance will always be 20 ohms.

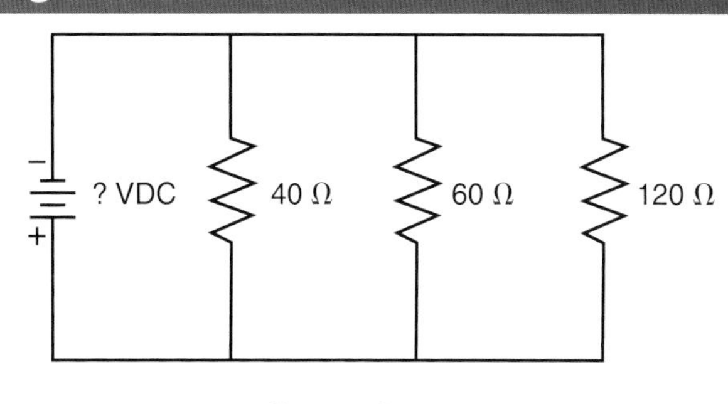

$R_T = 20\ \Omega$

Figure 5–15 Any resistor can have its total resistance calculated by using an assumed voltage.

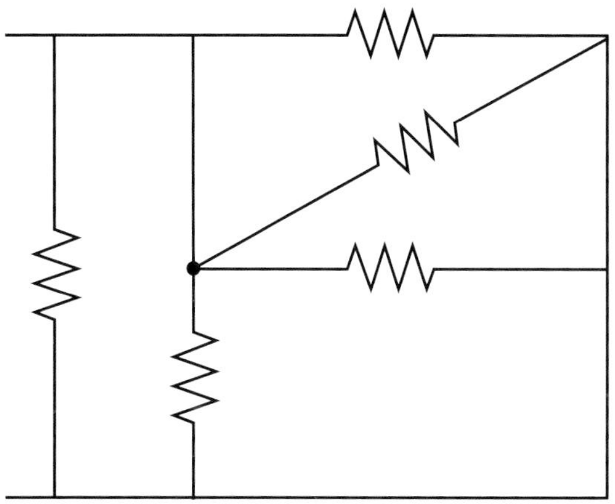

Figure 5–16 The convoluted pattern of resistors may not look like a parallel circuit but is one.

VOLTAGE SUPPLY

As we have seen throughout the study of parallel circuits, we have used the same voltage for each of the branch circuit voltages to determine branch currents. This is the **parallel circuit voltage rule**. The voltage across each branch of a parallel circuit is the same as any other branch in parallel. In fact, this rule helps us identify parallel circuits in wiring patterns in which the parallel configuration we are used to seeing is not as apparent. See Figure 5–16 to verify that the components are really in parallel, based on the rule for voltage in a parallel circuit. As you note, it might take some practice to "see" parallel paths in convoluted circuits.

MULTIPLE VOLTAGE SOURCES

Circuit branches are not the only components that may be connected in parallel. Voltage sources may also be connected this way.

One of the most common reasons for connecting voltage sources in parallel is to supply more current to a circuit. All real power supplies have power limits. Both generators and batteries have maximum amounts of current they can supply to a load.

If a circuit requires higher current but the same voltage, sources can be connected in parallel to combine their currents. Generators are connected in parallel all across a power system for this very reason. Figure 5–17 shows how batteries can be connected in parallel. Note that each battery will supply its rated current to the lights. Consequently, the total energy capability for the lights is tripled.

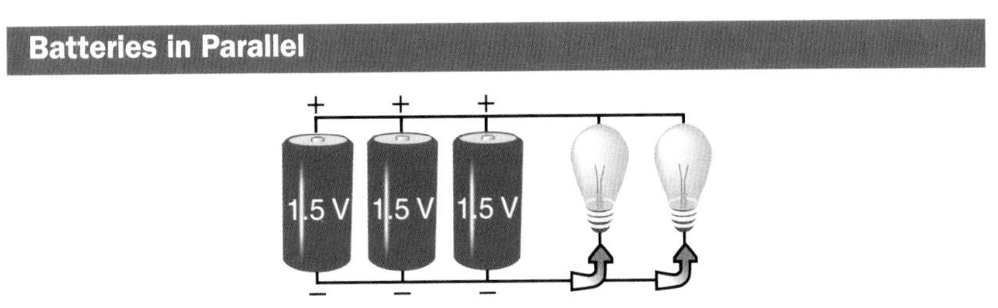

Figure 5–17 Batteries are connected in parallel to increase the current supply.

Figure 5–18 shows another example of voltage sources connected in parallel. Each voltage source in the circuit is a 12 V battery rated at 80 amp-hours. All negative ends are connected together, and all positive ends are connected together. The result is a source voltage of 12 VDC with a combined load capacity of 240 amp-hours.

The voltmeter is measuring all voltage sources. Because all terminals of the batteries are connected, the voltmeter senses only one voltage source in the circuit. In this circuit, the voltmeter reads 12 VDC. The voltage output from the voltage sources is equal to any individual battery. This would be true regardless of the number of batteries connected in the circuit.

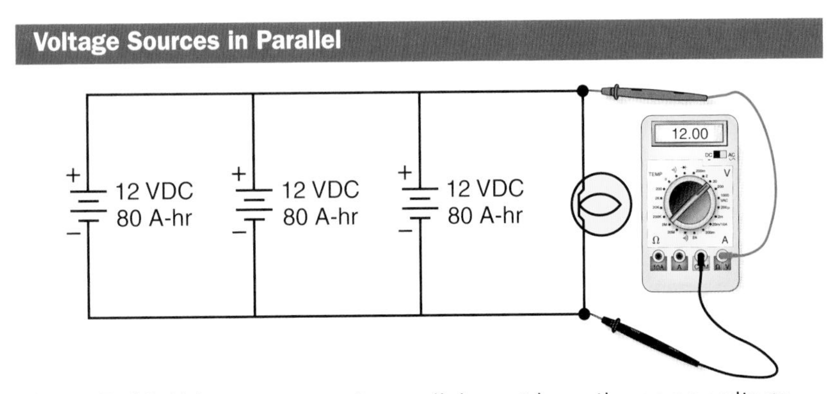

Figure 5–18 Voltage sources in parallel must have the same voltage and the same polarity.

Examine Figure 5–19. In this case, each power source provides a portion of the total circuit current. This configuration divides the current by the number of power sources connected in parallel. The advantage of placing power sources in parallel is an increased lifetime for each individual source. A single 12 V battery rated at 60 amp-hours would last twice as long if another 12 V 60 amp-hour battery were added in parallel. With two 12 V 60 amp-hour batteries in parallel, the total output increases to 120 amp-hours. Each battery will last twice as long in a circuit because each will be supplying half the required current.

Note that if the parallel sources are of equal power capacity (e.g., 60 amp-hours) the two will contribute equally to the total current. If they are not of equal power capacity (e.g., 30 amp-hours and 60 amp-hours), the larger one will generally contribute more of the circuit current.

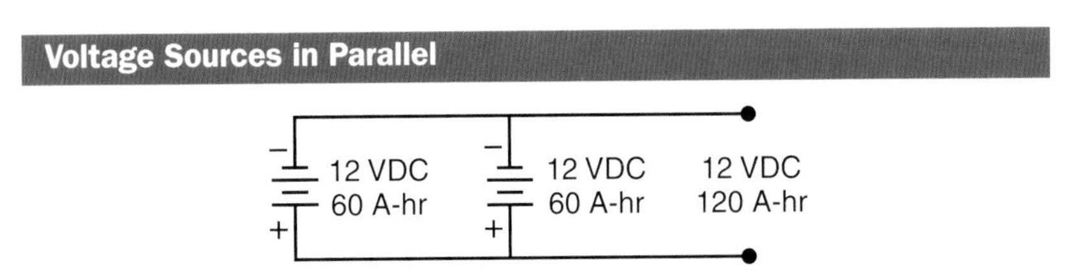

Figure 5–19 Voltage sources in parallel do not increase voltage but do increase current supply.

Wattage Formulas

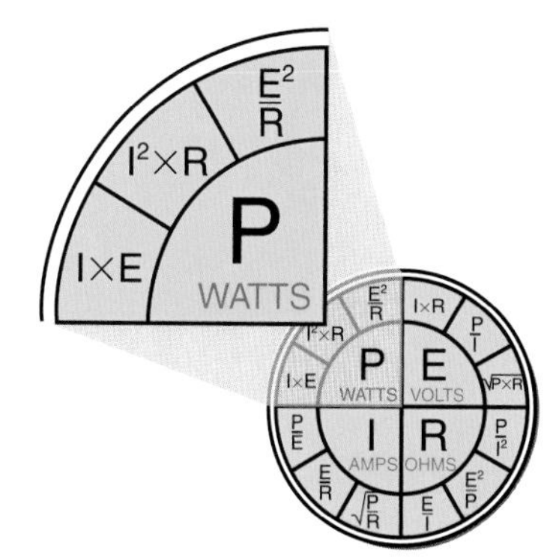

Figure 5–20 Wattage formulas are used with circuit values to determine unknowns.

Placing power supplies in parallel is common. In DC circuits, the power supply is often a battery source. As you may have done on your vehicle when your battery is weak or dead, you can jumper a battery from another vehicle to yours to provide the needed power source. Be sure to always connect the positive terminal to the positive terminal and the negative terminal to the negative terminal so that the voltages are the same polarity. Now the current sources are adding together and the voltage is the same, as in any typical parallel circuit.

CAUTION: Connecting voltage sources in parallel that do not have the same voltages and polarities is an unsafe practice and should never be done. Excessive currents could be produced that would damage the voltage sources and cause personal injury and explosions.

POWER CALCULATION IN PARALLEL CIRCUITS

The equations for calculating power in parallel are the same as those used in series circuits. The pie chart for Ohm's law (Figure 5–20) shows the three equations you may use. These equations are applied in the same manner as in a series circuit. The power for individual components can be calculated if any two of the three required values are known. The calculation for power of the total circuit is also the same.

There are two basic methods for determining total power: directly and indirectly. The general formula for calculating total power in a circuit is $P_T = P_1 + P_2 + P_3 + \ldots P_n$. Each wattage calculation is based on an individual component and we add them together up to n number of components to yield the total watts of the circuit.

THE DIRECT METHOD

Calculating power directly involves using one of three equations and replacing the symbols with total voltage (E_T), total current (I_T), or total resistance (R_T). The following examples use Figure 5–21 to show the direct method for calculating power.

Example

Calculate the total power in Figure 5–21.

Solution:

The values we have to calculate power are resistance and current. Thus, from Ohm's law:

$$P_T = I^2 \times R_T$$

Because the resistance in this equation is total resistance and this is a two-branch circuit, substitute the product over sum equation into the previous equation.

$$P_T = \frac{I^2(R_1 \times R_2)}{(R_1 + R_2)}$$

$$P_T = \frac{2^2 \times (150 \times 250)}{(150 + 250)}$$

$$P_T = \frac{4 \times (37{,}500)}{(400)}$$

$$P_T = 375 \text{ W}$$

Direct Formula

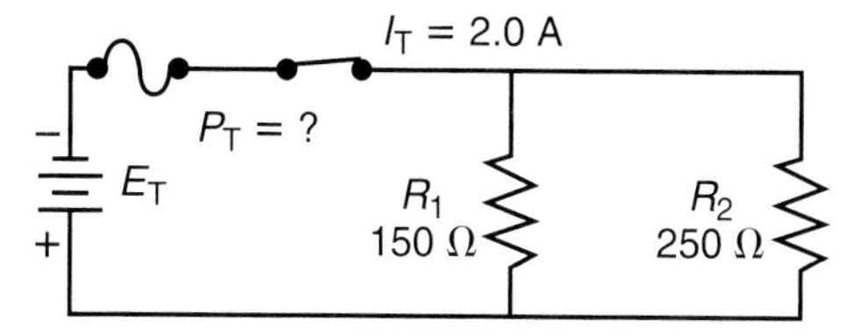

Figure 5–21 Using the direct formula, you can calculate the total wattage of the circuit by using circuit totals.

As previously mentioned, a fuse is used to limit the amount of current that flows through the wiring to protect the circuit from too much current. We need to know if the fuse for this circuit (Figure 5–22) is rated properly. If the circuit uses too much power, the current will be too high and the fuse will open, stopping the circuit current.

Solution: Because the total power and voltage are known, the basic equation can be transposed to:

$$P = I \times E \text{ transposed to } I = \frac{P}{E}$$

$$I = \frac{106.067\text{ W}}{100\text{ V}} = 1.067\text{ A}$$

These calculations show that the fuse will allow the normal load current to flow without opening the circuit (blowing). Of course, a 1.5 ampere fuse would also work. The exact one to choose is based on other issues.

Example

Determine the source voltage in Figure 5–21.

Solution:

Rearrange the equation for power that includes voltage:

$$P_T = I_T \times E_T$$

This transposes to:

$$E_T = \frac{P_T}{I_T}$$

$$E_T = \frac{375\text{ W}}{2\text{ A}} = 187.5\text{ V}$$

THE INDIRECT METHOD

The indirect method involves calculating the power for each individual component and then adding all results to determine total power.

Example

Use the indirect method and calculate total power for the circuit in Figure 5–22.

Solution:

The power for each component is determined using voltage and resistance.

$$P_1 = \frac{E_1^2}{R_1}$$

$$P_1 = \frac{100^2\text{ V}}{150\ \Omega} = 66.67\text{ W}$$

$$P_2 = \frac{E_2^2}{R_2}$$

$$P_2 = \frac{100^2\text{ V}}{250\ \Omega} = 40\text{ W}$$

Thus, the total power is:

$$P_T = P_1 + P_2$$

$$= 66.67\text{ W} + 40\text{ W} = 106.67\text{ W}$$

Indirect Formula

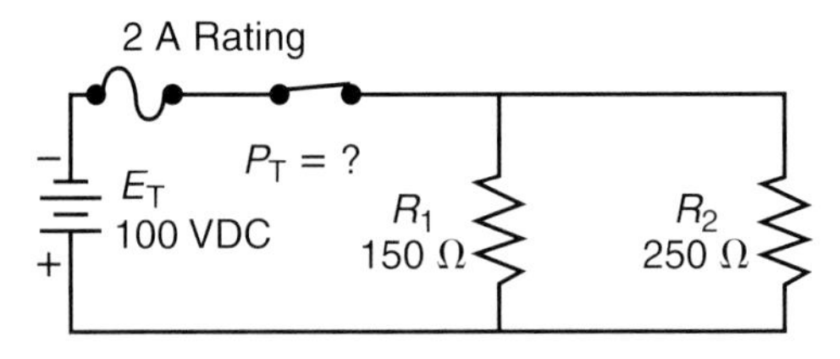

Figure 5–22 Using the indirect formula, you can calculate the total wattage of a circuit by using individual component values.

SUMMARY

The parallel circuit is the most used circuit in the electrical industry. An understanding of how the components appear and how to calculate the various quantities is essential. The basic rule for parallel circuits is that the voltage stays the same across all branches of the circuit. The branches consist of multiple current paths from the negative to the positive power supply terminals.

The total current is calculated by the sum of the individual branch currents. The effect of creating multiple (parallel) branches is that the total opposition to current flow in the circuit decreases with the addition of more branches. All branches in a parallel circuit contribute to the total current. The sum of the branch circuits is equal to the total current. The total current can be calculated in several ways. If the resistance and voltage drop of each branch are known, calculate each branch current and add the results:

$$I_T = I_1 + I_2 + I_3 + \ldots.\ I_n$$

Ohm's law for the circuit totals can be used to calculate total current if you know the source voltage and the circuit total resistance. The total circuit resistance for resistors in parallel is always less than the smallest resistor. Adding more resistors to branches in a parallel circuit causes the total resistance to decrease and the total current to increase. Total resistance of a parallel circuit can be calculated in different ways.

For any parallel circuit, the reciprocal equation can be used. For a two-branch circuit, the product over sum equation is useful. For circuits with equal values, the value of one resistor divided by the number of similar resistors can be used. Another method to find the total true resistance is to use an assumed voltage, calculate the assumed current, and then use these two values to find the actual resistance.

Voltage sources can also be connected in parallel. When voltage sources (batteries or generators) are connected in parallel, they will each contribute to the total current flow. Consequently, the two will supply a larger current than either one can individually. You should never connect sources of different voltage in parallel.

Power calculations in parallel circuits are very similar to series circuits. Each wattage load adds to the total watts of the circuit. We can use the total circuit values to find total watts, or find the individual component watts and add them for a total sum.

CHAPTER GLOSSARY

Assumed voltage method In a parallel circuit, the total resistance can be calculated by using an assumed voltage with the known resistive branches. This assumed voltage produces values of current, but will calculate actual resistance. The formula is:

$$\frac{E_{T\,assumed}}{I_{T\,assumed}} = R_{T\,actual}$$

Parallel circuit An electrical circuit that provides more than one possible path for current to flow from the source and back.

Parallel circuit voltage rule Each voltage across every branch of a parallel circuit is the same.

Parallel current rule In a parallel circuit, the total current is the sum of the individual branch currents.

Product over sum method (Product over sum equation) An equation that can be used to calculate the equivalent resistance of two resistors in parallel. It is a simplification of the reciprocal equation:

$$R_T = \frac{(R_T \times R_2)}{(R_T + R_2)}$$

Rule for parallel circuit resistance As resistance branches are added in parallel, the total resistance decreases and the total resistance for the circuit is smaller than the smallest branch resistor.

REVIEW QUESTIONS

1. A DC parallel circuit has a source voltage of 30 V, and there are two parallel branches. Branch 1 has a resistance of 6 ohms, and branch 2 has a resistance of 3 ohms.
   a. Draw the circuit.
   b. Find the current drawn across each load.
   c. Find the total current.
   d. Find the total resistance.
   e. Find the power across each load.
   f. Find the total power.
2. Using the circuit values from question 1:
   a. Solve for the power across each branch of the circuit.
   b. Solve for total power using the following formulas:
   $$P = \frac{E^2}{R} \text{ and } P = I^2 \times R$$
3. In a DC parallel circuit, if resistance is added, what happens to the total resistance?
4. If resistance is added in a DC parallel circuit, what happens to the total current?
5. What is the reciprocal formula for finding total resistance in a parallel DC circuit?
6. What is the product over sum equation for finding total resistance in a DC parallel circuit?
7. What is the equation for finding total resistance in a DC parallel circuit when all resistive loads are the same?
8. What is the relationship between the total resistance in a parallel circuit and the value of the smallest resistor in the parallel combination?
9. A DC circuit has the resistances connected in parallel. One branch is rated at 20 ohms, one branch at 40 ohms, and one branch at 60 ohms.
   a. Draw the circuit.
   b. Find the total resistance, using the reciprocal method.
10. A DC parallel circuit has two loads. Load 1 is rated at 10 Ω and load 2 is rated at 5 Ω. The circuit is supplied with 24 V.
    a. Draw the circuit.
    b. Calculate the total resistance of the parallel circuit.
    c. Calculate the current flow through each load.
    d. Calculate the voltage across each load.
11. A DC parallel circuit has three loads. Branch 1 is 360 Ω, branch 2 is 240 Ω, and branch 3 is 144 Ω. The circuit has a 120 V supply.
    a. Draw the circuit.
    b. Label each branch with its current.
    c. Calculate the total resistance.
12. The total power in a series circuit is equal to the __________ of the power in the individual components of that circuit.
13. The total power in a DC parallel circuit is equal to the __________ of the power consumed in the individual components of that circuit.
14. If the power is known across each branch of a parallel circuit, what is the formula for solving total power? Is this formula the same for a series DC circuit?
15. If total voltage and current are known, what is the formula for solving total power?
16. If total voltage and total resistance are known, what is the formula for solving total power?
17. If total current and total resistance are known, what is the formula for solving total power?

PRACTICE PROBLEMS

1. As shown in Figure PP5–1, a parallel DC circuit contains three branches. Fill in the missing information in the diagram.

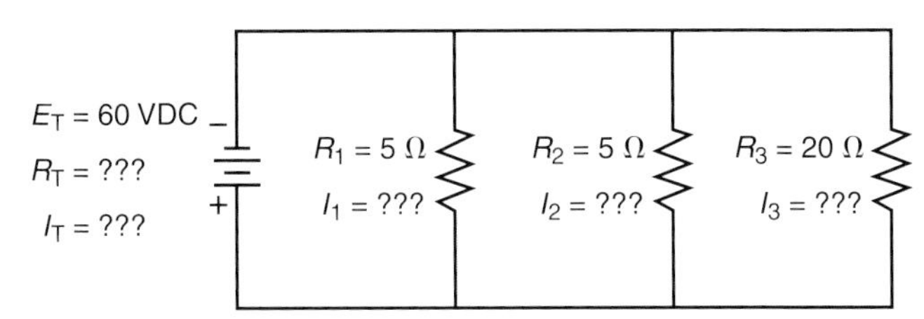

Figure PP5–1

2. As shown in Figure PP5–2, a parallel DC circuit contains three branches. Fill in the missing information in the diagram.

E_T = 120 VDC
R_T = ???
I_T = ???
P_T = ???

P_1 = 40 W
E_1 = ???
R_1 = ???
I_1 = ???

P_2 = 60 W
E_2 = ???
R_2 = ???
I_2 = ???

P_3 = 100 W
E_3 = ???
R_3 = ???
I_3 = ???

Figure PP5–2

6

DC Combination Circuits

TOPICS

- INTRODUCTION
- CURRENT IN COMBINATION CIRCUITS
- CIRCUIT SIMPLIFICATION USING EQUIVALENT COMPONENTS
- SOLVING COMPLEX CIRCUITS
- CALCULATING VOLTAGE DROPS IN COMBINATION CIRCUITS
- VOLTAGE ACROSS A SIMPLE COMBINATION CIRCUIT
- FINDING COMPONENT VALUES
- SOLVING CIRCUIT VALUES WITH PARTIAL INFORMATION
- POWER IN COMBINATION CIRCUITS

OVERVIEW

The chapters you have studied to this point were concerned with either a series circuit or a parallel circuit. This chapter introduces combination circuits. A combination circuit is one in which both series and parallel conditions exist. Some of the practical circuits you will encounter in your career will be purely series circuits or purely parallel circuits. Many of the electrical systems in use consist of combination circuits, which combine series and parallel components.

You will learn to follow currents through a combination circuit as the current divides and adds to flow through parallel branches or through series loads. In this chapter, we treat the loads as resistors to keep the circuit concepts less complicated. However, the loads may be any piece of electrical equipment that utilizes current and voltage.

In this chapter, you will learn how to reduce or combine resistances in combination circuits so that the resulting equivalent resistance appears to be nothing more than a single series resistor. An example of this is combining three parallel branch resistances into one equivalent resistance and adding this equivalent resistance to the rest of the circuit to calculate the total circuit resistance, as we have done in the previous chapter. This is illustrated in Figure 6–1, which shows one resistor (R_4) in series with the parallel combination of three other resistors (R_1, R_2, and R_3).

Voltage rules are explained as you learn to identify when and how to use series voltage rules and when to use parallel voltage rules. Finally, you will determine power calculations in a complex combination circuit.

OBJECTIVES

After completing this chapter, you will be able to:

- Identify alternative current paths in combination or series-parallel circuits
- Determine which components are in series and which are in parallel in combination circuits
- Apply series or parallel rules to determine the current through any branch or component of a combination circuit
- Reduce any combination circuit to an equivalent resistor
- Find voltage values for different parts of a combination circuit
- Determine each circuit component's current, ohmic, voltage, and power values
- Work with circuit components to find unknown component values elsewhere in the circuit
- Calculate circuit total current, resistance, voltage, and power values

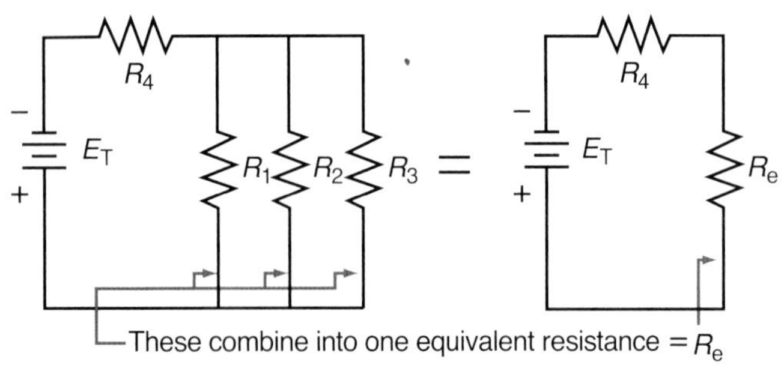

Figure 6–1 A parallel circuit in series with one resistor is depicted.

INTRODUCTION

The most basic circuit we construct as electrical professionals is the series circuit. This circuit has all needed components to allow electricity to work. The parallel circuit provides multiple paths for current to flow but maintains the voltage across the branches. As you know, many of the circuits we think of in a home are represented by parallel circuits. As we use common kitchen appliances such as a toaster, coffee maker, or waffle iron and various entertainment devices such as a DVD, television, VCR, or video game player, we plug them into a standard voltage of 120 V.

The circuitry that brings power to all of these pieces of equipment is typically some combination of series and parallel circuits. Some are simple combination circuits and some are complex. As you work through the examples in this chapter, remember that the resistor loads we use for examples can be any of the loads we mentioned. The process for determining circuit values is the intent of this chapter.

CURRENT IN COMBINATION CIRCUITS

Combination circuits are circuits in which both series and parallel conditions exist. A series circuit, as you know, is one in which there is only one path for current. Parallel circuits have multiple paths. To determine the type of circuit, trace the path for current through the components, using the schematic drawing. If the current can take more than one path, the circuit is not series.

When you studied current in earlier chapters, you learned that in purely series circuits, all components have the same amount of current. In parallel circuits, the total current is equal to the sum of all branch currents. The total current in a combination circuit can be calculated after each branch current of the parallel circuits is determined and the series circuits' currents are calculated (i.e., after the total equivalent resistance is known).

CURRENT FLOW THROUGH A SIMPLE COMBINATION CIRCUIT

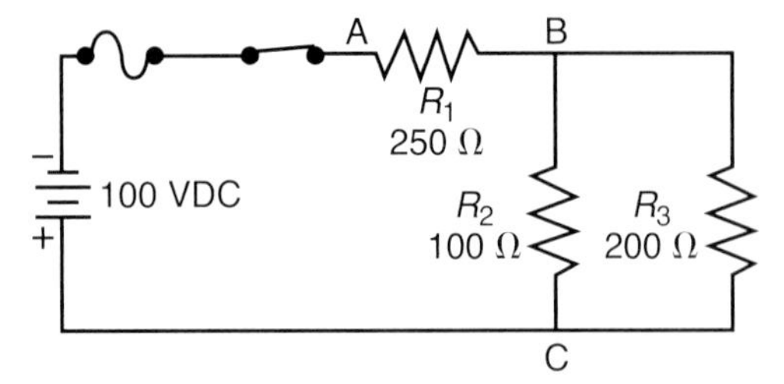

Figure 6–2 A combination circuit with a parallel component and series component.

Examine Figure 6–2. To determine the total current through the circuit, first determine the equivalent resistance of the parallel components consisting of R_2 and R_3. As you can see, there is more than one path to get from point B to point C. Therefore, this portion from B to C is a parallel circuit. The rules for parallel circuits apply to this part of the circuit. Use the parallel rules for parallel resistors to create an equivalent single resistor representing the combination of R_2 and R_3.

An easy method to use would be the product over sum equation:

$$R_{2,3(Equivalent)} = \frac{(R_2 \times R_3)}{(R_2 + R_3)}$$

$$R_{2,3(Equivalent)} = \frac{(100 \times 200)}{(100 + 200)}$$

$$R_{2,3(Equivalent)} = \frac{20{,}000}{300}$$

$$R_{2,3(Equivalent)} = 66.67\ \Omega$$

Now the figure with the equivalent resistance looks like that shown in Figure 6–3, which shows two resistors: R_1 in series with the combined equivalent resistor of $R_{2,3}$. This equivalent circuit now can be redrawn or "reduced to" the circuit shown in Figure 6–4. This circuit combines the series resistor R_1 and the equivalent resistor $R_{2,3}$ using the series circuit rules for series resistance, which is:

$$R_T = R_1 + R_{2,3}$$

$$R_T = 250\ \Omega + 66.67\ \Omega$$

$$R_T = 316.67\ \Omega$$

With this information, the total current can now be calculated with Ohm's law:

$$I_T = \frac{E_T}{R_T}$$

$$I_T = \frac{100\text{ V}}{316.67\ \Omega}$$

$$I_T = 0.316\text{ A}$$

This is the current flowing from the power supply and would be the current measured at point A in Figure 6–2. This is still the series circuit part of the combination circuit, and thus the series circuit rules for current apply. In other words, all of the current flows through R_1 to point B. Now the current has two paths to take.

By reviewing the rules for parallel current, we know the current will not split evenly. More current will flow through the smaller resistor. To be precise, we should calculate the amount of voltage available between point A and point B. Because all of the current flows through resistor R_1, we can determine how much voltage is dropped on R_1 and therefore how much is left for the parallel branches. From Ohm's law for R_1:

$$E_{R_1} = I_T \times R_1$$

$$E_{R_1} = 0.316\text{ A} \times 250\ \Omega$$

$$E_{R_1} = 79\text{ V dropped from point A to point B}$$

With a 100 V source applied, 79 V drops at R_1 and the series circuit rules for voltage drop determine that the remaining voltage is:

$$100\text{ V} - 79\text{ V} = 21\text{ V}$$

The voltage that appears across point B to C is 21 V. Now we go back to the parallel circuit rules in Figure 6–4. Looking at just R_2 and R_3 and using the voltage of 21 V, we can determine the current now through each resistor, as follows:

$$I_{R_2} = \frac{21\text{ V}}{100\ \Omega}$$

$$I_{R_2} = 0.21\text{ A}$$

R_1 Resistor in Series

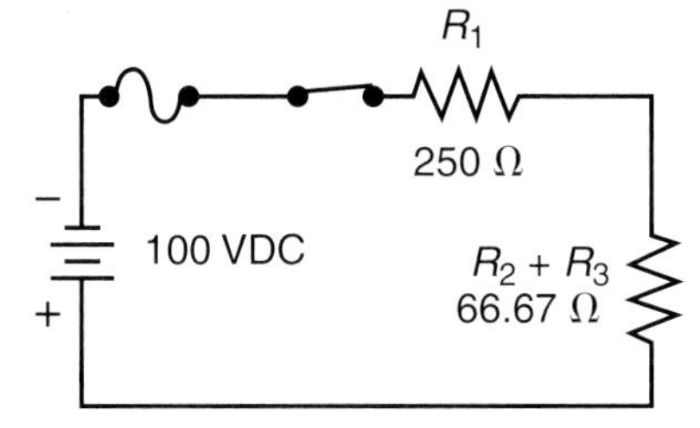

Figure 6–3 The R_1 resistor is in series with an equivalent resistor for R_2 and R_3.

Simplification Process

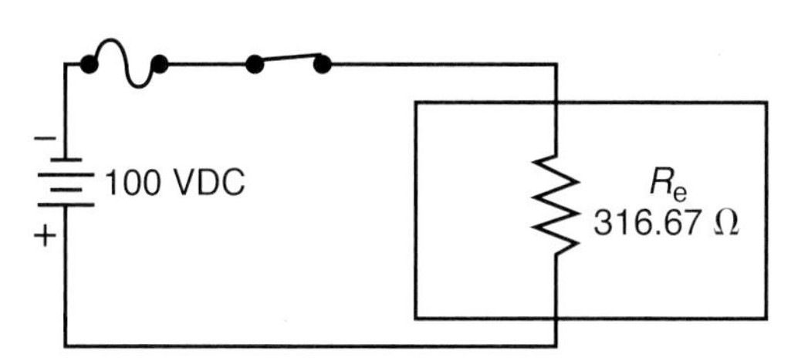

Figure 6–4 A circuit can be reduced to a single equivalent resistor.

and

$$I_{R_3} = \frac{21 \text{ V}}{200 \ \Omega}$$

$$I_{R_3} = 0.105 \text{ A}$$

To verify that this is correct, look at point C in Figure 6–2 to determine if the two parallel branch currents now add together to equal the original series current that split at the top of the parallel circuit (0.21 + 0.105 = 0.315 A). The total current flowing back to the source should be the same as the current flowing from the source. Note that they do not match exactly in many calculations because of rounding the decimal number to three significant digits. We say that these figures are close enough to prove the point.

If we wanted to calculate the values to many more significant digits, the answers would be precise. We are more interested in process here, rather than exact values. Another check to verify your calculation is to use the equivalent resistance ($R_{2,3}$). This resistance was found to be 66.67 Ω. The voltage drop on this part of the circuit was thought to be 21 V. If we use Ohm's law, $\frac{21 \text{ V}}{66.67 \ \Omega} = 0.315$ A. Again, this yields the circuit total current through the parallel equivalent branch of the total circuit.

REDUCING A COMPLEX COMBINATION CIRCUIT

The process we just used to solve for the current flow through a simple combination circuit is called simplification. Figure 6–5 is a simple combination circuit. The first step in analyzing this circuit is to reduce (simplify) the circuit as much as possible. Each section to be reduced will be a group of two or more resistors, with the equivalent resistance results taking the place of the group. The parallel parts of the circuit are analyzed first.

Resistors R_2 and R_3 are in parallel. First, find the equivalent resistance for R_2 and R_3 by using the product over sum equation, as described earlier, or by any appropriate method to determine the equivalent resistance of a parallel section of the circuit. With this result, redraw the circuit and replace the two parallel resistors with their "equivalent" single resistor, as shown in Figure 6–6.

Now the circuit in Figure 6–5 has been reduced to a simple series circuit with two resistors in series. In any combination circuit with multiple branches in parallel and with series circuits, we can simplify the circuit to a single resistor. This process can be time-consuming. It is best if you record each simplification in an intermediate diagram. As you solve for the totals, you will then rebuild the circuit from the simple back to the complex.

If you keep the intermediate circuits available, you can retrace your steps back to the original circuit. This first simplification was easy, going from three resistors to two resistors to one resistor. As you saw previously, we needed to go back from one (R-total) to two resistors and finally back to the original three resistors to solve the entire problem.

Single Equivalent Resistors

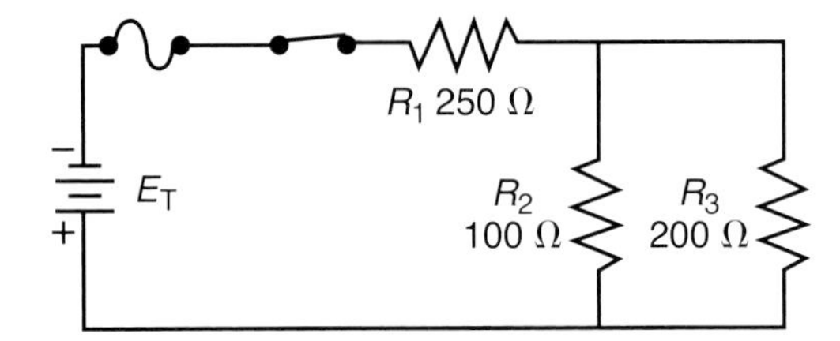

Figure 6–5 The process used to reduce the number of components is called simplification.

Combining Resistors

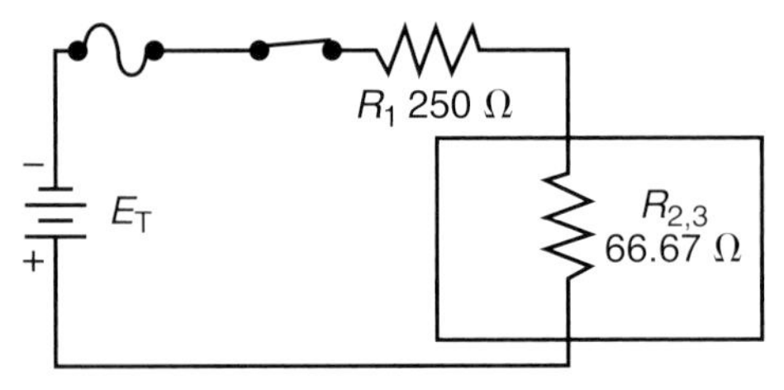

Figure 6–6 Resistors are combined to create equivalent resistors.

CIRCUIT SIMPLIFICATION USING EQUIVALENT COMPONENTS

We can take what looks like a difficult circuit and simplify it to a single resistor. We will use simple values for the first simplification of a larger circuit. See Figure 6–7 as the original circuit. Just by looking at it, we may not be able to tell how much current flows through each component or how much voltage is dropped across each component. By simplifying to a single resistor and then rebuilding back to the original, we can find all component values.

First decide what parts of the circuit are in parallel with each other and which components are in series with each other. In this example, we have a parallel circuit: R_1 and R_2, in series with R_3. Now the series circuit, consisting of the first three resistors, is in series with the next section of the parallel resistors R_4, R_5, R_6, and R_7. The way this is drawn, with all components aligned vertically, it is easier to see the relationships.

To simplify, combine the R_1 and R_2 into one equivalent resistor by using parallel circuit resistance rules. You may use products over sum, reciprocal, equal value resistor, or even assumed voltage if you wish. The two 50 Ω resistors combine to form one equivalent 25 Ω resistor (Figure 6–8).

It is often difficult to "see" the series and parallel relationships between components because they are physically connected in convoluted patterns. If you can redraw your more complex combination circuits so that the components are in vertical representations, it is sometimes easier to see the relationships. You may have to redraw several times to actually make the connecting lines look like the traditional series or parallel connections we see in schematic views of the diagram.

Complex Combination Circuit

Figure 6–7 A more complex combination circuit requires several steps to simplify.

First Reduction

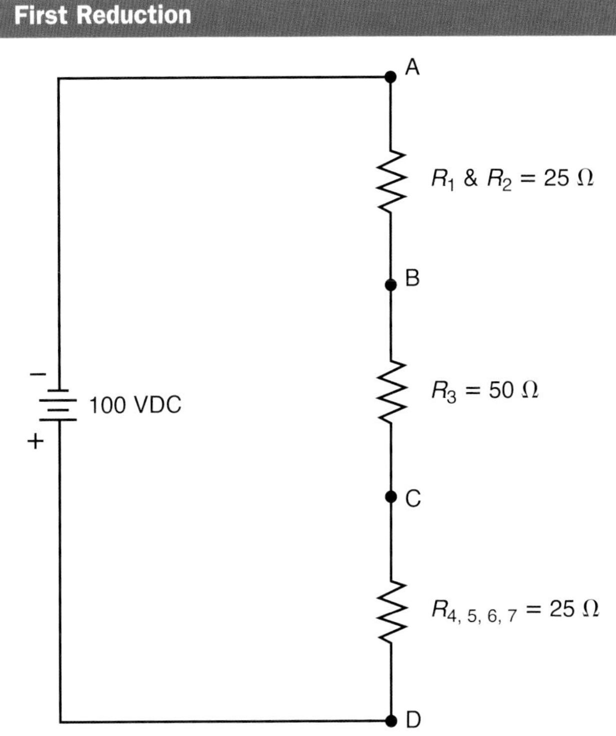

Figure 6–8 The first reduction of Figure 6–7 to an equivalent circuit is depicted.

Point A to point B would act like a 25 ohm resistor. Then combine all four resistors in parallel between points C and D. Again, use the parallel resistance rules to create one equivalent resistor of 25 ohms (Figure 6–8). Now we add all resistors using series resistors rules to yield one equivalent circuit total resistor (Figure 6–9). The circuit acts like a load of one 100 ohm resistor.

With the voltage of 100 VDC applied, the total current supplied to the circuit is 1 A (found by $\frac{E_T}{R_T} = I_T$ or $\frac{100\ \text{V}}{100\ \Omega} = 1\ \text{A}$). Now we can track the 1 A through the circuit. As the circuit current flow of 1 A gets to point A, it splits two ways: part goes through R_1 and part flows through R_2. How much goes each way? To decide, we go back to the equivalent 25 Ω resistor. Remember, the circuit acts like one resistor at this point.

The voltage across the equivalent resistor is found by $I \times R = E$, or $1\ \text{A} \times 25\ \Omega = 25\ \text{V}$. Therefore, 25 V drop across this part of the circuit A to B. Now that we know the voltage, we can determine how much current would flow through each branch of the parallel circuit:

$$\frac{E_{Component}}{R_{Component}} = I_{Component}$$

or

$$\frac{25\ \text{V}}{50\ \Omega} = 0.5\ \text{A}$$

through R_1 and the same through R_2, because it is the same resistance value.

At point B, the currents join to create the 1 A of total circuit again. The 1 A flows through the R_3 resistance of 50 Ω to create a voltage drop of 50 V. We now know that R_3 is 50 Ω, it has 1 A flowing through it, and it drops 50 V across it from point B to point C.

For the last section (point C to point D), we go back to the equivalent resistance (Figure 6–8). The four 100 Ω resistors create an equivalent resistor of 25 Ω. Therefore, the current through the equivalent 25 Ω drops 25 V across this section C to D. Again, if we know the voltage across the individual component (which is 25 V according to the parallel voltage rules) and if we know the component resistance, we can calculate the component current with

$$\frac{E_{Component}}{R_{Component}} = I_{Component}$$

For any resistor R_4, R_5, R_6, R_7, the component resistance is:

$$\frac{25\ \text{V}}{100\ \Omega} = 0.25\ \text{A}$$

The current splits four ways equally in this circuit because the resistors are all equal. If the resistors are not equal values, the current does not split equally. At point D, the currents all "collect" to return to the circuit current of 1 A that flows back to the source.

To double-check, we should be able to account for all of the voltage, as in Kirchhoff's laws. Referring to Figure 6–8, we can see that the circuit from A to B drops 25 V, B to C drops 50 V, and C to D drops 25 V, accounting for the total supply voltage.

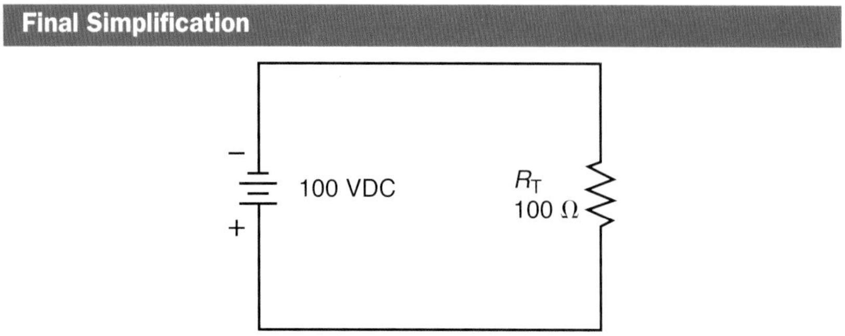

Final Simplification

Figure 6–9 Figure 6–7 is finally simplified to one total resistor.

SOLVING MORE COMPLEX COMBINATION CIRCUITS

Reducing a more complicated circuit to its equivalent series resistor is performed much the same way as previously described. The process can be structured to maintain a consistent approach. Table 6–1 outlines the general steps involved.

The circuit shown in Figure 6–10 will be reduced to its equivalent series resistor. Each section will be reduced individually and the circuit redrawn to show the equivalent circuit until only one resistor remains. The first task is to determine which series resistors may be combined. In this circuit, only R_4 and R_5 may be combined as series connected resistors. Again, the rule for series is that there is only one path for current to pass through the components. For the current to flow from point C to point F, it will flow through series components R_4 and R_5. The equivalent resistance of R_4 and R_5 is $R_{4,5}$.

$$R_{4,5} = R_4 + R_5$$

$$R_{4,5} = 85\ \Omega + 200\ \Omega = 285\ \Omega$$

The circuit may then be redrawn (Figure 6–11), with the equivalent resistance drawn in place of R_4 and R_5. The next step is to combine any parallel portions of the circuit. The parallel parts of the circuit are R_1 to R_2 and R_6 to R_7 (Figures 6–12 and 6–13). Using the product over sum equation:

$$R_{1,2(Equiv)} = \frac{(R_1 \times R_2)}{(R_1 + R_2)}$$

$$R_{1,2(Equiv)} = \frac{(25 \times 50)}{(25 + 50)}$$

$$R_{1,2(Equiv)} = 16.67\ \Omega$$

Table 6–1

Here are the steps to reduce a circuit to an equivalent resistance total.

Step	Description
1	Reduce one part of the circuit at a time.
2	Redraw the circuit, exchanging the equivalent circuit for the original resistors. This helps you visualize the next required step.
3	Ensure that all series resistors have been combined before a parallel portion is reduced.
4	Combine parallel portions to a single resistor.
5	Repeat, combining equivalent resistors until all portions are reduced to one equivalent resistance.

Complex Combination Circuit

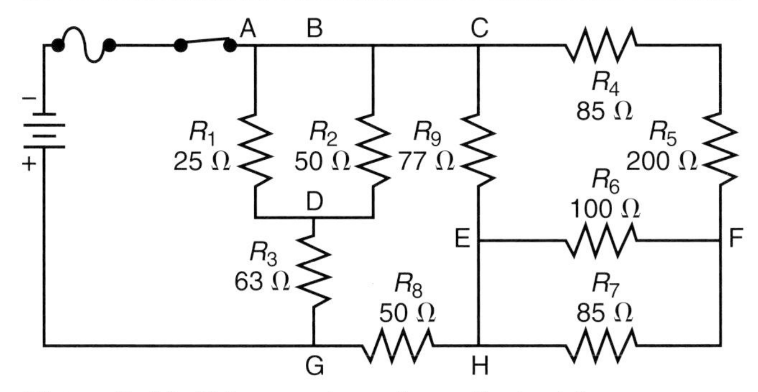

Figure 6–10 This complex schematic depicts a combination circuit.

Complex Combination Circuit

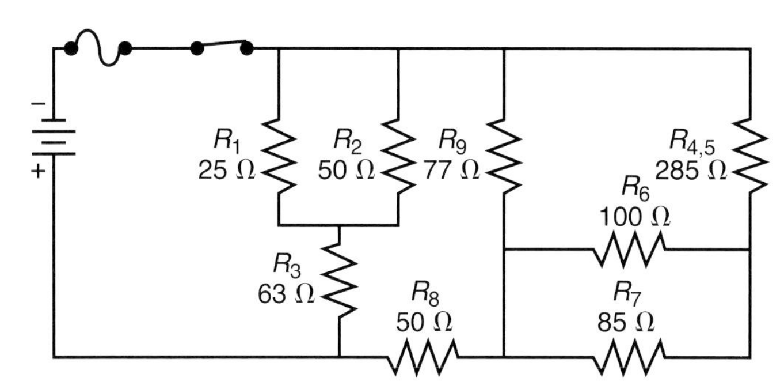

Figure 6–11 From Figure 6–10, R_4 and R_5 are combined.

Complex Combination Circuit

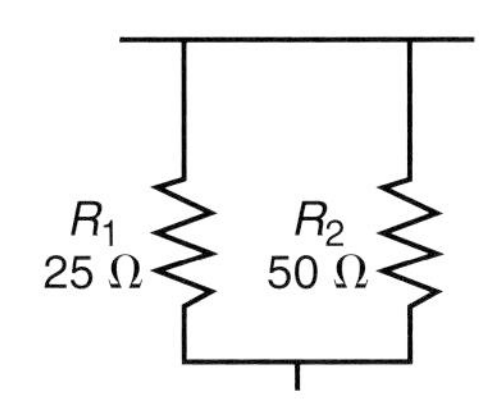

Figure 6–12 R_1 and R_2 are combined to an equivalent resistor.

Complex Combination Circuit

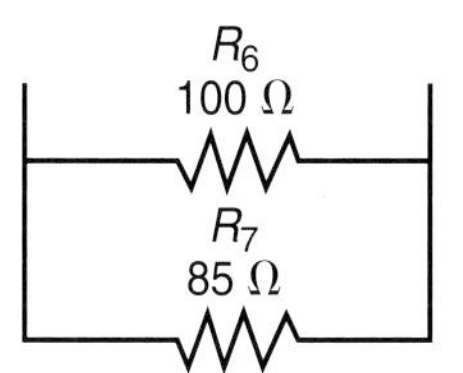

Figure 6–13 R_6 and R_7 are combined to an equivalent resistor as an intermediate step.

and

$$R_{6,7(Equiv)} = \frac{\text{Product}}{\text{Sum}}$$

$$R_{6,7(Equiv)} = \frac{(100 \times 85)}{(100 + 85)}$$

$$R_{6,7(Equiv)} = 45.95\ \Omega$$

Now redraw the circuit, combining the two parallel branch circuits into their respective equivalent resistances (Figure 6–14). Next, because $R_{1,2}$ and R_3 are in series in one branch and $R_{4,5}$ and $R_{6,7}$ are in series in another branch, the rules for series addition of resistors are applied (Figures 6–15 and 6–16). The result is:

$$R_{1,2,3} = R_{1,2} + R_3$$

$$R_{1,2,3} = 16.67 + 63 = 79.67\ \Omega$$

and

$$R_{4,5,6,7} = R_{4,5} + R_{6,7}$$

$$R_{4,5,6,7} = 285 + 45.95 = 330.95\ \Omega$$

The result is redrawn or simplified (Figure 6–17). Now, in this simplified schematic, it shows that $R_{4,5,6,7}$ and R_9 are in parallel (Figure 6–18). The product over sum equation can be used as follows:

$$R_{4,5,6,7,9} = \frac{(R_9 \times R_{4,5,6,7})}{(R_9 + R_{4,5,6,7})}$$

$$R_{4,5,6,7,9} = \frac{77\ \Omega \times 330.95\ \Omega}{77\ \Omega + 330.95\ \Omega}$$

$$R_{4,5,6,7,9} = 62.47\ \Omega$$

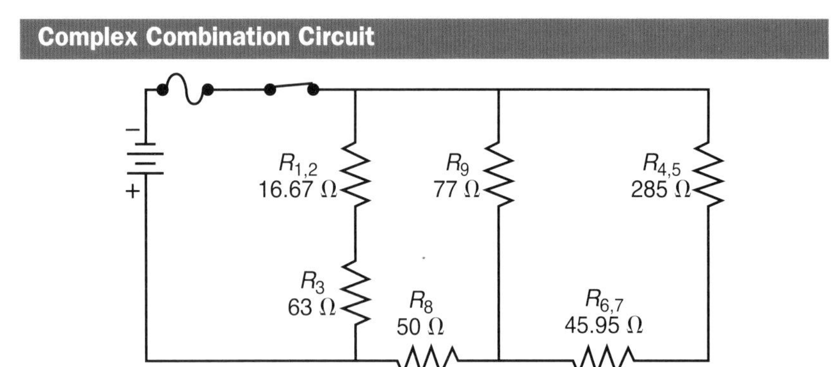

Figure 6–14 Figure 6–10 is redrawn with equivalent resistors to imitate the original circuit.

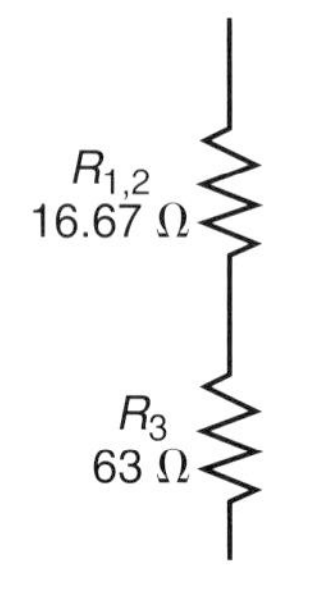

Figure 6–15 A further reduction of series components uses series resistance rules.

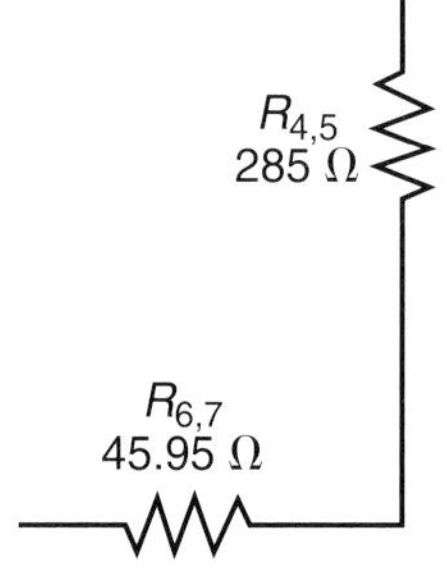

Figure 6–16 $R_{4,5,6,7}$ are all in series and can be combined.

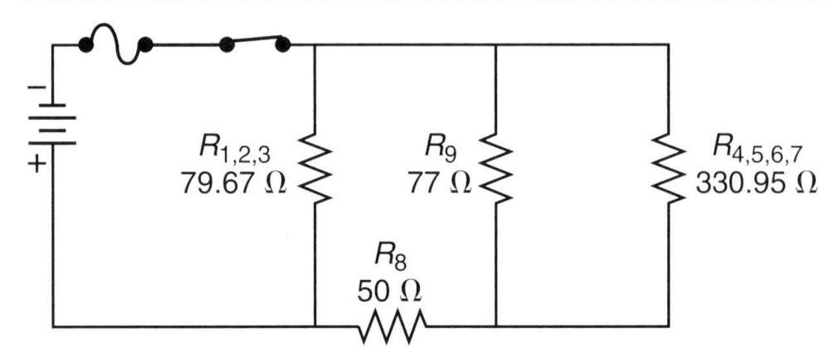

Figure 6–17 With all of the reductions to equivalent resistance, Figure 6–10 equates to this circuit.

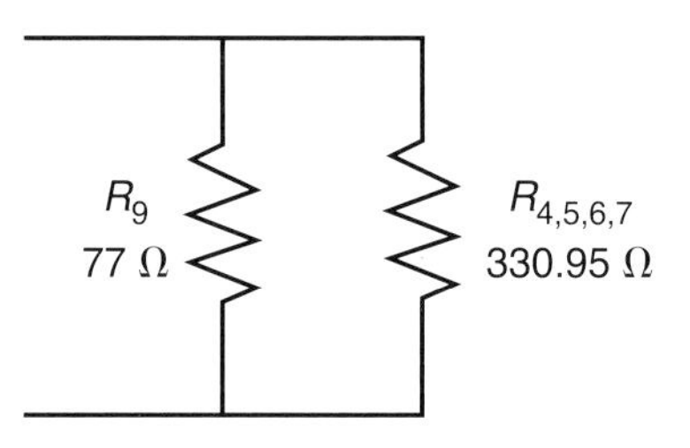

Figure 6–18 R_9 is combined with $R_{4,5,6,7}$.

The redrawn circuit is shown in Figure 6–19. Figure 6–20 shows the series combination of resistors $R_{4,5,6,7,9}$ in series with R_8.

$$R_{4,5,6,7,8,9} = R_8 + R_{4,5,6,7,9}$$

$$R_{4,5,6,7,8,9} = 50 + 62.47$$

$$R_{4,5,6,7,8,9} = 112.47\ \Omega$$

The result (Figure 6–21) is a simple parallel circuit with two resistors. The product over sum equation is used one final time to create the final result (Figure 6–22):

$$R_T = \frac{R_{1,2,3} \times R_{4,5,6,7,8,9}}{R_{1,2,3} + R_{4,5,6,7,8,9}}$$

$$R_T = \frac{79.67\ \Omega \times 112.47\ \Omega}{79.67\ \Omega + 112.47\ \Omega}$$

$$R_T = \frac{8{,}960.48\ \Omega}{192.14\ \Omega}$$

$$R_T = 46.64\ \Omega$$

This is the equivalent resistance from Figure 6–10. The nine resistors from the original circuit appear as one single resistive source in Figure 6–22.

TechTip!

When calculating resistance, current, or voltages about a complex network, errors can accumulate to such an extent that results are not sufficiently accurate. A good way to overcome this pitfall is to avoid rounding intermediate results. Rounding *should* be performed, however, upon the final result. The final result is only as accurate as the least precise factor used in a calculation. For example, the sum of 1.5 Ω + 19.99 Ω is 21.5 Ω, *not* 21.49 Ω. Rounding in this example is not only convenient, but also important.

This simplification process allows us to break very complex circuits down to manageable levels. At this point, we do not know what the voltages are or what the current is through each component, but the process allows us to analyze this one step at a time. We will now reconstruct the original circuit and apply the appropriate series and parallel circuit rules.

Complex Combination Circuit

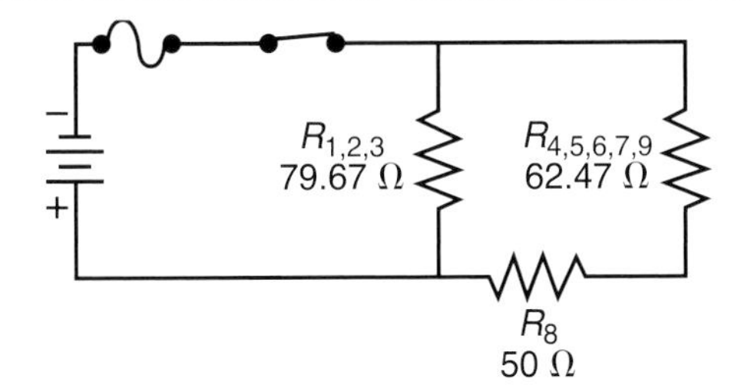

Figure 6–19 The next intermediate step toward simplification reduces the original circuit to three resistors.

Complex Combination Circuit

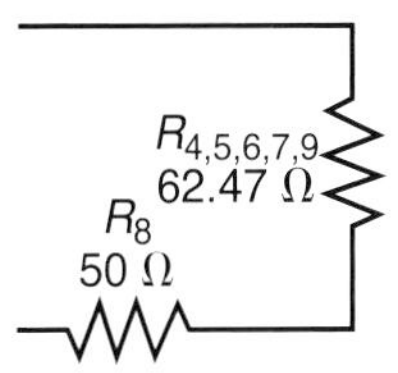

Figure 6–20 Add R_8 to $R_{4,5,6,7,9}$ as a series resistor.

Complex Combination Circuit

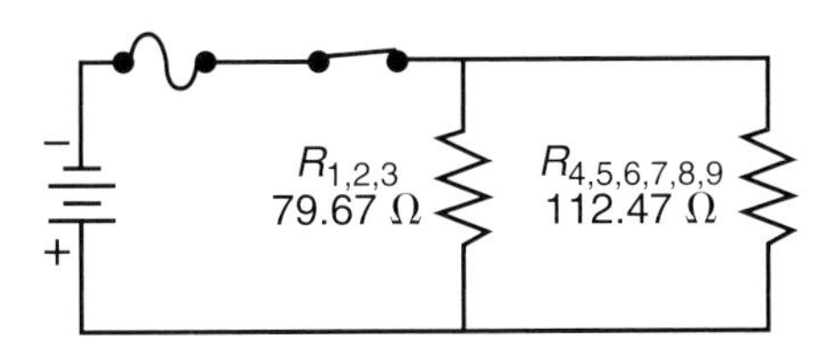

Figure 6–21 The combination of $R_{1,2,3}$ and $R_{4,5,6,7,8,9}$ in parallel will yield a final single equivalent resistor.

Complex Combination Circuit

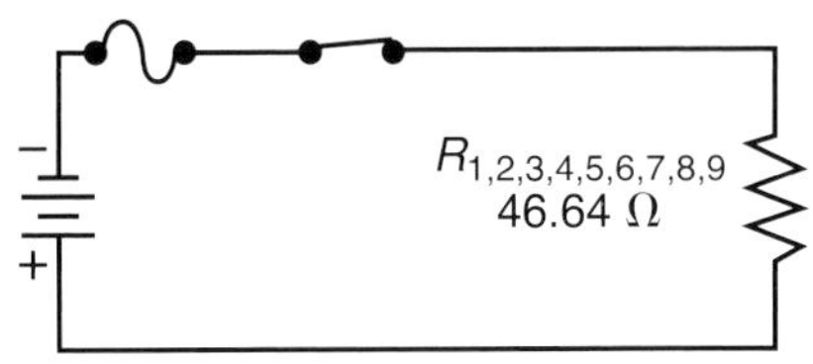

Figure 6–22 Figure 6–10 is reduced to a single resistor (R_T).

SOLVING COMPLEX CIRCUITS

Circuits that are more complex and have many branches in combination must be solved with Ohm's law. Figure 6–23 provides a more complex circuit to calculate the currents through each individual resistor.

Example

What is the total current in Figure 6–23?

Solution:

From working with this problem in a previous lesson we know that the total voltage $E_T = 200$ V and the total resistance $R_T = 46.64\ \Omega$ (Figure 6–22). By knowing these two variables, solve for the total current (I_T) using Ohm's law. The total equivalent resistance circuit for Figure 6–23 can be seen in Figure 6–22.

$$I_T = \frac{E_T}{R_T}$$

$$I_T = \frac{200\text{ V}}{46.64\ \Omega}$$

$$I_T = 4.29\text{ A}$$

Complex Combination Circuit

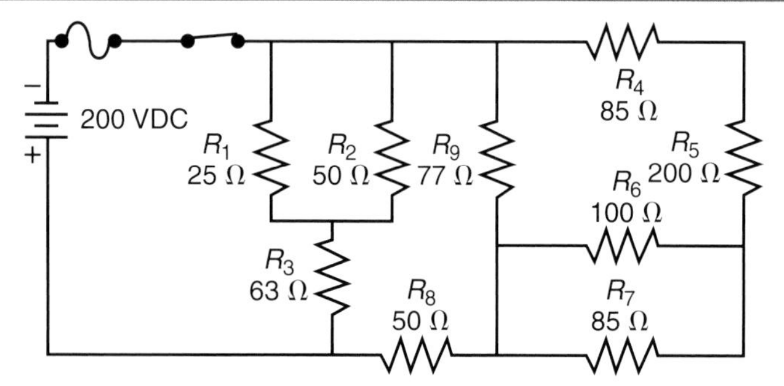

Figure 6–23 The same complex combination circuit as in Figure 6–10 is shown, with applied voltage.

Rebuilding the Circuit

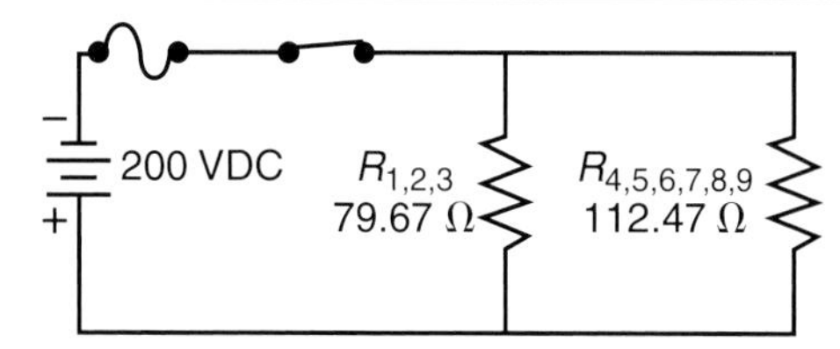

Figure 6–24 Rebuilding toward the original circuit looks the same as Figure 6–21, now with voltage applied.

Example

What are the current values through each resistor 1 to 9?

Solution:

The value of currents through each respective resistor must also be calculated with Ohm's law. Work backward from the total equivalent resistance, $R_T = 46.64\ \Omega$ (Figure 6–22) and calculate the current through each branch.

Figure 6–24 shows an expansion of Figure 6–22, indicating how the single resistance breaks into a simple parallel circuit with two branches. The current can be calculated directly with Ohm's law because the voltage source is impressed across both branches.

$$I_{1,2,3} = \frac{E_T}{R_{1,2,3}}$$

$$I_{1,2,3} = \frac{200\text{ V}}{79.67\ \Omega}$$

$$I_{1,2,3} = 2.51\text{ A}$$

$$I_{4,5,6,7,8,9} = \frac{E_T}{R_{4,5,6,7,8,9}}$$

$$I_{4,5,6,7,8,9} = \frac{200\text{ V}}{112.47\ \Omega}$$

$$I_{4,5,6,7,8,9} = 1.78\text{ A}$$

Checking the results, you see that the sum of the two branch currents is equal to the total current.

$$I_T = I_{1,2,3} + I_{4,5,6,7,8,9}$$

$$= 2.51 + 1.78 = 4.29\text{ A}$$

Separating $R_{1,2,3}$ back into its individual resistors results in the branch shown in Figure 6–25. From the previous calculation, 2.51 A are flowing though this branch. The 2.51 amps are shared by resistors R_1 and R_2 because they are in parallel. The total 2.51 A flow through R_3 because it is in series with the parallel branch $R_{1,2}$. Because R_1 and R_2 are a current divider, the **law of proportionality** equation can be used.

THE LAW OF PROPORTIONALITY FOR CURRENT

The law of proportionality tells us that if we take the total current through a parallel set of components we can determine how the current splits up according to the proportions of the resistance to the total equivalent resistance. Actually, the current splits so that the most current goes through the least resistance. For instance, if a 40 Ω and a 10 Ω resistor were in parallel with 40 V applied, 1 A would flow through the 40 Ω resistor and 4 A would flow through the 10 Ω resistor.

The 40 Ω resistor has four times the resistance of the 10 Ω resistor and the inverse (¼) of the current of the 10 Ω resistor. Current and resistance are inversely proportional. The total current in the example is 5 A through an equivalent resistance of 8 Ω for the two parallel resistors. If we use the proportion of the single component of 10 Ω resistance divided by the total for the circuit equivalent of 8 Ω, we have:

$$\left[\frac{10}{(8)}\right] = 1.25.$$

The inverse of 1.25 is 0.8 as applied to the total current. Therefore, 0.8 × 5 A = 4 A (flows through the 10 Ω resistor). The current flow through the parallel resistor of 40 Ω compared to the total of 8 is $\frac{40}{8} = 5$. The inverse of 5 is 0.2.

Therefore, 0.2 × 5 A = 1 A (of current flows through the 40 Ω resistor). For our circuit, the law of proportionality is determined per the following calculations:

$$I_1 = I_{1,2,3} \times \left[\frac{1}{\left(\frac{R_1}{R_{1,2(Equiv)}}\right)}\right]$$

$$= 2.51\text{ A} \times \left[\frac{1}{\left(\frac{25}{16.7}\right)}\right] = 1.67\text{ A}$$

$$I_2 = I_{1,2,3} \times \left[\frac{1}{\left(\frac{R_2}{R_{1,2(Equiv)}}\right)}\right]$$

$$= 2.51\text{ A} \times \left[\frac{1}{\left(\frac{50}{16.7}\right)}\right] = 0.84\text{ A}$$

$$I_1 + I_2 = 1.67\text{ A} + 0.84\text{ A} = 2.51\text{ A}$$

Rebuilding the Circuit

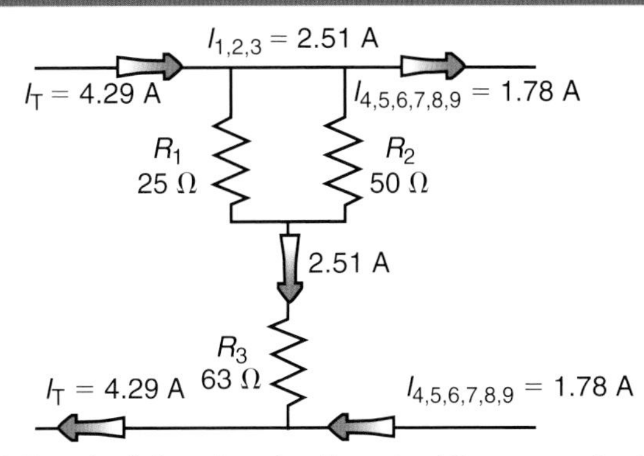

Figure 6–25 By rebuilding the circuit and adding currents, the circuit is built back to the original.

The current through R_3 is simply the branch current because it is a series resistor in the branch.

$$I_3 = 2.51\text{ A}$$

At this point, we can verify that the circuit values are correct using another method and the voltage rules for parallel and series circuits. As noted in Figure 6–24, the equivalent resistor for $R_{1,2,3}$ is represented by a resistor of 79.67 ohms. That resistor would have 200 V across it. If we use the series voltage rules, part of the 200 V drops across the series R_3 resistance.

This can be calculated by $E = I \times R$ or $E = 2.51\text{ A} \times 63\ \Omega = 158.13\text{ V}$. The rest of the 200 V applied must drop across the other series component, made up of the parallel branches of R_1 and R_2. Consequently, 200 V – 158.13 V = 41.87 V across the branches of R_1 and R_2. Now we can use that branch voltage and the individual resistance to calculate each branch current.

$$\text{Branch 1} = \frac{E}{R_1} = I_1 = \frac{41.87\text{ V}}{25\ \Omega}$$

$$I_1 = 1.67\text{ A}$$

$$\text{Branch 2} = \frac{E}{R_2} = I_2 = \frac{41.87\text{ V}}{50\ \Omega}$$

$$I_2 = 0.84\text{ A}$$

$$I_3 = 2.51\text{ A}$$

These figures correspond to the figures calculated by the law of proportionality. For the next step, you see that $R_{4,5,6,7,8,9}$ can be drawn as two parallel resistors in series with one resistor (Figure 6–26). The current through R_8 can be stated directly because it is a series resistor in the branch.

$$I_8 = 1.78 \text{ A}$$

The current through R_9 is calculated with the law of proportionality.

$$I_9 = I_{4,5,6,7,8,9} \times \left[\frac{1}{\left(\frac{R_9}{62.59}\right)}\right]$$

$$= 1.78 \times \left[\frac{1}{\left(\frac{77}{62.59}\right)}\right] = 1.45 \text{ A}$$

Expanding the Circuit

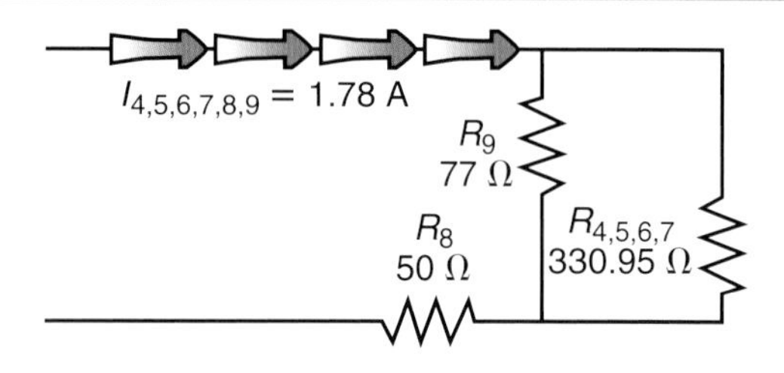

Figure 6–26 Expanding the circuit and adding currents help track the circuit concepts.

Recreating the Circuit

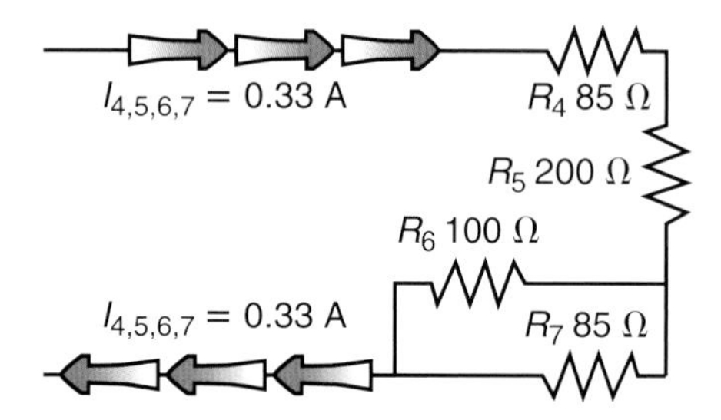

Figure 6–27 Expanding or recreating the circuit helps establish current flows.

This means that the remainder of the 1.78 A flows through $R_{4,5,6,7}$.

$$I_{4,5,6,7} = I_{4,5,6,7,8,9} - I_9$$

$$I_{4,5,6,7} = 1.78 - 1.45 = 0.33 \text{ A}$$

From Figure 6–27, you can see that R_4 and R_5 are in series with the branch and thus have the same current value as the entire branch.

$$I_4 = I_5 = 0.33 \text{ A}$$

R_6 and R_7 are parallel resistors, and thus the law of proportionality applies.

$$I_6 = I_{4,5,6,7} \times \left[\frac{1}{\left(\frac{R_6}{R_{6,7(Equiv)}}\right)}\right]$$

$$= 0.33 \text{ A} \times \left[\frac{1}{\left(\frac{100}{46.0}\right)}\right] = 0.152 \text{ A}$$

and

$$I_7 = I_{4,5,6,7} \times \left[\frac{1}{\left(\frac{R_7}{R_{6,7(Equiv)}}\right)}\right]$$

$$= 0.33 \text{ A} \times \left[\frac{1}{\left(\frac{85}{46}\right)}\right] = 0.179 \text{ A}$$

Thus, the currents through all resistors are as follows:

$$I_1 = 1.67$$

$$I_2 = 0.84$$

$$I_3 = 2.51$$

$$I_4 = 0.34$$

$$I_5 = 0.34$$

$$I_6 = 0.152$$

$$I_7 = 0.18$$

$$I_8 = 1.79$$

$$I_9 = 1.45$$

Note that adding the currents of all resistors does not give you the total current (I_T) for the circuit. This is because some of the resistors are in parallel and some are in series.

CALCULATING VOLTAGE DROPS IN COMBINATION CIRCUITS

The previous sections provided the beginning steps in determining circuit voltages:

1. Calculate the equivalent resistance.
2. Calculate circuit currents.
3. Calculate voltage drops using Ohm's law.

As you study and practice the problems in this chapter, remember that actual application goes well beyond the diagrams shown. Always make every effort to relate the lesson material to what you have experienced or observed on the job. For example, consider three lamps connected in parallel to a 120 V supply. Your first thought relative to this circuit might be that it resembles the physical circuit shown in Figure 6–28. A simple schematic equivalent circuit for the three lamps is shown in Figure 6–29.

In fact, if you consider the voltage drop of the conductors, then Figure 6–29 is not entirely accurate. Earlier in this book, you learned that the conductors (wires) would also have resistance and a subsequent voltage drop. We have ignored those drops before. However, the diagram should really look more like that shown in Figure 6–30. This more accurate diagram is in fact a combination circuit, with series and parallel components.

VOLTAGE ACROSS A SIMPLE COMBINATION CIRCUIT

Using Figure 6–30 as a practical example of wire resistance, we will calculate the voltage available to the load with a particular current draw. For this example, we will assign the wattage of the three lamps in parallel as 60 W each.

Voltage drop, or line drop, affects other utilization equipment on the same voltage source. You may have noticed that when a large appliance or motor starts in your home, the lights in the room dim for an instant. This is a result of the series and parallel circuit produced by the motor of the appliance, and the lights in the room actually being in parallel with one another. The circuit that supplies these products in the house also has a series resistance. As the motor starts, it typically draws six times as much current as it does when it is running.

This large amount of current is also traveling through the line conductors, which have conductor resistance. The current and the resistance cause a voltage drop in the line, and consequently there is less voltage at the loads; thus, the light receives less voltage momentarily and the lights get dimmer. As the motor current goes back to normal, the line current and the line drop decrease, allowing the voltage at the lights to go back to normal.

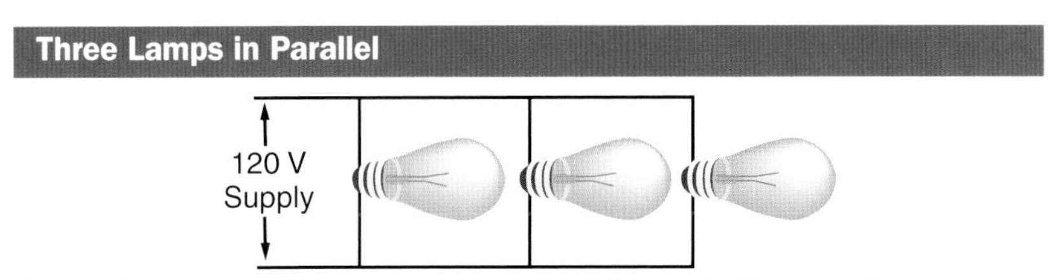

Figure 6–28 Three lamps are connected in parallel in this circuit.

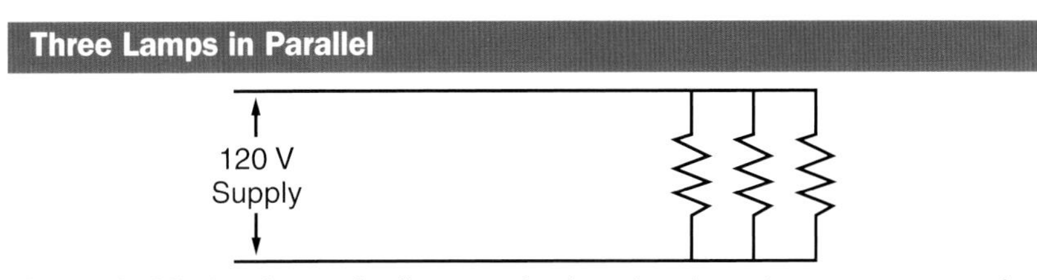

Figure 6–29 A schematic diagram depicts the three lamps connected in parallel.

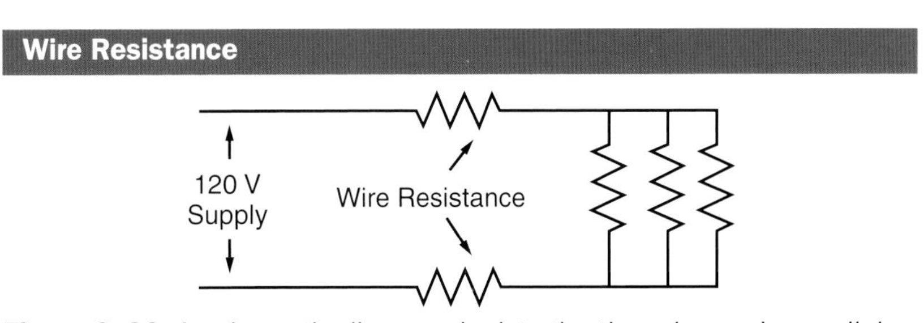

Figure 6–30 A schematic diagram depicts the three lamps in parallel and series wire resistance.

Using Watt's law we can find the current through each parallel branch by $I = \frac{P}{E}$ or $I = \frac{60\text{ W}}{120\text{ V}} = 0.5\text{ A}$ for each lamp. The total load of the three lamps can be represented as one load with 1.5 A at 120 V (Figure 6–31). For this example we will assume the wire resistance of each wire is 1 ohm (Figure 6–31).

Now we can use the series rules for voltage drop and calculate the voltage drop in the wires to and from the load. The volt drop formula is really only an adaptation of Ohm's law (Chapters 1 and 3). We know that the load current of 1.5 A flows through the wire resistance of 1 Ω and that therefore the volt drop in each line wire is 1.5 V in each direction, giving a total of 3 V dropped in getting the current to and from the load. Kirchhoff's law, which you will study in more depth in the next chapter, requires that we account for all of the source voltage of 120 V. We know that 3 V are lost in line drop, and thus the voltage remaining across the parallel branches of the load is 120 V – 3 V = 117 V.

Calculating Voltage

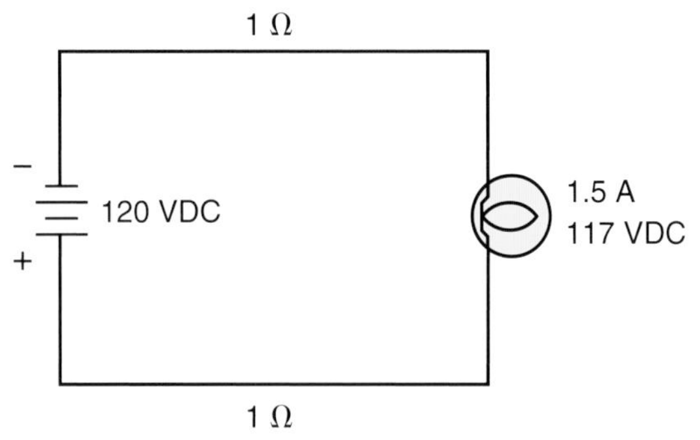

1 Ω

– 120 VDC +

1.5 A
117 VDC

1 Ω

Figure 6–31 Three resistive elements and series circuit wire resistance are depicted.

SOLVING COMPLEX CIRCUITS

Previously you used the circuit shown in Figure 6–23 and derived the current through each resistor. You reduced the resistance to an equivalent resistance (Figure 6–22) and calculated the total current as follows:

$$I_T = \frac{E_T}{R_T}$$

$$I_T = \frac{200\text{ V}}{46.64\ \Omega} = 4.29\text{ A}$$

Working backward, you expanded the equivalent resistances into their individual circuit component values and stopped when you had series equivalent resistors for which the current could be calculated. In the first step, you expanded the circuit into two parallel resistors (Figure 6–24) and found the current through each to be as indicated in Table 6–2.

Using the equivalent circuits and the expanded circuits, we can assign each component a value rating to use in Ohm's law and Watt's law to find any individual or combination value in the circuit. The last column of the table uses Watt's law to find the wattage dissipated by each piece of **utilization equipment** (resistor at this point). We can use the formula I^2R or $E \times I$ or $\frac{E^2}{R}$ to solve for P.

Table 6–2

Values are shown for the individual measures in the combination circuit.

Resistor	Resistance	Current	Voltage	Wattage
R_1	25	1.67	41.75	69.7
R_2	50	0.84	41.9	35.2
R_3	63	2.51	158.13	396.9
R_4	85	0.34	28.90	9.83
R_5	200	0.34	68.00	23.1
R_6	100	0.1562	15.62	2.44
R_7	85	0.1837	15.62	2.87
R_8	50	1.78	89.00	158.4
R_9	77	1.44	110.88	159.7

FINDING COMPONENT VALUES

As with all other types of circuits, the power utilized in a combination circuit is the sum of the power consumed by each of the individual components. Each component adds a portion to the total power.

You have worked in previous lessons to systematically reduce the various components in a combination circuit down to an "equivalent resistance" representative of the various components, regardless of whether they are connected in series or parallel in the circuit.

This equivalent resistance can also be used to calculate the total power of the combination circuit. When the total equivalent resistance is used to find the power, the value will be equal to the sum of the power consumed by the individual components. You will continue to develop additional skills in analyzing combination circuits and specifically in determining the power used in these circuits. We now explore how determining the power requirements of a combination circuit can be used to your advantage.

In addition, you will continue to practice the same systematic approach to solving for the various unknown values by analyzing the circuit and working with given information. These values, when inserted into the various formulas, will provide insight into all possible circuit parameters. In addition, you will also be using known power values to calculate for voltage and current, working backward to solve for the unknown. Again, the key to solving all combination circuits is building confidence in analyzing the circuit for what is known and then systematically solving for the unknowns by reducing the various resistances to equivalent resistance values.

SOLVING CIRCUIT VALUES WITH PARTIAL INFORMATION

Often we know only parts and pieces of the combination circuit puzzle, but because we are trained professional electricians, we are expected to be able to decipher all unknown quantities. You will solve for the unknown values and create a table (as in Table 6–2) from a new circuit (Figure 6–32).

The easiest way to unravel the mystery is to fill in what you know and look for a part of the circuit where you can identify two of the three factors needed in Ohm's law. Using the figure, build a table (e.g., Table 6–3) with all the values you can determine from what you are able to directly measure or collect as information.

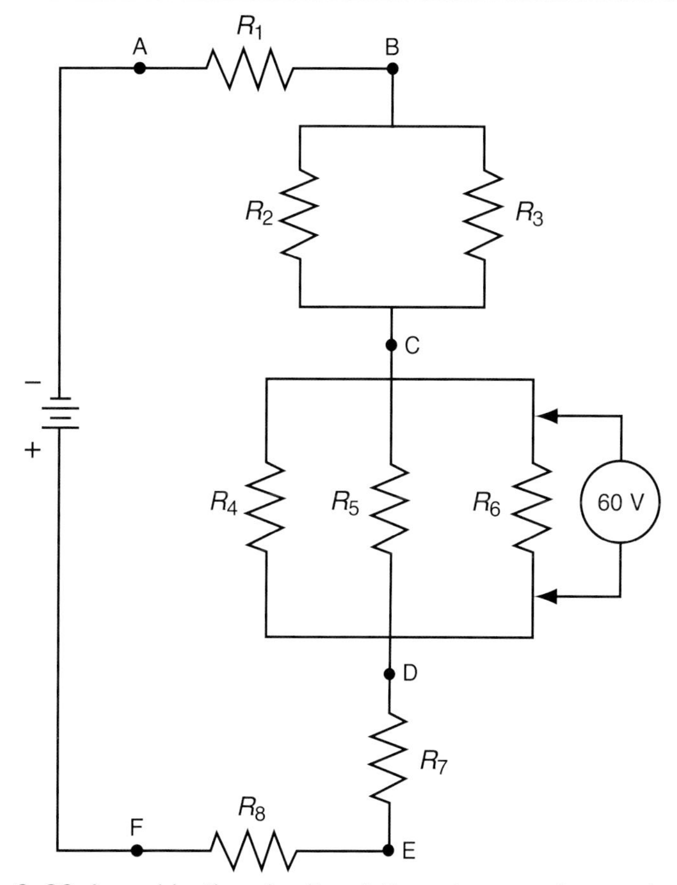

Figure 6–32 A combination circuit variation, where you know values inside the circuit, is depicted.

Table 6–3

A table is started to identify the known circuit variables.

Load	Resistance (Ω)	Current (A)	Voltage (V)	Power (W)
R_1	10	—	—	—
R_2	—	2.5	—	—
R_3	—	—	10	—
R_4	—	1.0	—	—
R_5	30	—	—	—
R_6	—	2.0	60	—
R_7	—	—	100	—
R_8	30	—	—	—

Now you can begin to fill in the unknowns. Begin at R_6. You have identified R_6 as having 60 V across it and 2 A through it.

$$R_6 = \frac{60\text{ V}}{2\text{ A}}$$

$$R_6 = 30\ \Omega$$

R_5 also has 60 V across it because of parallel voltage rules.

$$I_5 = \frac{60\text{ V}}{30\ \Omega}$$

$$I_5 = 2\text{ A}$$

R_4 has 60 V across it and 1 A.

$$R_4 = \frac{60\text{ V}}{1\text{ A}}$$

$$R_4 = 60\ \Omega$$

By reviewing the series current rules, we know that R_7 and R_8 are in series with the parallel branch points C to D. Therefore, the total current of the parallel branches added together at point D is 5 A. You already know that the voltage across R_7 from point D to E is 100 V and thus can calculate the other values needed.

$$R_7 = \frac{100\text{ V}}{5\text{ A}}$$

$$R_7 = 20\ \Omega$$

You know that 5 A continues flowing from point E to F through the 30 Ω resistor. You can now calculate the voltage across R_8.

$$E_8 = 5\text{ A} \times 30\ \Omega$$

$$E_8 = 150\text{ V}$$

Now we can begin working back toward the source from point C. We know there is 5 A total in the parallel branches from point C to point D. This means that there must have been 5 A coming through the parallel branches from point B to C. With this knowledge and the information that $I_2 = 2.5$ A, we deduce that R_3 also must have the other 2.5 A to make the total of 5 A. Therefore, R_3 had 2.5 A and 10 V.

$$R_3 = \frac{10\text{ V}}{2.5\text{ A}}$$

$$R_3 = 4\ \Omega$$

R_2 had 2.5 A and the same parallel voltage of 10 V, and thus we calculate:

$$R_2 = \frac{10\text{ V}}{2.5\text{ A}}$$

$$R_2 = 4\ \Omega$$

You could also assume that half the current went through one resistor and the other half through the second (same-size) parallel resistor. The last resistor to solve for is R_1. R_1 is in series with the rest of the circuit. All of the current must pass through R_1, and thus 5 A passes through R_1 with a resistance of 10 Ω.

$$E_1 = 5\text{ A} \times 10\ \Omega$$

$$E_1 = 50\text{ V}$$

Fill in all values in the table (Table 6–4) and mark them on the schematic diagram of Figure 6–32.

Fill in individual power consumption values for each piece of resistive equipment (Table 6–4). To verify these calculations, you can use the circuit totals ($I_T = 5$ A). The equivalent R_T after circuit simplification is 74 Ω. The total voltage should be:

$$E_T = I_T \times R_T$$

$$E_T = 5\text{ A} \times 74$$

$$E_T = 370\text{ V}$$

Table 6–4

A table can be used to track the calculated values as they are found.

Load	Resistance (Ω)	Current (A)	Voltage (V)	Power (W)
R_1	10	5.0	50	250
R_2	4	2.5	10	25
R_3	4	2.5	10	25
R_4	60	1.0	60	60
R_5	30	2.0	60	120
R_6	30	2.0	60	120
R_7	20	5.0	100	500
R_8	30	5.0	150	750
Total	**74**	**5.0**	**370**	**1,850**

Use your chart and schematic diagram voltages to verify the correct total voltage. The voltages are:

- A–B = 50 V
- B–C = 10 V
- C–D = 60 V
- D–E = 100 V
- E–F = 150 V

Total voltage is the sum of each series segment (in this case, **370 V**). The last column is the wattage, which is the total of each component's dissipated wattage. The sum total is 1,850 W. Verify this several ways.

$$P_T = E_T \times I_T$$

$$P_T = 370 \times 5$$

$$P_T = 1{,}850 \text{ W}$$

or

$$P_T = I^2 \times R$$

$$P_T = 5^2 \times 74$$

$$P_T = 1{,}850 \text{ W}$$

or

$$P_T = \frac{E^2}{R}$$

$$P_T = \frac{370^2}{74}$$

$$P_T = 1{,}850 \text{ W}$$

POWER IN COMBINATION CIRCUITS

As you have previously calculated, the total power consumed by a series or parallel circuit is equal to the sum of the power consumed by each of the circuit's components. The same is true of the combination circuit. Often only one parameter (such as the resistance, the voltage, or the current) will be given. Your job is to reduce and combine the various resistances to find the second parameter.

Often this requires the systematic reduction of the circuit to a single equivalent resistance value. Then you work back in the opposite direction to provide the missing values for the various components, using Ohm's law. Remember, you cannot find the power until you have two of the three parameters for each of the components. The total power is derived by adding all individual powers from each of the components.

$$P_T = P_1 + P_2 + P_3 + \ldots P_n$$

Calculate the power dissipated by each component and the total power of the combination circuit in Figure 6–33. The first step is to review the circuit, analyzing for what is known and what is not known and comparing what you see to what you are solving for. In Figure 6–33, there are no components that have two values from which you can work. This means that you will have to begin by combining circuit components to solve for the necessary second variable.

You could use the equation $P_T = E_T \times I_T$ to calculate the total power. However, Figure 6–33 does not have the value for I_T. The first step is therefore to find I_T. To calculate the value for I_T, you need the value for the total resistance (R_T). Knowing R_T allows you to calculate I_T with the following formula:

$$I_T = \frac{E_T}{R_T}$$

To calculate R_T in this circuit, you start by combining resistors R_2 and R_3:

$$R_{2,3} = \frac{(R_2 \times R_3)}{(R_2 + R_3)}$$

$$= \frac{(100 \times 200)}{(100 + 200)} = 66.67\ \Omega$$

Just for validation, we could use an assumed voltage for the 100 Ω and the 200 Ω resistors in parallel. Assume 400 V as the parallel voltage, as an example.

Calculating Wattage

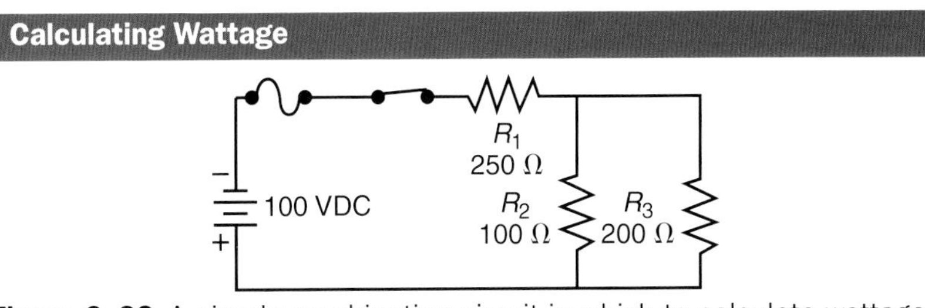

Figure 6–33 A simple combination circuit in which to calculate wattage is depicted.

The 100 Ω resistor would allow 4 A and the 200 Ω resistor would allow 2 A, for a total of 6 A. This 6 A would be the total current with 400 V applied. Thus, $\frac{E}{I} = R$ actual or $\frac{400\text{ V}}{6\text{ A}} = 66.67\ \Omega$.

The second step is to reduce series resistor R_1 and equivalent resistor $R_{2,3}$ to a single value. This gives R_T. Because this is a series circuit, R_T is calculated by adding the series elements:

$$R_T = R_1 + R_{2,3}$$
$$= 250\ \Omega + 66.67\ \Omega = 316.67\ \Omega$$

Now that you have the source voltage $E_T = 100$ V and the total resistance R_T, you can solve for the total current (I_T), using the following formula:

$$I_T = \frac{E_T}{R_T} = \frac{100\text{ V}}{316.67\ \Omega} = 0.316\text{ A}$$

You now have all three parameters for the equivalent circuit. With these values, you can find the total power for the circuit P_T and the power consumed by each of the circuit components R_1, R_2, and R_3. First you solve for the total circuit power (P_T), using the following formula:

$$P_T = E_T \times I_T$$
$$= 100\text{ V} \times 0.316\text{ A} = 31.6\text{ W}$$

Now you have the needed information to solve for the individual power dissipations, starting with R_1. Because in a series circuit the current is the same throughout the circuit, and because you know the total current I_T, the current flow in R_1 and $R_{2,3}$ can be calculated as:

$$I_T = I_1 = I_{2,3} = 0.316\text{ A}$$

Now you have the current and the resistance for R_1, and thus you can solve for the power consumed by R_1 using the following formula:

$$P_1 = I^2 \times R_1 = (0.316)^2 \times 250\ \Omega$$
$$= 0.099856 \times 250 = 24.964\text{ W}$$

There are several ways you can calculate the power dissipated by R_2 and R_3. Because you know the current through the parallel branch $R_{2,3}$ and the equivalent resistance of $R_{2,3}$ (calculated earlier in the problem), you can calculate the voltage across this branch, using Ohm's law. Use the following formula:

$$E_{2,3} = I_{2,3} \times R_{2,3}$$
$$= 0.316\text{ A} \times 66.67\ \Omega = 21.07\text{ V}$$

Because this is a parallel branch, you know that the voltage is the same across both R_2 and R_3:

$$E_{2,3} = E_2 = E_3 = 21.07\text{ V}$$

Now you have the voltage drop and resistance of R_2 and R_3 and can calculate the power dissipated by each of those components, using the following formulas:

$$P_2 = \frac{E^2}{R_2} = \frac{21.07^2}{100} = 4.439\text{ W}$$

$$P_3 = \frac{E^2}{R_3} = \frac{21.07^2}{200} = 2.219\text{ W}$$

You have completed all of the calculations required to find the total power and the power consumed by the individual components that made up the circuit in Figure 6–33. You should check your calculations by adding the individual power dissipations and comparing them to the total power calculated previously:

$$P_T = P_1 + P_2 + P_3$$
$$= 24.964\text{ W} + 4.439\text{ W} + 2.219\text{ W}$$
$$= 31.622\text{ W}$$

Previously you calculated P_T as $0.316 \times 100 = 31.6$ W. The slight error is caused by rounding off decimals during the calculations.

CIRCUITS WITH KNOWN POWER USAGE

The circuits you will work on in your job will often have known values of power and voltage. One of the aspects you will have to consider is the real power requirement for circuits with components that are already rated. For example, examine Figure 6–34. In this circuit, three lamps are connected to the power source with 300 feet of #16 American wire gauge (AWG) copper wire.

You can see from the figure that each lamp is rated at 25 W and that the lamps are connected in parallel.

When you look at a circuit like this one, you must be sure to take all circuit parameters into account. The electrician who installed the lamps calculated that the lamps would draw 0.5 A each with 50 V applied, making the total circuit current the sum of the current in the individual branches (i.e., 1.5 A). The following formulas support the electrician's calculations. To calculate the current of the individual lamps (the total current is the sum of all three lamps):

$$I_{Total} = I_{Lamp\ 1} + I_{Lamp\ 2} + I_{Lamp\ 3}$$

$$= 0.5\ \text{A} + 0.5\ \text{A} + 0.5\ \text{A} = 1.5\ \text{A}$$

This all seems to check out fine. However, look at the circuit again. The load, all three lamps, is 300 feet from the source voltage. You know from earlier chapters that all conductors, even copper, have some resistance. You also know that the longer the conductor, the higher the resistance: #16 AWG copper has a resistance of 4.99 Ω per 1,000 feet.

This information comes from the *NEC*® and applies to seven-stranded conductor at 75°C. Note that this value differs slightly from the value given in an earlier chapter. This is because the value in that table is for solid conductor wire, not stranded wire.

Review other sizes of conductor to see how much resistance they have per thousand feet of conductor. Note that the larger the conductor, the lower the resistance value. The fact that the load is 300 feet from the power source means that the resistance of the copper #16 conductor, going to and returning from the lamps, must be considered in the calculations.

Because resistance exists in the conductors supplying the lighting load, the calculations for this circuit have other factors to consider. Examine Figure 6–35. This circuit was redrawn to represent all resistances that must be considered in the circuit when performing the calculations.

Combination Series-Parallel Circuit

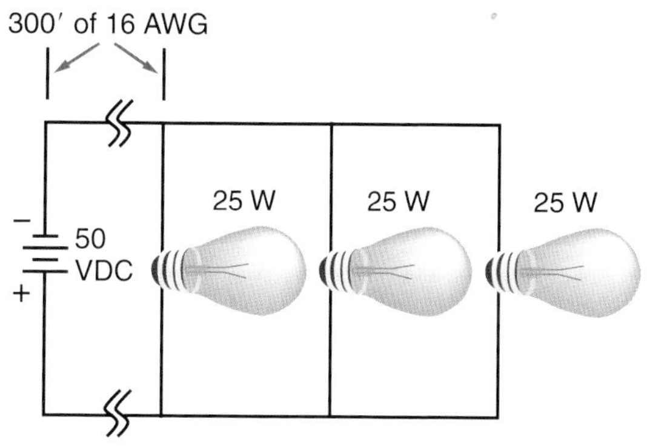

Figure 6–34 A simple parallel wiring of three lamps is really a combination series-parallel circuit.

Parallel Simple Lamp Circuit

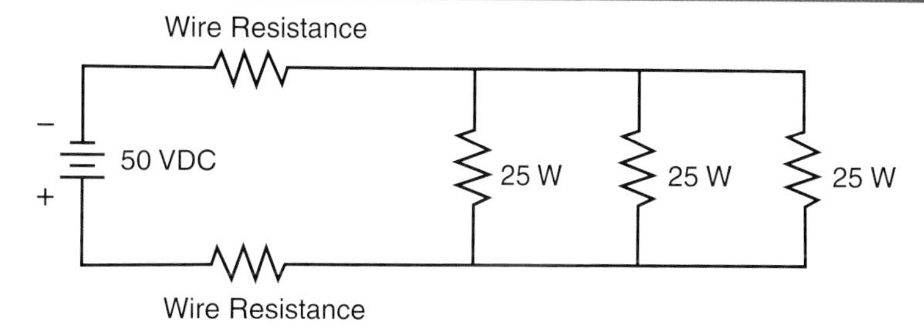

Figure 6–35 This schematic diagram shows a series-parallel simple lamp circuit.

The power rating of a lamp is the amount of power the lamp will dissipate with rated voltage applied. Because rated voltage will not be applied because of the voltage drop across the wire, you need to know the lamp resistances so that you can solve this circuit as a combination circuit, using the methods you have learned in this and previous chapters. Because the lamps are rated at 25 W at 50 V, you can calculate the resistance value of the lamp as follows:

$$R_{Lamp} = \frac{E_{Lamp}^{\ 2}}{P_{Lamp}}$$

$$R_{Lamp} = \frac{50^2}{25\ \text{W}}$$

$$R_{Lamp} = 100\ \Omega$$

Note that this is the resistance of each individual lamp. To completely analyze this circuit, you also need to know resistance of the wire feeding the lamps. Based on the resistive value of #16 AWG wire being 4.99 ohms per 1,000 feet of wire, you can calculate the total resistance as:

$$R_{Wire} = \left(\frac{4.99}{1{,}000 \text{ ft}}\right) \times 300 \text{ ft} = 1.497\ \Omega$$

Remember, this circuit has 300 feet of wire going to the lamps and 300 feet of wire coming back from the lamps. Both distances must be taken into account. Figure 6–36 shows the complete circuit with all resistances included. This is a combination circuit that can be reduced with the methods you have learned. Start with calculation of the resistance of the three lamps in parallel, $R_{All\ lamps}$:

$$R_{All\ lamps} = \frac{R_{One\ lamp}}{\text{number of lamps}}$$

$$R_{Total\ lamps} = \frac{100}{3}$$

$$R_{Total\ lamps} = 33.33\ \Omega$$

The circuit has now been reduced to a simple series circuit with the two lengths of copper wire and the parallel combination resistance of the lamps. The total resistance is the sum of the three series elements:

$$R_T = R_{Wire\ 1} + R_{Wire\ 2} + R_{All\ lamps}$$
$$= 1.497 + 1.497 + 33.33$$
$$= 36.324\ \Omega$$

Parallel Connected Load

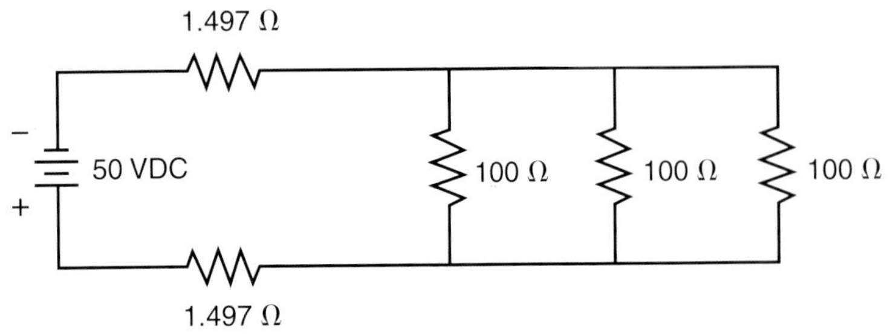

Figure 6–36 A schematic diagram illustrates wire resistance in series with parallel connected loads.

The total current is calculated as:

$$I_T = \frac{E_T}{R_T}$$

$$I_T = \frac{50}{36.32\ \Omega} = 1.38 \text{ A}$$

Note that the value calculated by the electrician installing this project (1.5 A) is different from the value calculated previously. The reduction in current is caused by the added resistance of the wire. Now we will find out how much voltage is actually supplied to the lamps. We know that the total current for this circuit is actually 1.38 A, and we know that the equivalent resistance for the lamps in series with the resistance equals 33.33 Ω. Using Ohm's law, we can easily calculate the voltage across the lamps as follows:

$$E_{Lamps} = I_{Lamps} \times R_{Lamps}$$
$$= 1.38 \text{ A} \times 33.33\ \Omega = 46 \text{ V}$$

This means that we have lost (dropped) 4 V on the wires. In percentage, this is:

$$\%\text{Drop} = \frac{(E_{Rated} - E_{Actual})}{E_{Rated}} \times 100$$

$$\%\text{Drop} = \frac{(50 - 46)}{50} \times 100$$

$$= 0.08 \times 100 = 8\%$$

This much drop is excessive. Maximum voltage drops should normally be limited to no more than 3%. The lamps are not dissipating the amount of power for which they were designed. In other words, instead of the lamps actually providing light based on their design parameters, the lights will now be operating according to the actual installation parameters. Let's see what effect the installation parameters actually have on the lamps. Because we now know that the installation is being supplied by 46 V to the lamps, we can recalculate the actual wattage of the lamps as follows:

$$P_{Lamp} = \frac{E_{Lamp}^{\ 2}}{R_{Lamp}} = \frac{46^2}{100} = 21.16 \text{ W}$$

This circuit, by taking into account the resistance of the conductors, in reality has lost approximately 15% of the lamp output. This is:

$$\%\text{Loss} = \frac{(P_{Rated} - P_{Actual})}{P_{Rated}} \times 100$$

$$= \frac{(25 - 21.16)}{25} \times 100$$

$$= \frac{3.84}{25} \times 100 = 15.36\%$$

You can see that the entire circuit must be taken into account when designing an installation. All of the resistance components in a combination circuit consume power and must be taken into account. Take a look at the resistance of the conductors and see how much power their resistance consumes. Because you have already calculated the current and the resistance of the conductors, you can easily calculate the power consumed by the conductors from the following formula:

$$P_{Wires} = 2 \times (1.38^2 \times 1.497) = 5.7 \text{ W}$$

Note the multiplier of 2. There are two conductors, and you must consider both of them. What the previous calculation tells you is that out of all power consumed by the installation, a portion of that power (5.69 W) is consumed by the conductors and wasted as heat. By not taking into account the power lost in the conductors, the actual circuit did not perform to optimum design efficiency, the actual lamp output was reduced, and the conductors consumed power that was wasted as heat. All of this may have been avoided or reduced by selecting a larger conductor size.

In series and parallel circuits, especially circuits with long runs of conductor or small-diameter conductor (such as fire alarms or speaker systems), you must be aware of voltage drop and of wattage loss due to conductor resistance.

FieldNote!

Another example of a line drop and line loss problem is in the use of fire alarm circuits. In monitor circuits where there are many heat sensors or smoke sensors in a zone of protection, the two conductors that go from sensor to sensor will have line drop. In notification circuits (Figure 6–37), the sensing elements are all in parallel across the two parallel conductors. Normally, the sensors present an open connection to the parallel circuit.

The monitoring of the circuit is accomplished by placing a resistor across the line conductors at the very end of the circuit. This resistor is called the end-of-line resistor. Its purpose is to allow a small amount of current to flow all along the negative conductor to the end, through the resistor and all the way back to the source.

If there is a break anywhere along the circuit (open), the line current goes to zero and the monitoring electronics indicate a circuit problem. If any sensor senses heat or smoke, it creates a closed circuit (short) from one conductor to the other. The monitoring electronic circuit senses a large increase in current and the building fire alarm is activated.

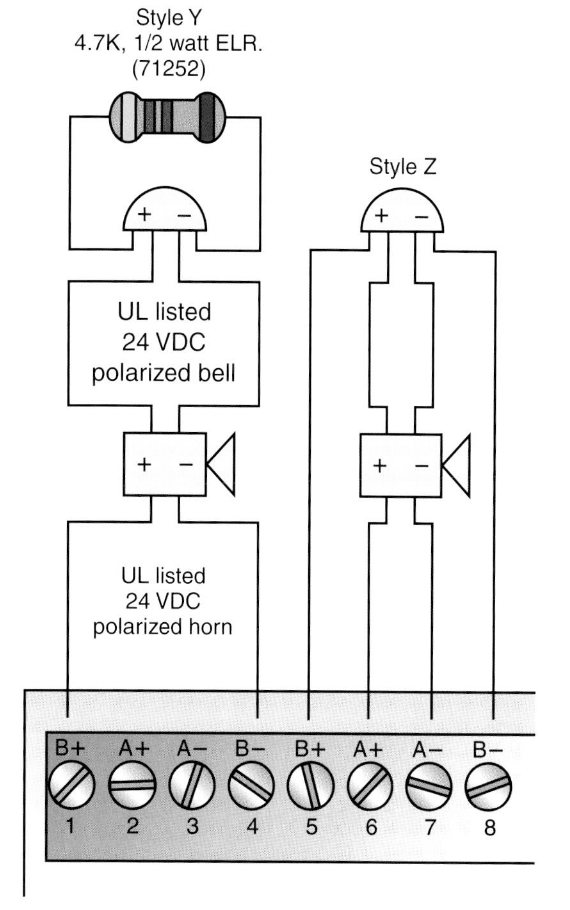

Figure 6–37 This typical monitored fire alarm system is known as "Style Y" in a supervised power-limited monitoring system.

SUMMARY

This chapter dealt with the combination circuit. This circuit is very common in the electrical industry. First we traced current flow through the circuit to determine if the components were in series or parallel. When we were able to identify the type of circuit, we began to reduce or simplify the circuit so that we could calculate the specific values for each component. After the current paths were established, we found the correct rules to use in each portion of the circuit to find R_T (the single resistor equivalent) for the entire circuit.

The voltage rules and methods were explained to allow you to see what voltages dropped at various parts of the combinations of resistors. Then we were able to verify the component values compared to the total value. Several different circuits were explained to show that there are various ways of arriving at the same conclusion. Finally, the wattage calculations for different circuits were calculated and confirmed as correct. The series parallel combination of circuit segments was used to return to the effects of current on a circuit where wire resistance is a factor.

CHAPTER GLOSSARY

Combination circuit A circuit consisting of both parallel and series connected components. Combination circuits are sometimes referred to as series-parallel circuits.

Law of proportionality Current and resistance are inversely proportional for a series circuit, and voltage drop and resistance are directly proportional in a series circuit. The voltage of a parallel circuit is constant, and current flows as an inverse proportion to the resistance values compared to the total equivalent resistance.

Utilization equipment As defined by the *National Electrical Code*® (*NEC*®), utilization equipment is equipment that utilizes (uses) electric energy for electronic, electromechanical, chemical, heating, lighting, or similar purpose.

REVIEW QUESTIONS

1. Define a combination circuit.
2. Why would you solve for total resistance in a combination circuit?
3. Describe the general process for simplifying a combination circuit to one equivalent resistor.
4. Explain how to use an assumed voltage to find a parallel equivalent resistance.
5. Give an example of how to use the law of proportionality when finding the currents in parallel branches.
6. Explain why you would need to rebuild the circuit after simplifying it.
7. List the formulas for solving power in a combination circuit.
8. What effect does the length of the circuit conductor have when calculating power or voltage drop? How can this effect be modified or improved?
9. Where do you find the resistance rating for wire?
10. What is the maximum voltage drop recommended on a branch circuit?
11. Explain why you cannot just add the voltages of each component to get a sum for the total circuit voltage.
12. Explain why you cannot just add the component currents in a combination circuit to get the total current.

PRACTICE PROBLEMS

1. Solve for I_T, R_T, and P_T in Figure PP6–1.
2. Solve for R_T, I_T, and E_T in Figure PP6–2.
3. Solve for E_T, I_T, R_T, and P_T in Figure PP6–3.
4. Using the circuit in Figure PP6–4, calculate the currents and voltages in each circuit element.
5. A commercial warehouse is 1,000 feet long. Each end of the warehouse has two 400 W 120 V HPS wall pack–type light fixtures mounted on it. The supply panel is a 120 V panel and is located in the middle of the warehouse. There are two circuits, one for each end of the warehouse. The wall packs are fed with #12 stranded copper wire.
   a. Draw the schematic diagram for this scenario.
   b. What is the total current of each circuit?
   c. What is the power consumed by each element of the circuits?
   d. What is the total power consumed by the circuits?

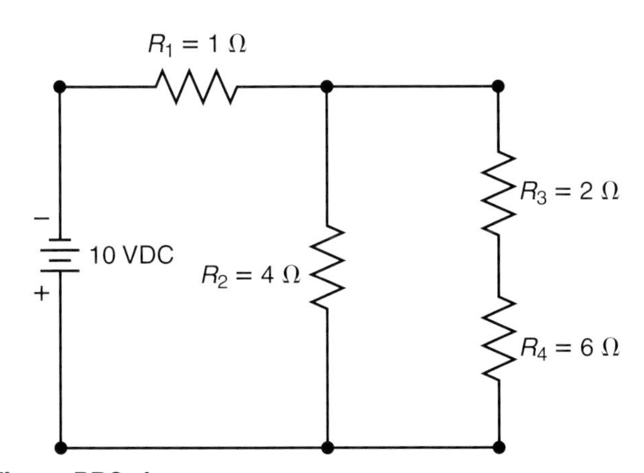

Figure PP6–1

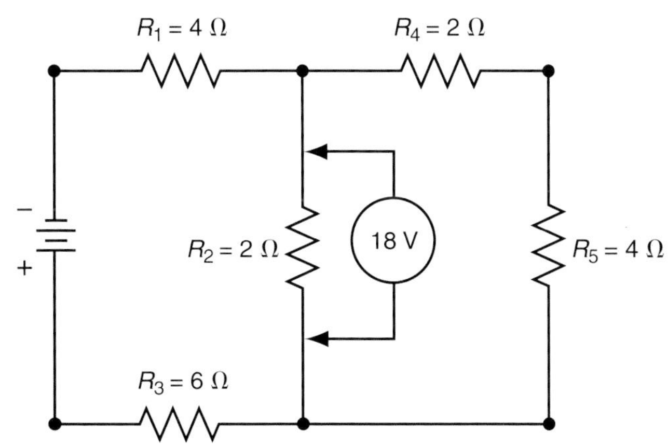

Figure PP6–2

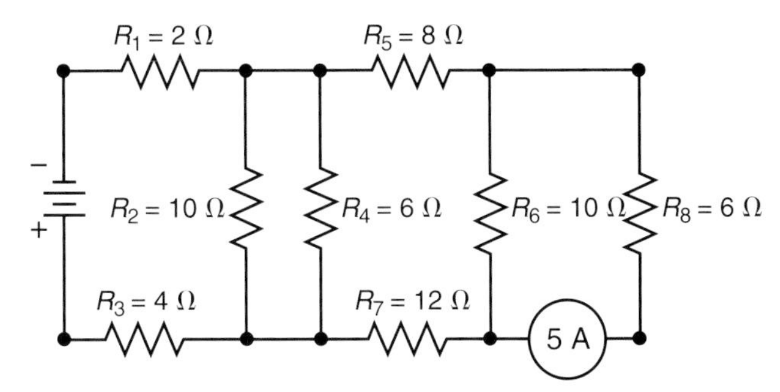

Figure PP6–3

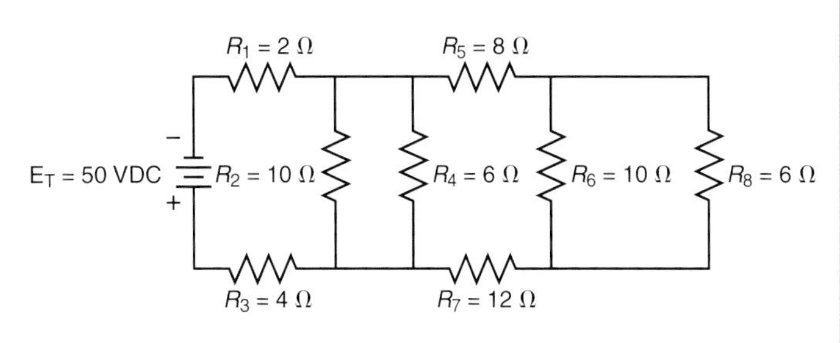

Figure PP6–4

7

DC Circuit Analysis

TOPICS

- INTRODUCTION
- VOLTAGE DIVIDERS
- LAW OF PROPORTIONALITY FOR VOLTAGE
- CURRENT DIVIDERS
- THE SUPERPOSITION THEOREM
- KIRCHHOFF'S LAWS
- THEVENIN'S THEOREM
- NORTON'S THEOREM
- LIMITING CIRCUITS BY WATTAGE RATING

OVERVIEW

In this chapter, you will learn some of the most useful circuit concepts in electricity. These theorems and processes allow you to analyze circuits and find solutions for various circuit configurations. Voltage dividers will enable you to take a voltage supply and create multiple voltages using only resistors. The voltage divider is a frequently used circuit, whether you are looking for an extra voltage to supply power, analyzing a circuit to troubleshoot, or modifying a circuit so that it works more appropriately for the task at hand.

Previously you learned that voltage distributes across resistors in a series circuit. The voltage division in a voltage divider circuit is proportional to the resistance of each individual resistor. Current in a parallel circuit divides. You will learn how to calculate the current through branches of a current divider using simple formulas, which help determine how to provide multiple currents to complex circuits.

The superposition theorem is one of the many useful tools that can be used in circuit analysis. It is so useful because it applies Ohm's law to circuits that have more than one voltage source. In simple terms, we can calculate the effect of one voltage source at a time and then superimpose the results of all sources. You can also use the superposition theorem to calculate the current flow through any circuit branch containing more than one voltage source.

Kirchhoff's laws are needed to help analyze more complex circuits. These laws allow for the mathematical justification of voltages and polarities when several sources are used. We will study them in more depth in this chapter.

Thevenin's and Norton's theorems can save you time in recalculating entire circuits for changing loads. They are used extensively in electronic circuit applications and have application in many electrical power system applications. They can be especially useful when dealing with many unknowns in complex circuits, such as those encountered during network analysis in a power system. In this chapter, you will learn about both of the theorems and how to use them.

The last concept explored is that of limiters where circuit conditions dictate which components will work in some series or parallel circuits but may be destroyed in other similar-looking circuits. You will determine what the limits are for circuit parameters.

OBJECTIVES

After completing this chapter, you will be able to:

- State the law of proportion for series circuits
- Describe voltage dividers in series circuits
- Explain the difference between directly proportional and inversely proportional relationships
- State the law of proportionality as it applies to series and parallel circuits
- Solve problems involving resistors in parallel, using the law of proportionality
- State the steps necessary to apply the principle of superposition
- Apply the theory of superposition to solve for multiple voltage source circuits
- Apply Kirchhoff's laws to solve for circuit variables in complex circuits
- Explain Thevenin's and Norton's theorems
- Apply Thevenin's and Norton's theorems to solve for circuit unknowns
- Solve circuit problems in terms of open-circuit voltages and maximum circuit currents
- Determine limiting components in series and parallel circuits

INTRODUCTION

In previous chapters, Ohm's law was used exclusively to calculate resistance, current, and voltage values for resistors in series DC circuits. The law of proportionality, discussed in this chapter, also uses Ohm's law to find the voltage across a single resistor in a series of resistors. This law describes the relationship between resistors and circuit-resistive elements in a circuit. Using this law, you can calculate the voltage across any resistor or combination of resistors in a series circuit without knowing the circuit current.

This chapter also covers the method used to separate circuit voltages into different values. These different voltages may be required by other parts of the same circuit or by other circuits. Circuits of this type, where the voltage is divided between two or more resistors, are called **voltage dividers**. They "divide" the voltage into as many different values as are needed for different applications. See Figure 7–1 as an example of uses for a voltage divider circuit where a load uses only part of the applied voltage. Note that the lower resistor is a variable resistor, which allows the voltage divider to vary the divided voltage.

Parallel circuits by definition have more than one current path. By taking a total current and dividing it among the branches, the current can be used for different applications in each branch. As you can see in Figure 7–2, the current divides into two paths and reunites to again form the total current. By using the current divider concepts, each branch current can be calculated as a percentage of the total current even though the exact voltage might not be known yet.

The purpose of the superposition theorem is to provide actual calculated values when more than one voltage source or **current source** is acting on a circuit. The definition of the superposition theorem states that the voltage or current in any element resulting from several sources acting together is the sum of the voltages or currents resulting from each source acting alone.

Gustav R. Kirchhoff devised a mathematical rationale for determining voltages and correct polarities for loads in complex circuits and presented it about 1845. His ideas are particularly helpful when multiple voltage sources are used in the same circuit.

Thevenin's theorem creates an equivalent series circuit with respect to any pair of terminals in the network.

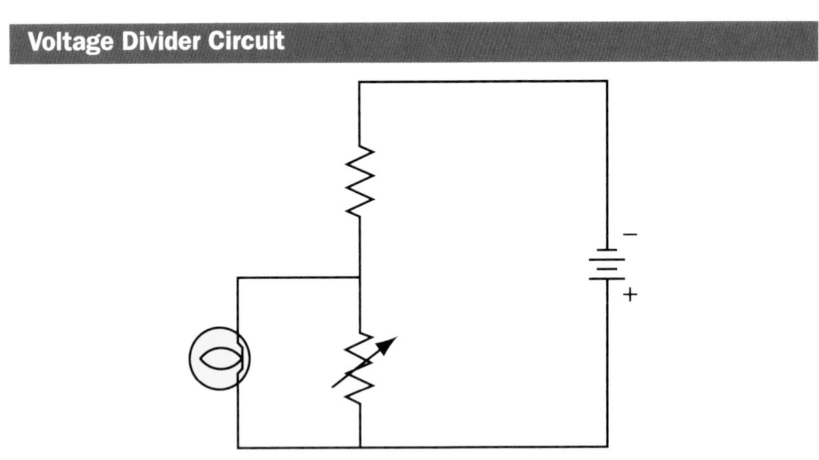

Figure 7–1 This voltage divider circuit has a variable resistor at the voltage selection point.

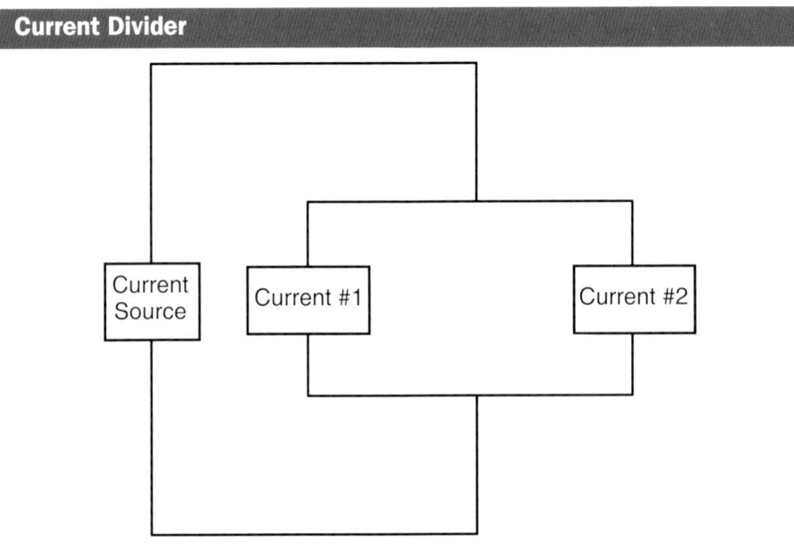

Figure 7–2 Current flow divides between different branches to create a current divider.

This can be accomplished regardless of how many voltage sources are involved and no matter what the circuit configuration of the original complex circuit is. Thevenin allows us to simplify complex circuits that deliver electrical energy to a load and quickly calculate the effects of the circuit. Thevenin's theorem is related to Norton's theorem. Norton uses a current source and parallel resistors to calculate similar effects in a complex system.

In many cases, you will be called on to design a circuit that works for various loads with different voltage and current parameters. To do this you must know the limiting factors as you combine various rated devices or utilization equipment in series circuits or in parallel. You need to determine what the limits of current and voltage are before you build and test the circuit. This chapter introduces you to these limiting factors.

VOLTAGE DIVIDERS

Refer to the simple voltage divider circuit shown in Figure 7–3. This circuit divides the applied voltage into two measurable voltages. This is a voltage divider with two resistors (R_1 and R_2) and a DC voltage source. The voltage applied is 30 VDC. Assume that the resistance values of R_1 and R_2 are equal ($R_1 = R_2$). You know from previous lessons that the voltage will distribute the drops into proportional values. To review, the process works according to the following principles:

- The current through a DC series circuit is the same value through all components.
- The total resistance value of a DC series circuit is the sum of all resistance values in the circuit.
- The sum of all voltage drops in a DC series circuit is equal to the source voltage.

Because R_1 and R_2 are of equal resistances, the voltage drop across each will be the same number. This is shown in the following equations:

$$\text{Because } R_1 = R_2 \text{ (given)}$$

$$\text{and } I_1 = I_2 = I_T \text{ (series circuit),}$$

$$\text{then } I_1 \times R_1 = I_2 \times R_2$$

$$\text{and thus } E_1 = E_2 \text{ (Ohm's law)}$$

In Chapter 4, you learned that the sum of the voltages in a series circuit is equal to the voltage source (E_T). Because $E_T = E_1 + E_2$ in Figure 7–3 and $E_1 = E_2$, then E_1 and E_2 are each half the amount of E_T. The voltage divider in Figure 7–3 divides the total voltage into two equal parts. The voltage divider in Figure 7–3 is relatively simple because both resistors are of equal value. The following description covers a circuit whose resistors are not equal. R_1 in Figure 7–4 is 10 Ω, and R_2 is 5 Ω; R_1 is twice the value of R_2.

Voltage Divider

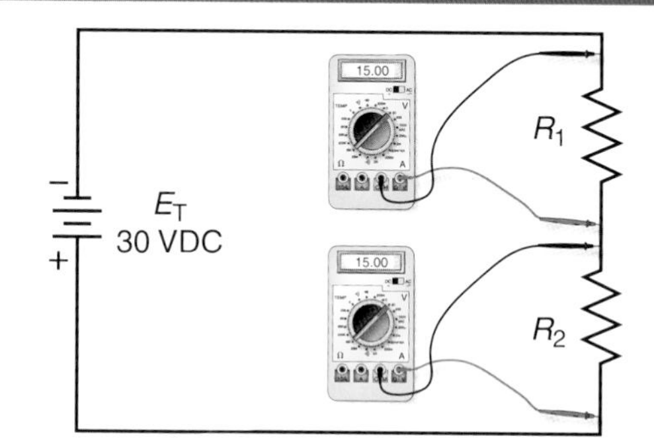

Figure 7–3 The voltage is divided by two series resistors to create a voltage divider.

Utilizing Voltages

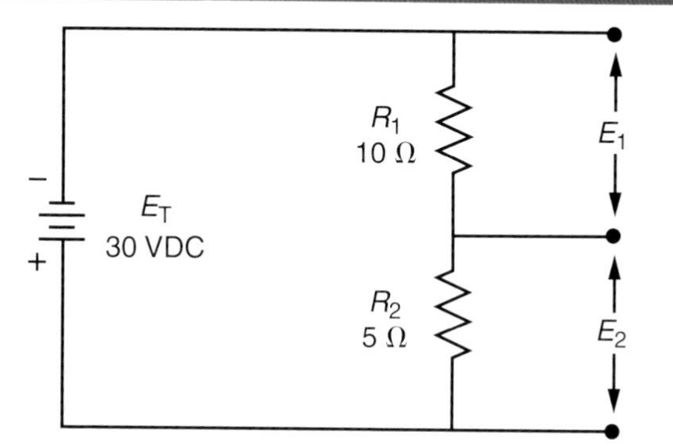

Figure 7–4 The voltages can be utilized as if they were a source at E_1 or E_2 or both.

Using the previous statements and Ohm's law, we can show how the voltage drops across the resistors compare with each other.

Because $R_1 = 2R_2$ (given)

and $I_1 = I_2$ (series circuit),

$$\frac{E}{R} = I$$

and

$$\frac{E_1}{R_1} = \frac{E_2}{R_2} \text{ (Ohm's law)}$$

$$E_1 = \frac{(E_2 \times R_1)}{R_2}$$

and because $R_1 = 2R_2$

$$\text{then } E_1 = \frac{(E_2 \times 2R_2)}{R_2 \text{ (substitution)}}$$

and thus $E_1 = 2E_2$ (canceling the R_2)

The sum of the voltages across all resistors is equal to the source voltage (E_T). Thus, according to Kirchhoff's law:

$$E_T = E_1 + E_2$$

Because $E_1 = 2E_2$,

then $E_T = 2E_2 + E_2$

$$E_T = 3E_2$$

$$\text{and thus } E_2 = \frac{1}{3E_T}$$

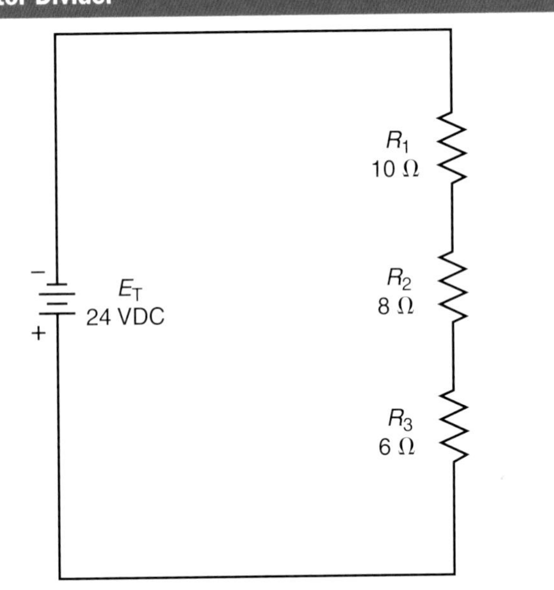

Figure 7–5 With a three-resistor divider, the voltages can be determined proportionally.

Comparing this to the resistances:

$$R_T = R_1 + R_2$$

$$R_T = 2R_2 + R_2$$

$$R_T = 3R_2 \text{ (substitution)}$$

$$R_2 = \frac{1}{3R_T}$$

This example illustrates the point that the fractional equivalent of the voltage dropped on any resistor is the same as the fractional equivalent that each resistor represents. Because R_2 is one third of the total resistance, it will drop one third of the total voltage.

LAW OF PROPORTIONALITY FOR VOLTAGE

Because the current through a series circuit is the same at any point, the voltage across any resistor is:

$$E_R = I_T \times R_R$$

Here, E_R is the voltage across the resistor being measured, I_T is the total current, and R_R is the resistance of the resistor being measured. The total current in the circuit is represented by:

$$I_T = \frac{E_T}{R_T}$$

Now, if the equation for total current is placed in the previous equation for the voltage across any resistor, the previous equation becomes:

$$E_R = E_T\left(\frac{R_R}{R_T}\right)$$

The last equation is known as the law of proportionality in a series circuit, as we have mentioned earlier. It says that the voltage dropped across any resistor in a series circuit is equal to the total voltage multiplied by the fraction of the value of the resistor in question, divided by the total resistance.

The law of proportionality can be extended to calculating the voltage drop across more than one resistor. Consider Figure 7–5, for example.

If you wish to calculate the total voltage drop across resistors R_2 and R_3, you can do so by substituting into the law of proportionality equation the sum $R_R = R_2 + R_3$. This means that the total voltage drop across R_2 and R_3 can be calculated as:

$$E_{(R_2+R_3)} = E_T\left[\frac{(8+6)}{(8+6+10)}\right]$$

$$= 24 \times \left(\frac{14}{24}\right) = 14 \text{ V}$$

You can verify this by inspection of Figure 7–5. The proportion of the two resistors $R_2 + R_3$ compared to the total resistance of $R_1 + R_2 + R_3$ is the same proportion of the total voltage. With this knowledge, we can divide the source voltages into voltage we need to operate sub-circuits (as in Figure 7–6). Note that the percentage of the total resistance will allow you to calculate the percentage of the total voltage dropped. If R_2 is 25% of the total resistance, it will have 25% of the total voltage, whatever the source voltage might be. In this case, the source voltage is 100 VDC and the voltage across R_2 is 25 V.

We can now use the 25 V across R_2 as a source of power for another circuit operating at 25 V. As the current load is added to the 25 V divided voltage, it tends to create a combination circuit (as you have just studied). The current through the 25 V load is actually drawn from the original 100 V source, through R_1. This will actually start to change the voltage divider voltage ratios.

A rule of thumb for the 25 V added load is to keep the resistive load value high (at least 10 times the divider resistor) to limit the amount it "loads down" the circuit. If the load draws only 1/10 of the divider current, the original circuit will not be altered a great deal.

Output and Input Voltage

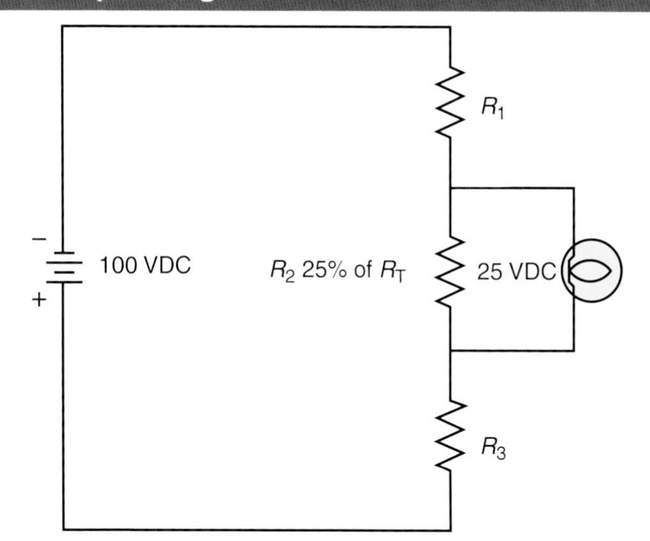

Figure 7–6 The output voltage is proportional to the input voltage divided proportionally among the resistors.

CURRENT DIVIDERS

In series circuits, the voltage drop across a resistor is related to the value of the resistance. As the resistance increases, the voltage drop increases and the current through the same resistor decreases, its value changing in the opposite direction. If resistance increases, current decreases. We say that values that change in the same direction are directly proportional. Values that change in the opposite direction are inversely proportional.

Because the parallel circuit voltage is equal across the branches in parallel, the branches in a parallel circuit respond similarly for current. As the resistance in a branch increases, the current decreases. Remember, though, that the voltage across the branch remains the same as long as the source voltage does not change.

So far, you have seen parallel circuits simply as separate parallel paths for current, all returning to the same power source.

Current Divider Circuit

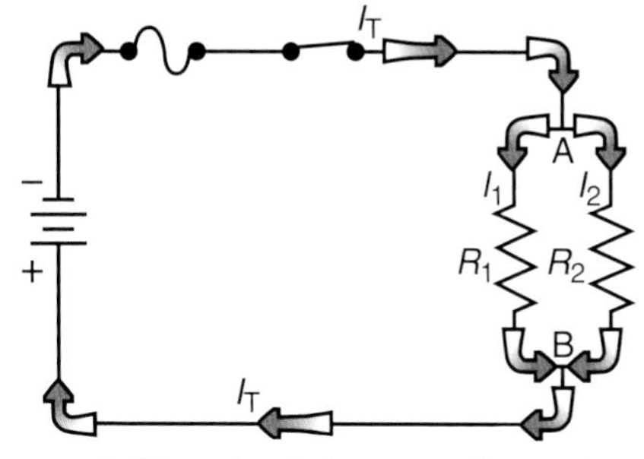

Figure 7–7 A current divider circuit has parallel paths for current.

Let's look at a two-branch parallel circuit in a different way. Figure 7–7 shows a circuit with two paths for current. You can see that the current divides at point A, travels through the two paths, and recombines at point B before returning to the source.

If the total current for this circuit were known, but not the voltage, calculating the current through one of the resistors would be a bit involved. With the **current divider** concept, you can find all the values of the circuit without knowing the circuit totals. The current divider will tell you how the current will divide, even if you do not know the applied voltage in the circuit. By simplifying the equations and relating one branch current to the total circuit current, you make the calculation easier.

For example, the current through any branch of a parallel circuit is:

$$I_1 = \frac{E_T}{R_1}$$

In addition, the total voltage in any circuit is:

$$E_T = I_T \times R_T$$

Substituting total voltage into the current equation:

$$I_1 = (I_T \left(\frac{R_T}{R_1}\right))$$

This is the reciprocal formula, explained in Chapter 5. The current splits in the inverse proportion of the resistance of the branch compared to the total resistance of the parallel circuit. The general form of this equation is:

$$I_x = (I_T \left(\frac{R_T}{R_x}\right))$$

This equation can be used to calculate the current through any circuit branch by first calculating the total resistance (R_T) of the branch using the parallel resistance formulas you learned earlier and then using the values in the previous equation.

Example

What are the currents through R_1 and R_2 in Figure 7–8?

Solution:

Using the current divider equation with a known I_T, and finding the R_T as 17.53 Ω:

$$I_1 = (I_T \left(\frac{R_T}{R_1}\right)) = 1.5 \times \left(\frac{17.53}{27}\right) = 0.974 \text{ A}$$

$$I_2 = (I_T \left(\frac{R_T}{R_2}\right)) = 1.5 \times \left(\frac{17.53}{50}\right) = 0.526 \text{ A}$$

Now check the result using Ohm's law.

$$E_T = I_T \times R_T = 1.5 \times 17.53 = 26.295 \text{ V}$$

Because the voltage in a parallel circuit is the same across all branches, solve for the current through R_1 and R_2 and see if the total current I_T equals the sum of the current through each of the resistances. The calculations are as follows:

$$I_1 = \frac{E_T}{R_1} = \frac{26.295}{27} = 0.974 \text{ A}$$

$$I_2 = \frac{E_T}{R_2} = \frac{26.295}{50} = 0.526 \text{ A}$$

The checked values show that the original calculation was correct. The total current, I_T, can be found by adding the currents from each of the branches as follows:

$I_T = I_1 + I_2 = 0.974 \text{ A} + 0.526 \text{ A} = 1.5 \text{ A}$ (as was given in the original circuit).

As with voltage dividers, there is a law of proportionality that applies to current dividers. This law states that the amount of current passing through a branch of a current divider is proportional to the resistance of the other branch, divided by the sum of the two branches, times the total current. The following is a mathematical description of this law.

Parallel Circuit Rules

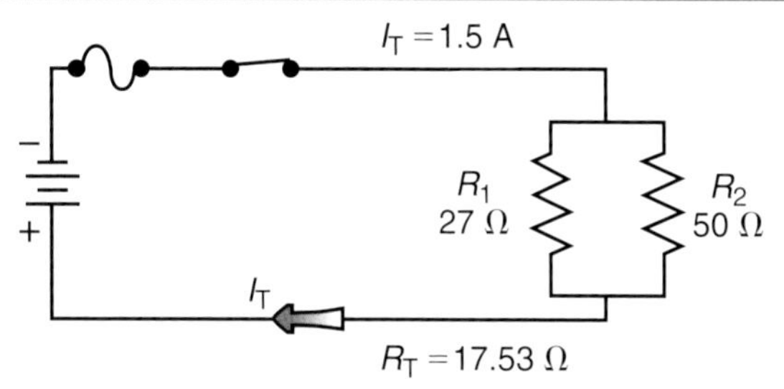

Figure 7–8 Use parallel circuit rules to calculate the current divider currents.

Example

Calculate the currents through both branches of the circuit in Figure 7–8, using the law of proportionality.

$$I_1 = (I_T\left(\frac{R_2}{(R_1 + R_2)}\right))$$

$$= 1.5 \times \left(\frac{50}{27 + 50)}\right) = 0.974 \text{ A}$$

and

$$I_2 = (I_T\left(\frac{R_1}{(R_1 + R_2)}\right))$$

$$= 1.5 \times \left(\frac{27}{(27 + 50)}\right) = 0.526 \text{ A}$$

Note that both methods (current divider and law of proportionality) produced the same result as using Ohm's law.

THE SUPERPOSITION THEOREM

DEFINITION

The voltage or current in any element resulting from several sources acting together is the sum of the voltages or currents resulting from each source acting alone. To produce the effects for a single source at a time, all other sources must be turned down to zero while the first source effects are calculated. This means disabling the other sources so that they cannot generate voltage or create current. This has to be done without changing the resistance of the circuit. You can do this by assuming that the voltage sources not being measured are shorted across their terminals.

SUPERPOSITION THEOREM REQUIREMENTS

There are two basic circuit requirements necessary for the use of the superposition theorem.

1. All of the components must be linear and bilateral. Linear means that the current is proportional to the applied voltage. Bilateral means that the calculated amount of the current through a component or circuit would be the same even if the voltage source applied were connected in reverse polarity.
2. The circuit components are usually passive. Passive components do not amplify or rectify. Examples of linear and bilateral components include capacitors, resistors, and air-core inductors. Examples of active components include transistors, semiconductor diodes, and tubes.

You may find it necessary to verify the operation of a circuit having multiple sources. Before getting bogged-down, do the following: assure that all the source voltages (or currents) are energized; verify the accuracy of each of the voltage (or current) sources; and confirm all switches/relays are properly configured.

Finally, check voltages/currents of all elements of the circuit, including switches/relays.

You have worked quite a bit by now with voltage sources (batteries). In the following pages of this chapter, you will be exposed to current sources as power supplies. You will be able to recognize a current source by it's unique symbol: (↓) (↑)

Using Superposition

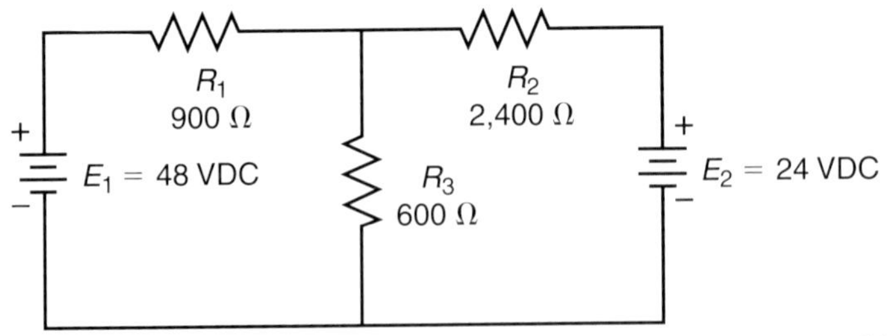

Figure 7–9 For a circuit with two sources, you may use superposition to solve for circuit values.

SUPERPOSITION THEOREM STEPS

The following example takes a step-by-step approach to using the superposition theorem when solving for the effect of multiple voltage sources in a circuit (Figure 7–9).

Example

Find the voltage drop and current flow of R_3 in Figure 7–9.

Solution:

Step 1

Reduce all but one **voltage source** to zero. In this example (as seen in Figure 7–9), you will remove the E_2 voltage source and replace it with a jumper (Figure 7–10). By inserting the jumper in place of the E_2 source, you have removed the effects of the E_2 source on the circuit. Essentially, you have turned the source to zero by putting a short circuit across it.

If this were a **current source**, you would reduce the current to zero by creating an open circuit at the current source. To create the open circuit at the current source, the current source would be removed from the circuit. However, this time you would not install the jumper in place of the current source.

Next, calculate the total resistance of the new circuit (Figure 7–10) using Ohm's law. First, reduce resistors R_2 and R_3 (which are now in parallel) to an equivalent resistance that is called $R_{2,3}$.

$$R_{2,3} = \frac{(R_2 \times R_3)}{(R_2 + R_3)} = \frac{(2{,}400\ \Omega \times 600\ \Omega)}{(2{,}400\ \Omega + 600\ \Omega)} = 480\ \Omega$$

Now calculate the equivalent resistance for resistors R_1 and $R_{2,3}$. Because they are in series, you know that:

$$R_T = R_1 + R_{2,3} = 900\ \Omega + 480\ \Omega = 1{,}380\ \Omega$$

Now that the total resistance is known, calculate the total current for the circuit.

$$I_T = \frac{E_T}{R_T} = \frac{48\ \text{V}}{1{,}380\ \Omega} = 0.0348\ \text{A}$$

Because the total current and resistance are known, the voltage drop across the parallel resistors R_2 and R_3 can be found.

$$E_{2,3} = I_T \times R_{2,3} = 0.0348\ \text{A} \times 480\ \Omega = 16.7\ \text{V}$$

Finally, calculate the current through R_3.

$$I_{R_3} = \frac{E_{R2,3}}{600\ \Omega} = \frac{16.7\ \text{V}}{600\ \Omega} = 0.0278\ \text{A}$$

Examine Figure 7–11. Note the polarity of the voltage drop and the direction of the current flow for R_3.

Step 2

The next step is to repeat the previous steps, using E_2 as the only voltage source. To do this, remove the jumper and reinstall E_2 into the circuit. Then remove E_1 and replace it with a jumper, just as you did in step 1. Figure 7–12 shows the newly configured circuit. Now calculate the total resistance in the redrawn circuit: $R_T = R_1$ and R_3 in parallel for 360 Ω, which is in series with R_2. The total circuit resistance is 2,760 Ω. Now that the total resistance is known, calculate the total current for the circuit.

$$I_T = \frac{E_T}{R_T} = \frac{24 \text{ V}}{2{,}740 \ \Omega} = 0.0087 \text{ A}$$

Because the total current and resistance are known, the voltage drop across the parallel resistors R_1 and R_3 can be found.

$$E_{1,3} = I_T \times R_{1,3} = 0.0087 \text{ A} \times 360 \ \Omega = 3.13 \text{ V}$$

Finally, the current through R_3 can now be calculated.

$$E_{1,3} = 3.13 \text{ V}$$

$$I_{R_3} = \frac{E_{1,3}}{R_3} = \frac{3.13 \text{ V}}{600 \ \Omega} = 0.0052 \text{ A}$$

Examine Figure 7–13. Note that the polarity of the voltage drop and the direction of the current flow for R_3 are the same as in Figure 7–11 after step 1.

(continued)

Voltage Drop and Current Flow

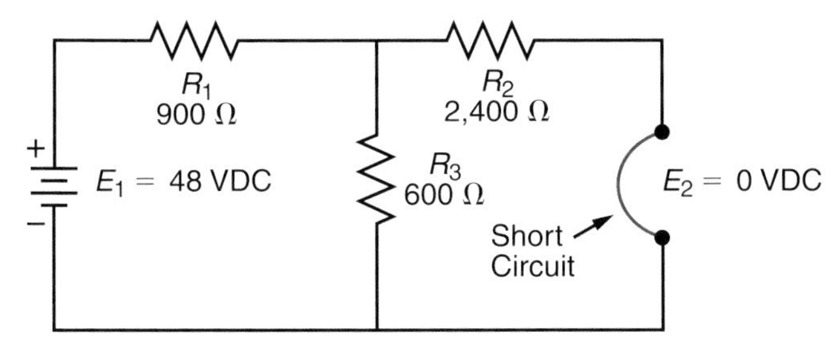

Figure 7–10 Step 1 in the process is to short out (on paper) one voltage source.

Voltage Drop and Current Flow

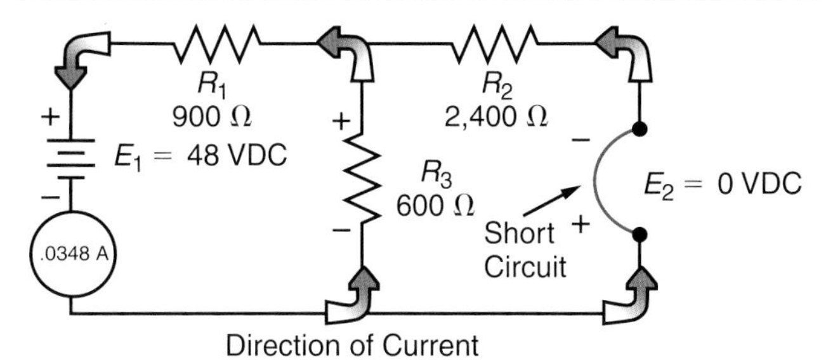

Figure 7–11 Solve circuit parameters with one voltage source first.

Voltage Drop and Current Flow

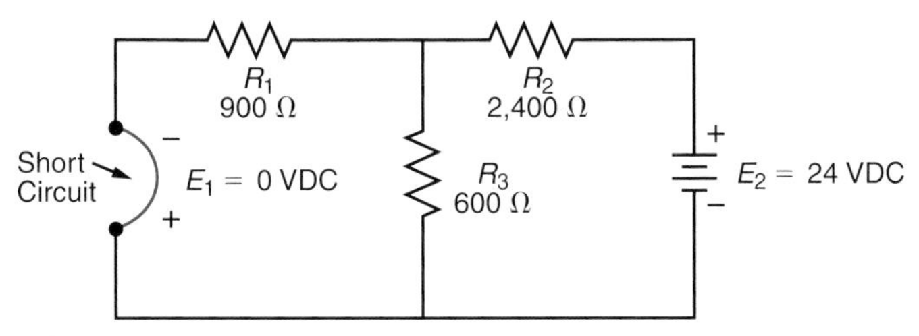

Figure 7–12 Remove the first source and add the second source to solve the circuit again.

Voltage Drop and Current Flow

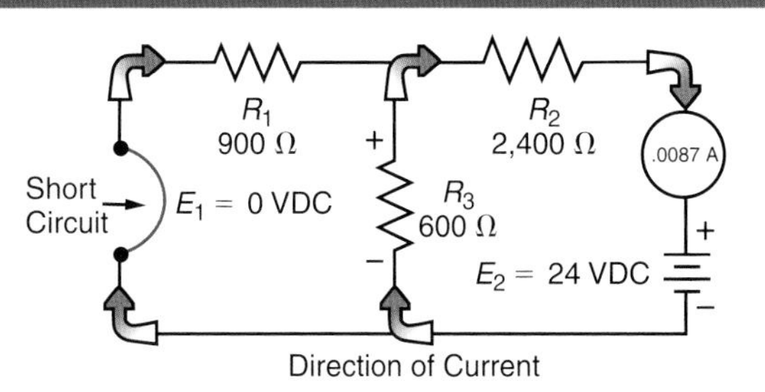

Figure 7–13 Step 2 requires new current and voltage solutions with a second source.

Example (continued)

Step 3

The last step is to find the algebraic sum of the two currents and voltage drops in R_3. Examine Figures 7–11 and 7–13. Because both currents are flowing in the same direction and both voltage drops have the same polarity, the voltages and currents are simply added.

$$I_{R_3} = I_{R_3\text{-step1}} + I_{R_3\text{-step2}} = 0.0052\text{ A} + 0.0278\text{ A} = 0.0330\text{ A}$$

$$E_{R_3} = E_{R_3\text{-step1}} + E_{R_3\text{-step2}} = 3.13\text{ V} + 16.7\text{ V} = 19.83\text{ V}$$

E_{R_1} is calculated as in step 1 as $I_T \times R_1$ or $0.0348\text{ A} \times 900\ \Omega = 31.3\text{ V}$ because the current from E_2 does not affect it. E_{R_2} is calculated in step 2 as $I_T \times R_2$ or $0.0087\text{ A} \times 2{,}400\ \Omega = 20.87\text{ V}$ because the current from E_1 does not affect R_2. We now have a circuit as originally drawn with $E_{R_1} = 31.3$ V, $E_{R_2} = 20.87$ V, $E_{R_3} = 19.83$ V. This adds up to the original two voltage sources' total input of 72 V. We also know the current in any part of the circuit.

Example

Using Figure 7–14, solve for the total current flow through R_3. Note that this circuit has both a voltage source and a current source.

Solution:

Step 1

Reduce the current source to zero by opening the circuit at the terminals of the current source. Figure 7–15 shows the redrawn circuit. Now calculate the total resistance in the redrawn circuit.

$$R_T = R_1 + R_3 = 900\ \Omega + 600\ \Omega = 1{,}500\ \Omega$$

Note that R_2 did not add to the total resistance seen by E_1 because no current could flow through the open part of the circuit. Now that the total resistance is known, calculate the total current for the circuit.

$$I_T = \frac{E_T}{R_T} = \frac{48\text{ V}}{1{,}500\ \Omega} = 0.032\text{ A}$$

Because the total current and resistance are known, the voltage drop across the resistors R_1 and R_3 can be found.

$$E_1 = I_T \times R_1 = 0.032\text{ A} \times 900\ \Omega = 28.8\text{ V}$$

$$E_3 = I_T \times R_3 = 0.032\text{ A} \times 600\ \Omega = 19.2\text{ V}$$

Step 2

Now reverse the process for I_1. Short out E_1 and reduce its voltage to zero. Figure 7–16 shows how this will affect the circuit. Reduce the voltage source to zero. Replacing the voltage source with a short circuit does this (Figure 7–16). Now calculate the total resistance in the redrawn circuit. First, using Ohm's law, solve for the parallel branch made up of resistors R_1 and R_3.

R_1 and R_3 are in parallel to create an equivalent resistance of 360 Ω.

Because the new $R_{1,3}$ is in series with R_2, you can easily calculate the total resistance R_T value.

$$R_T = R_{1,3} + R_2 = 360\ \Omega + 2{,}400\ \Omega = 2{,}760\ \Omega$$

Now that the total resistance is known, calculate the total voltage for the circuit.

$$E_T = I_1 \times R_T = 0.04\ \text{A} \times 2{,}760\ \Omega = 110.4\ \text{V}$$

Because the total current and resistance are known, the voltage drop across the parallel resistors R_1 and R_3 can be found.

$$E_{1,3} = I_T \times R_{1,3} = 0.04\ \text{A} \times 360\ \Omega = 14.4\ \text{V}$$

Finally, the current through R_1 and R_3 can now be calculated.

$$I_1 = \left(\frac{E_{1,3}}{R_1}\right) = \left(\frac{14.4\ \text{V}}{900\ \Omega}\right) = 0.016\ \text{A}$$

$$I_3 = \left(\frac{E_{1,3}}{R_3}\right) = \left(\frac{14.4\ \text{V}}{600\ \Omega}\right) = 0.024\ \text{A}$$

Examine Figure 7–17. Note the polarity of the voltage drop and the direction of the current flow for R_1 and R_3.

(continued)

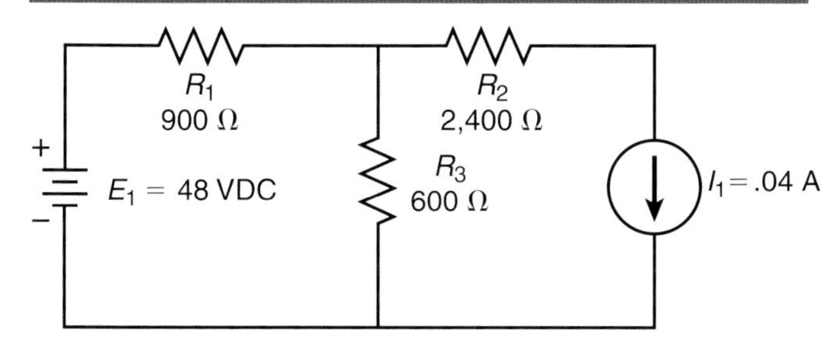

Figure 7–14 The superposition theorem is used for voltage or current sources or both.

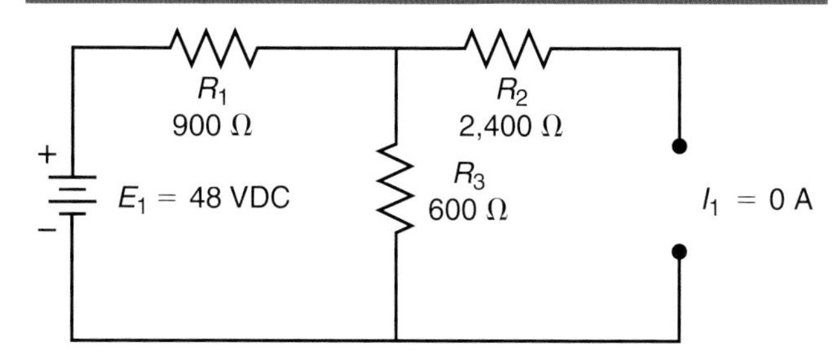

Figure 7–15 Step 1 is to open circuit the current source and solve for circuit values.

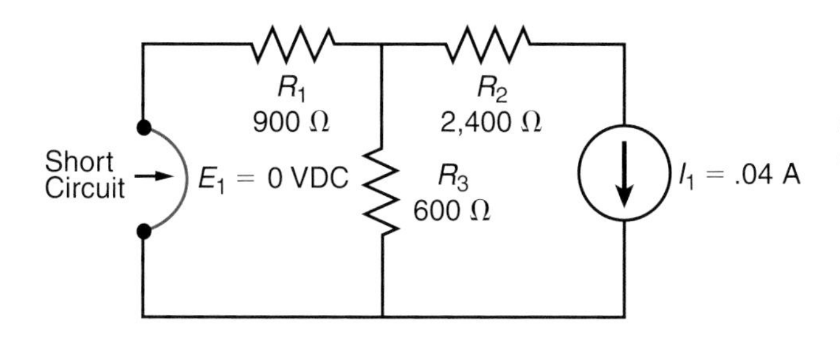

Figure 7–16 Step 2 in superposition is to short the voltage source (on paper).

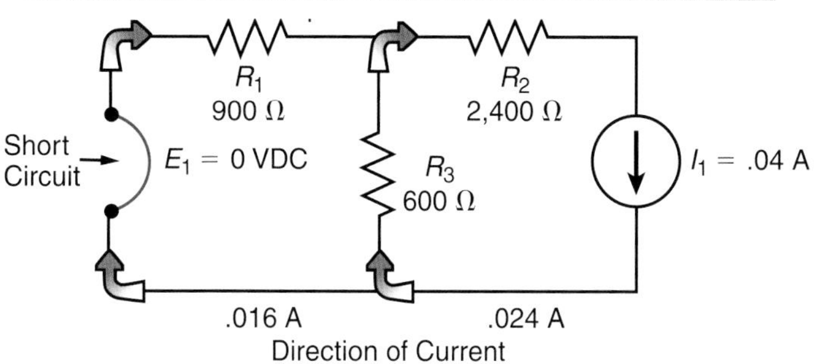

Figure 7–17 Calculate circuit parameters and note polarities.

Example (continued)

Step 3

The last step is to find the algebraic sum of the two currents and voltage drops in R_1 and R_3. Examine Figures 7–15 and 7–17. Because the two currents that flow through R_1 are in opposite directions and the voltage drops have the opposite polarities, the voltages and currents are algebraically added together (subtracted) and the larger of the currents and voltage drops determines the direction of polarity and current flow for R_1.

$$I_{1\ total} = I_{T\ (with\ current\ source\ open)} - I_{1\ (with\ E1\ shorted)}$$

$$I_{1\ total} = 0.032\ \text{A} - 0.016\ \text{A} = 0.016\ \text{A}$$

$$E_{1\ total} = E_{1\ (with\ current\ source\ open)} - E_{comb\ (with\ V1\ shorted)}$$

$$E_{1\ total} = 28.8\ \text{V} - 14.4\ \text{V} = 14.4\ \text{V}$$

Note that the current flow and voltage drops for I_3 are in the same direction and of the same polarity. This allows us to add the currents and voltage drops for the total effect on R_3.

$$I_{3\ total} = I_{3\ (with\ current\ source\ open)} + I_3 R = 0.032 + 0.024 = 0.056\ \text{A}$$

$$E_{3\ total} = E_{3\ (with\ current\ source\ open)} + E_{1,3\ (with\ V1\ shorted)}$$

$$E_{3\ total} = 19.2\ \text{V} + 14.4\ \text{V} = 33.6\ \text{V}$$

$$E_{3\ total} = I_T \times R_3 = 0.056\ \text{A} \times 600\ \Omega = 33.6\ \text{V}$$

$$E_{2\ total} = 0.04\ \text{A} \times 2{,}400\ \Omega = 96\ \text{V}$$

In this example, the voltage totals cannot be algebraically added because the totals reflect opposing voltages being added.

KIRCHHOFF'S LAWS

Kirchhoff's two laws are the following:

1. The algebraic sum (Σ) of the currents entering and leaving any node (junction point) is zero. This can be expressed mathematically as:

   $$0 = \Sigma I_{IN} - \Sigma I_{OUT}$$

   or

   $$\Sigma I_{IN} = \Sigma I_{OUT}$$

   Where the Greek symbol Σ (sigma) means "the sum of."

2. The algebraic sum (Σ) of the voltages around any closed path is zero. This can be expressed mathematically as:

   $$0 = \Sigma E_{source} - \Sigma(I \times R)$$

   or

   $$\Sigma E_{source} = \Sigma(I \times R)$$

In many cases, the use of Kirchhoff's laws is more direct than use of the superposition theorem.

KIRCHHOFF'S CURRENT LAW

Kirchhoff's current law states that the sum of any currents entering and leaving any given point in a circuit must be equal to zero. When the values of current entering and leaving a point within the circuit are summed, the currents entering the point are positive and the currents leaving the point are negative. In reference to Figure 7–18, Kirchhoff's current law can be written algebraically as:

$$I_{R_1} + I_{R_2} + I_{R_3} = 0$$

As you examine Figure 7–18, you might realize that the current flow through R_3 is actually flowing away from node E, not into it as shown.

This is the beauty of both of Kirchhoff's laws: As long as you are consistent with your assumptions, the direction of current flow will work out correctly when you do the algebraic solution. This means that if you always assume your currents to be into each node, their actual direction will be sorted out by the algebra.

For example, in Figure 7–18 you have assumed that the current through R_3 is flowing into node E. When this circuit is solved, I_{R_3} will have a negative sign showing that you assumed the wrong direction. Of the two laws, you will probably use Kirchhoff's voltage law more often than Kirchhoff's current law. The same principle applies in the voltage law, as you will see in the next section.

KIRCHHOFF'S VOLTAGE LAW

Note: A small "i" is often used in the formulas in Kirchhoff's laws. The small "i" is used because the actual currents are unknown and formulas are used to express assumed current paths. If a capital "I" were used, it may be confused with actual current flow in the circuit.

Refer to Figure 7–18 again. The circuit has three possible loops (closed paths): A-B-C-F-E-D, A-B-E-D, and C-B-E-F. In Kirchhoff's voltage law, you need to select loops so that every component is included in at least one. Consequently, you need only the last two (A-B-E-D and C-B-E-F) to analyze the entire circuit. Remember that you need to assume only as many loops as you need to include every component. In some cases, one or more of the components will be included in more than one of the loops, as was the case in superposition theorems. Figure 7–19 shows that resistor R_3 is included in both the i_1 loop and the i_2 loop.

Note that as each current flows around its own loop it will create a voltage drop across the resistors. In resistor R_3, the voltage drop will be caused by both i_1 and i_2. Although the choice of polarities for the voltage drops is not critical (as stated earlier),

Identifying Points

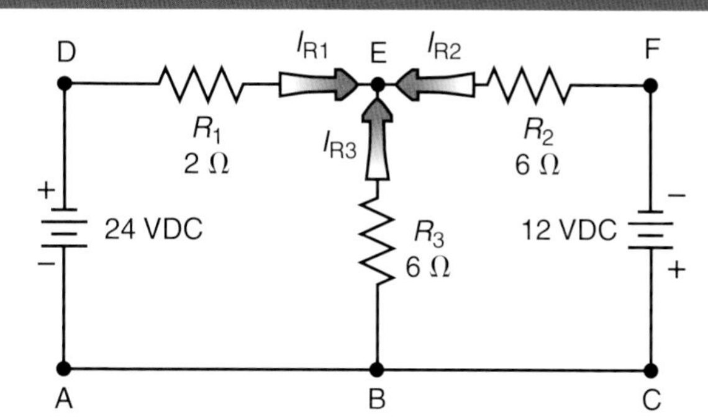

Figure 7–18 Kirchhoff's laws of current use nodes (identifying points).

Assumed Current Loops

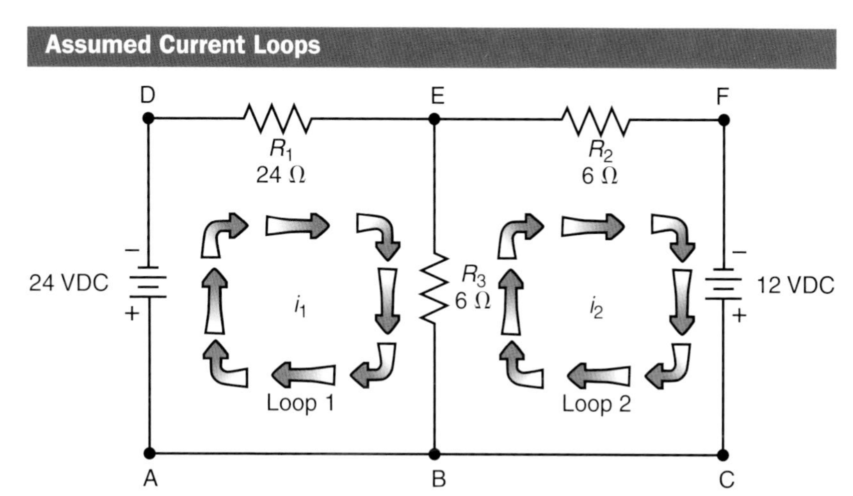

Figure 7–19 Kirchhoff's voltage laws use assumed current loops.

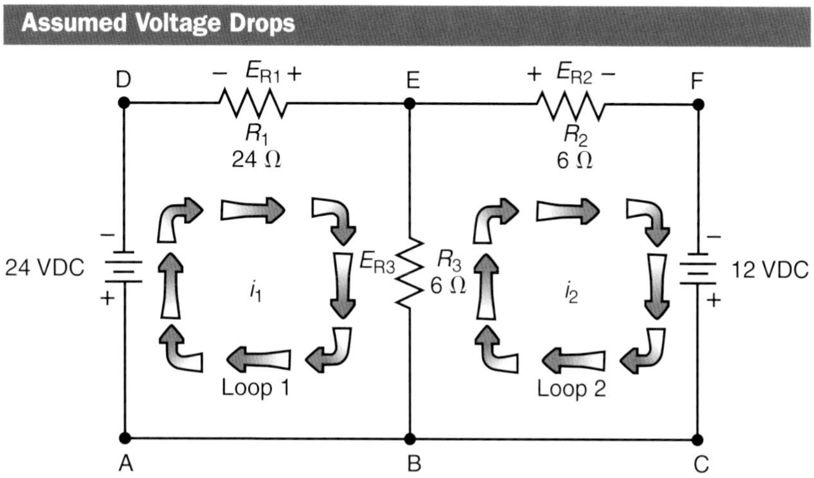

Assumed Voltage Drops

Figure 7–20 Kirchhoff used assumed voltage drops in his calculations, as shown.

it makes sense to select them logically. In Figure 7–20, each voltage drop is labeled and assigned a polarity. The polarities are chosen as follows:

1. E_{R_1} is chosen with its negative end located where the i_1 current flows in.

This is a reasonable guess and is the way most of the polarities should be chosen.

2. E_{R_2} is chosen with its positive end located where the i_2 current flows in. This is opposite the way you would normally choose it. The polarity is chosen for *this example only* and will help illustrate that the algebra will eventually sort out the correct polarity.
3. E_{R_3} is shared between two loops. Therefore, its polarity cannot be assumed. In such cases, the polarity is often picked on the basis of the first loop you encounter as you analyze. In this case, it assumes the polarity of loop 1 (i_1).

Because we have assumed clockwise rotation for i_1 and i_2, the Kirchhoff voltage equations can be developed by starting at the source and adding voltages in a clockwise direction for each loop. Be careful to observe the signs you have assumed.

- For loop 1, starting 24 V in sequence, noting the polarity of the beginning of each voltage encountered: $+24 - E_{R_1} - E_{R_3} = 0$
- For loop 2, starting 12 V in sequence, noting the beginning polarity of each voltage encountered: $-12 + E_{R_3} + E_{R_2} = 0$

Note that for each loop you start with the source. This is not necessary, but it is easy to remember. If there is more than one source, you can start with either one. As you go around the loop in the same direction as the current, you add or subtract each voltage depending on the sign you encounter.

In the equation for loop 1, you are moving clockwise and you encounter the positive sign for the 24 V battery first. Therefore, it is listed as positive. The other voltages are identified in the same way. Loop equations 1 and 2 can be simplified even more by moving the sources to the right side of the equation and multiplying equation 1.

$$+24 - E_{R_1} - E_{R_3} = 0, \text{ by } -1$$

This gives:

$$E_{R_1} + E_{R_3} = 24$$

And loop 2 times –1 becomes:

$$-12 + E_{R_3} + E_{R_2} = 0$$

$$E_{R_3} + E_{R_2} = 12$$

APPLYING KIRCHHOFF'S VOLTAGE LAW

Example

Equations for loops 1 and 2 define the voltages. However, even these equations are not yet enough. Note that you have three unknowns (E_{R_1}, E_{R_2}, and E_{R_3}) and only two equations. Fortunately, you can simplify by substituting the Ohm's law equivalents for the three unknowns. From Ohm's law, you know that:

$$E_{R_1} = i_1 \times R_1$$

$$E_{R_2} = -i_2 \times R_2$$

$$E_{R_3} = i_1 R_3 - i_2 R_3$$

Note that the value for E_{R_2} is negative. This is because the current i_2 is flowing into the positive end of R_2 as we have defined it. This means that the actual value will be minus the current times the resistance as shown earlier. Note also that because i_1 flows into the negative end of R_3 and i_2 flows into the positive end, the total voltage will be as shown. Substituting these values into the loop equations gives:

$$i_1 R_1 + (i_1 - i_2) \times R_3 = 24$$

and

$$(i_1 - i_2) \times R_3 - i_2 R_2 = 12$$

Simplifying and collecting terms gives:

$$i_1 (R_1 + R_3) - i_2 (R_3) = 24$$

and

$$i_1R_3 - i_2 (R_2 + R_3) = 12$$

Substituting the known values for the three resistors gives:

$$30i_1 - 6i_2 = 24$$

and

$$6i_1 - 12i_2 = 12$$

Note that the last two equations can be divided by 6, which leaves:

$$5i_1 - i_2 = 4$$

and

$$i_1 - 2i_2 = 2$$

You now have two equations with two unknowns to solve for, i_1 and i_2. There are several ways to solve for these two unknowns. The easiest would be to use a scientific calculator, computer spreadsheet, or other software that has the ability to solve them. In fact, when working on very complicated circuits with three or more loops, this may be the most practical approach.

Solution:
In this simple example, you can use the fairly simple approach of eliminating one of the unknowns and solving for the other. First, multiply $5i_1 - i_2 = 4$ by 2 to produce:

$$2 \times (5i_1 - i_2 = 4) = 10i_1 - 2i_2 = 8$$

You can then subtract $i_1 - 2i_2 = 2$ from the last equation to get:

$$(i_1 - 2i_2 = 2) - (10i_1 - 2i_2 = 8)$$

Expanding and collecting terms gives:

$$(i_1 - 10i_1 - 2i_2 + 2i_2) = -6$$

This further simplifies to:

$$-9i_1 = -6 \Rightarrow i_1 = \frac{6}{9} = 0.67 \text{ A}$$

Now substitute the value for i_1 into $i_1 - 2i_2 = 2$:

$$0.67 - 2i_2 = 2 \Rightarrow 1 - i_2 = 1.33 \Rightarrow i_2 = -0.67 \text{ A}$$

The result for i_2 is very important. Because it came out negative, your original assumption that i_2 flows clockwise is incorrect. This will work in this manner every time. Regardless of what you assume, as long as you are consistent the signs will work out at the end. Now you can calculate the voltage drops on each resistor and then add the results appropriately to check the answer. Earlier you found the equations of interest, and thus you can now calculate the results.

$$E_{R_1} = i_1 \times R_1 \Rightarrow 0.67 \times 24 = 16 \text{ V}$$

$$E_{R_2} = -i_2 \times R_2 \Rightarrow -(-0.67) \times 6 = 4 \text{ V}$$

$$E_{R_3} = i_1R_3 - i_2R_3 \Rightarrow 0.67 \times 6 - (-0.67 \times 6) = 4 + 4 = 8 \text{ V}$$

Example

Comparing these to the two loops, you see that 24 = $E_{R_1} + E_{R_3}$ = 16 + 8 = 24 V and that 12 = $E_{R_2} + E_{R_3}$ = 6 + 6 = 12 V. Note also that the voltage E_{R_2} came out positive, which means that although the original choice was not consistent with the assumed direction of i_2, it is consistent with the actual direction of i_2. Figure 7–21 shows the second example. Although it is considerably more complex, the approach is exactly the same. Table 7–1 outlines the actual steps.

Solution:

Step 1
This problem requires that at least four loops be chosen. The best approach is to select the smallest number of loops possible.

Step 2
Figure 7–22 shows the four loops and the labels chosen for this problem. Note that every component is included in at least one loop.

Step 3
Figure 7–23 shows the voltage labels for each component and their polarities, based on assumed direction of the loop currents.

(continued)

Four-Loop Problem

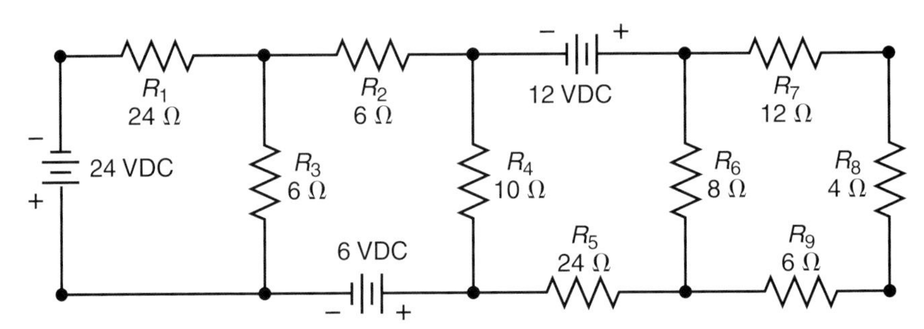

Figure 7–21 This complex four-loop problem uses Kirchhoff's laws.

Four-Loop Problem

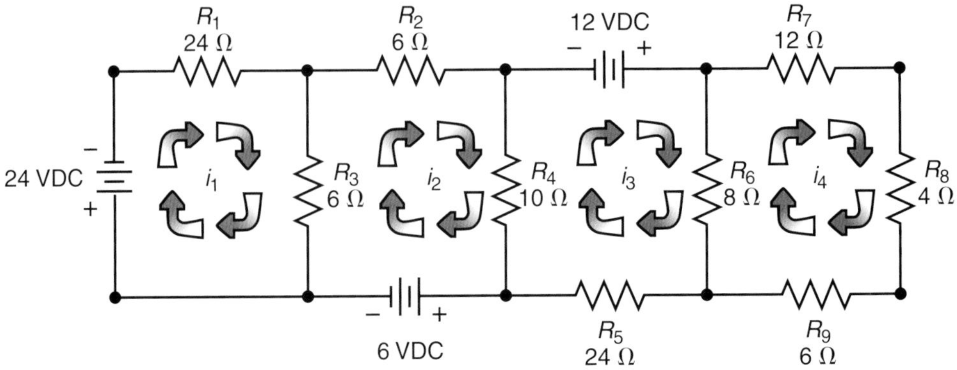

Figure 7–22 The complex diagram of Figure 7–21 can be broken down into four current loops.

Table 7–1

Eight steps are used in the application of Kirchhoff's Laws.

Step	Description	Comments
1	Using the schematic diagram, select the loops you will use for Kirchhoff's voltage law.	You must have enough loops to ensure that all components are included in at least one loop.
2	Draw and name each loop current. You may assume any direction for the loop currents, but usually being consistent is best.	This text always uses clockwise current flow. However, you may prefer counter-clockwise or even a combination.
3	Label each component with its voltage drop.	1. Usually assume that the end of the component into which the current flows is negative. 2. If a component has current from more than one loop, you may use whichever one you wish. Usually, the first current you consider determines polarity.
4	Write Kirchhoff's voltage law by summing all currents around each loop and setting the sum to zero.	Starting with any component, sum its voltage (subtract if you encounter a negative sign first) and then move to the next one in the direction of the assumed current flow. When you have summed all of the elements, set them to zero. Repeat this process for each loop.
5	Using Ohm's law, create a table of the values of each voltage drop.	For example, in Figure 7–20 the voltage drop $E_{R_1} = i_1 \times R_1$.
6	Substitute the Ohm's law values into the loop equations you developed in step 4.	
7	Solve the resulting equations for each loop current.	This can be done by using the rules of algebra on a scientific calculator.
8	Using the currents calculated in step 7 and the table created in step 5, calculate the voltage drops for each component.	

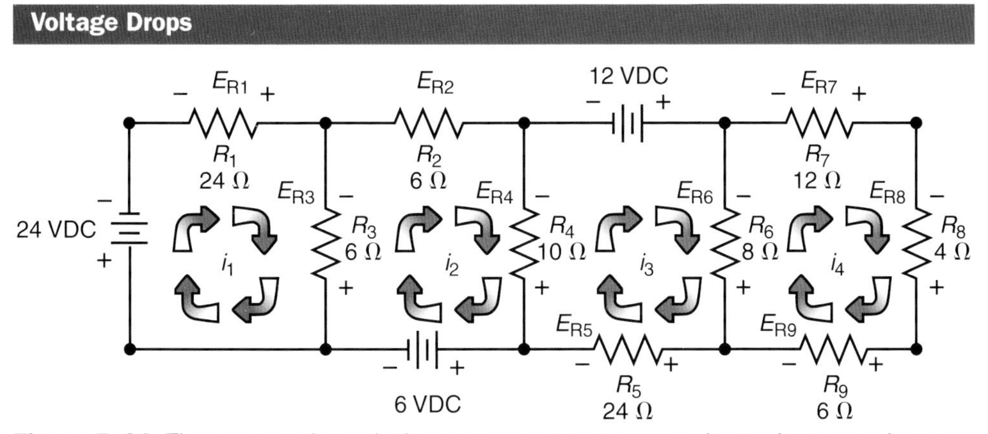

Figure 7–23 The current through the components creates voltage drops as shown.

Example (continued)

Step 4

You can now write the four loop equations, starting with i_1 and progressing to i_4.

$$\text{For } i_1\text{: Start} = +24 - E_{R_1} - E_{R_3} = 0 \Rightarrow E_{R_1} + E_{R_3} = 24$$

$$\text{For } i_2\text{: Start} = +6 + E_{R_3} - E_{R_2} - E_{R_4} = 0 \Rightarrow E_{R_2} - E_{R_3} + E_{R_4} = 6$$

$$\text{For } i_3\text{: Start} = +12 - E_{R_6} - E_{R_5} + E_{R_4} = 0 \Rightarrow E_{R_4} - E_{R_5} - E_{R_6} = 12$$

$$\text{For } i_4\text{: Start} = -E_{R_7} - E_{R_8} + E_{R_9} + E_{R_6} = 0 \Rightarrow E_{R_6} - E_{R_7} - E_{R_8} - E_{R_9} = 0$$

Step 5

Next, using Ohm's law, create a table of values for each voltage drop.

$$E_{R_1} = i_1R_1 = 24i_1$$

$$E_{R_2} = i_2R_2 = 6i_2$$

$$E_{R_3} = (i_1 - i_2)\, R_3 = 6(i - i_2)$$

$$E_{R_4} = (i_2 - i_3)\, R_4 = 10(i_2 - i_3)$$

$$E_{R_5} = i_3R_5 = 24i_3$$

$$E_{R_6} = (i_3 - i_4)\, R_6 = 8(i_3 - i_4)$$

$$E_{R_7} = i_4R_7 = 12i_4$$

$$E_{R_8} = i_4R_8 = 4i_4$$

$$E_{R_9} = i_4R_9 = 6i_4$$

Step 6

Substituting the values into the four loop equations gives:

$$24i_1 + 6(i_1 - i_2) = 24 \Rightarrow 30i_1 - 6i_2 = 24 \Rightarrow 5i_1 - i_2 = 4$$

$$6i_2 - 6(i_1 - i_2) + 10(i_2 - i_3) = 6 \Rightarrow -6i_1 + 22i_2 - 10i_3 = 6$$

$$10(i_2 - i_3) - 24i_3 - 8(i_3 - i_4) = 12 \Rightarrow 10i_2 - 42i_3 + 8i_4 = 12$$

$$8(i_3 - i_4) - 12i_4 - 4i_4 - 6i_4 = 0 \Rightarrow 8i_3 - 30i_4 = 0$$

Step 7

Solving the equations can be done in a number of ways, but with this system it is easiest to solve them with a handheld calculator. The results are:

$$i_1 = 0.885 \text{ A}$$

$$i_2 = 0.426 \text{ A}$$

$$i_3 = -0.194 \text{ A}$$

$$i_4 = -0.0518 \text{ A}$$

Step 8

The actual voltage drops can now be calculated as follows:

$$E_{R_1} = i_1R_1 = 24i_1 = 24 \times 0.885 = 21.2 \text{ V}$$

$$E_{R_2} = i_2R_2 = 6i_2 = 6 \times 0.426 = 2.56 \text{ V}$$

$$E_{R_3} = (i_1 - i_2)\, R_3 = 6(i_1 - i_2) = 6 \times (0.885 - 0.4262) = 2.754 \text{ V}$$

$$E_{R_4} = (i_2 - i_3)\, R_4 = 10(i_2 - i_3) = 10 \times (0.426 + 0.194) = 6.2 \text{ V}$$

$E_{R_5} = i_3R_5 = 24i_3 = 24 \times -0.194 = -4.66$ V

$E_{R_6} = (i_3 - i_4)\ R_6 = 8(i_3 - i_4) = 8 \times (0.194 + 0.0518) = -1.14$ V

$E_{R_7} = i_4R_7 = 12i_4 = 12 \times -0.0518 = -0.622$ V

$E_{R_8} = i_4R_8 = 4i_4 = 4 \times -0.0518 = -0.2072$ V

$E_{R_9} = i_4R_9 = 6i_4 = 6 \times -0.0518 = -0.311$ V

If you add the voltages around any loop shown in Figure 7–23, you will see that they add to zero, allowing for rounding errors.

THEVENIN'S THEOREM

Assume that the block on the left in Figure 7–24 contains a network connected to terminals A and B. Thevenin's theorem states that the entire network connected to A and B can be replaced by a single voltage source (E_{TH}) in series with a single resistance (R_{TH}) connected to the same terminals. This allows us to take any complex circuit, with any number of voltage or current sources, and reduce it to one source and one equivalent resistance. The definitions of these two values are as follows:

- E_{TH} is the open-circuit voltage measured at terminals A and B.
- The Thevenin resistance (R_{TH}) is open-circuit voltage (E_{TH}) divided by the short-circuit current (I_{SC}) that would occur between terminals A and B.

The following sections illustrate methods that can be used to determine and check a Thevenin equivalent circuit.

A French engineer, M. L. Thevenin, devised his theorem to simplify voltages in a network. This procedure was first discussed by a German scientist named Helmholtz and appeared in a different form. This theorem is sometimes referred to as the Helmholtz-Thevenin theorem. Paraphrased, Thevenin's theorem states: No matter how many sources and components and no matter how they are interconnected, they can all be represented by an equivalent series circuit with respect to any pair of terminals in the network.

Thevenin Analysis

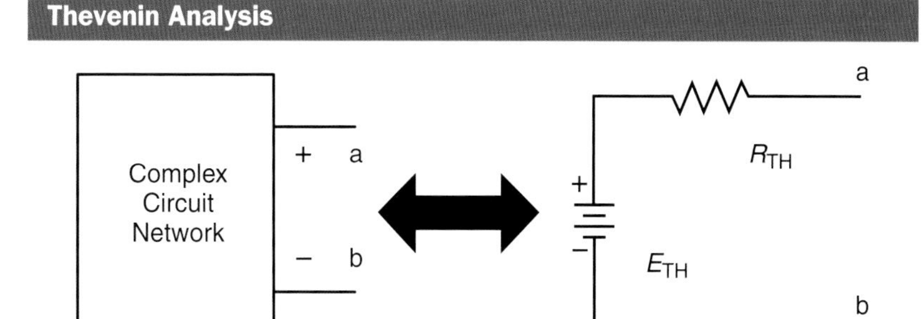

Figure 7–24 This basic circuit can be used for analysis of Thevenin's assumption.

CREATING A THEVENIN EQUIVALENT CIRCUIT

Let us assume that you wish to create a Thevenin equivalent circuit for terminals A and B of Figure 7–25, so that any load attached to terminals A and B on the Thevenin circuit will result in the same voltage and current as if the same load were attached to the original circuit. The following explanation shows how to find the Thevenin equivalent values.

Thevenin's Theorem

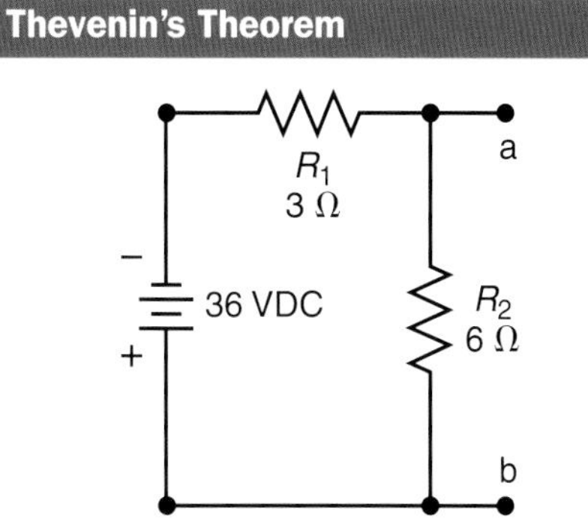

Figure 7–25 A sample circuit for explanation of Thevenin's theorem is shown above.

FINDING E_{TH}

E_{TH} is the open-circuit voltage at terminals A and B. This is clearly equal to the voltage drop across resistor R_2, which can be calculated by using the voltage divider formula and proportional voltage drops as follows:

$$E_{TH} = E_{R_2} = 36 \times \left[\frac{R_2}{(R_1 + R_2)}\right]$$

$$= 36 \times \left[\frac{6}{9}\right] = 24 \text{ V}$$

FINDING R_{TH}

R_{TH} can be found by first placing a short circuit at points A and B and then measuring or calculating the current that flows between A and B. From Figure 7–26, I_{SC} can be calculated as:

$$I_T = \frac{E_T}{R_2} \text{ with } R_2 \text{ shorted}$$

$$I_{SCTH} = \frac{36}{3} = 12 \text{ A}$$

Short Circuit Current

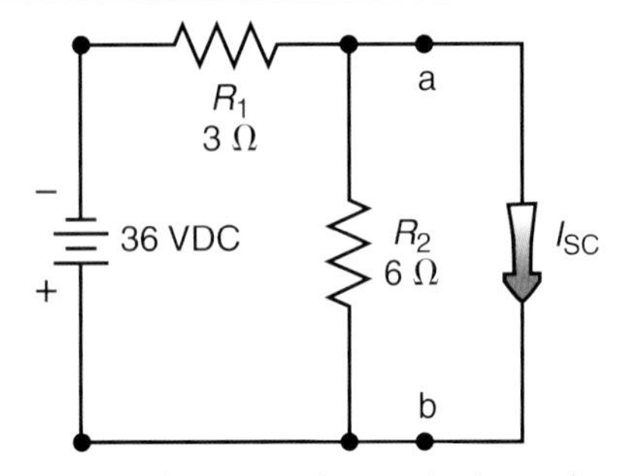

Figure 7–26 Short out R_2 (on paper) to calculate short-circuit current.

R_{TH} is now calculated as follows:

$$R_{TH} = \frac{E_{TH}}{I_{SC}} = \frac{24}{12} = 2\ \Omega$$

An alternative method of calculating R_{TH} is to short-circuit the voltage source (mathematically) and then calculate the equivalent resistance between terminals A and B. In this case, the equivalent resistance of R_1 and R_2 in parallel is 2 Ω. The complete Thevenin equivalent circuit is shown in Figure 7–27.

CHECKING RESULTS

If your calculations have been accurate, any load placed at terminals A and B in Figure 7–25 should create the same load voltage (E_L) and load current (I_L). First check the original circuit with an 8 ohm load connected, as shown in Figure 7–28. The voltage across the load can be calculated by first realizing that R_L and R_2 are connected in parallel. The equivalent resistance of these two is:

$$R_{L,2} = 3.43\ \Omega$$

Now the voltage drop across R_L can be calculated by using the voltage divider formula.

$$E_{L,2} = 36 \times \left[\frac{R_{1,2}}{R_T}\right]$$

$$= 36 \times \frac{3.43}{6.43} = 19.2 \text{ V}$$

Because the voltage across $E_{L,2}$ is the same as the voltage across E_L, $E_L = 19.2$ V.

Load Connections

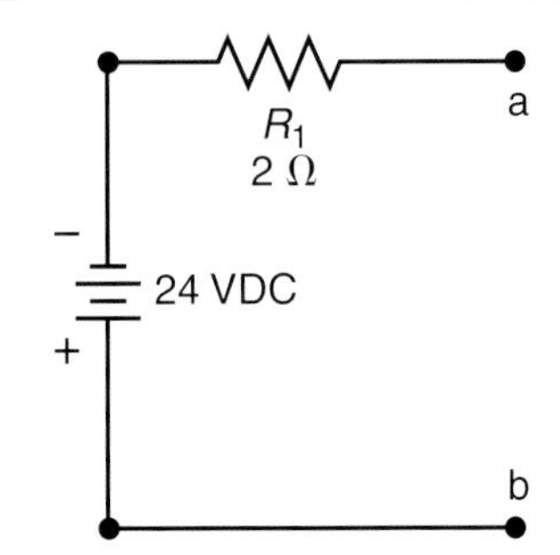

Figure 7–27 The Thevenin circuit equivalent is waiting for load connections to a and b.

Load Resistance

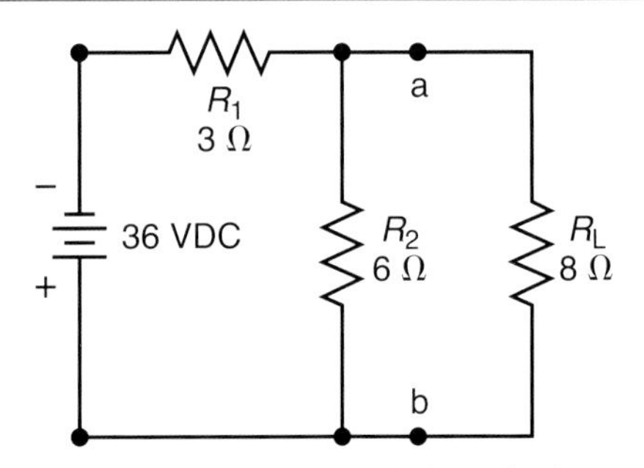

Figure 7–28 This Thevenin circuit has 8 ohms of load resistance connected.

The next check is to verify the voltage drop across an 8-ohm load connected to terminals A and B of the Thevenin equivalent circuit (as shown in Figure 7–29). The simple approach here is to use the voltage divider again.

$$E_L = 24 \times \left(\frac{8}{10}\right) = 19.2 \text{ V}$$

Clearly, the Thevenin equivalent circuit of Figure 7–27 performs exactly as the original circuit in Figure 7–25. Again, this allows us to take any circuit that is supplying power to a load and quickly find the effects of connecting various resistive loads to the source. We will use this process to analyze circuits to transfer the greatest amount of power to the load as we look at electronic circuits that supply power.

NORTON'S THEOREM

The concept is that the load you place on terminals A and B will create the same voltage and current. The two quantities needed are defined as follows.

1. I_N is the measured current between terminals A and B when a short circuit is applied.
2. R_N is calculated the same as R_{TH}. It is the open-circuit voltage (E_{ab}) divided by the short-circuit current (I_{SC}) that would occur between terminals A and B. As with Thevenin's theorem, it can also be calculated by turning all sources to zero and calculating the resistance between terminals A and B.

CURRENT SOURCE

The Norton circuit is especially useful when analyzing circuits that have parallel resistances. The series resistance equivalent of a Thevenin circuit would register varying voltages on parallel loads based on the different current through the resistances. Norton's theorem allows a constant current across the varying loads and thus easier voltage calculations.

An American, Edward Lawry Norton (who worked for Bell Telephone Laboratories), introduced his theorem in 1926. It was published by Hans Mayer, a Seimens researcher, in 1926. Norton's theorem is an extension of Thevenin's theorem. This theorem simplifies complex circuit networks using current sources instead of voltage. This circuit replaces the Thevenin circuit voltage source and series resistor with a current source and parallel resistor. Figure 7–30 shows the comparison between Thevenin and Norton circuits.

Load Connected

R_{TH} 2 Ω — a — 24 VDC — R_L 8 Ω — b

Figure 7–29 An equivalent Thevenin circuit with load connected is illustrated.

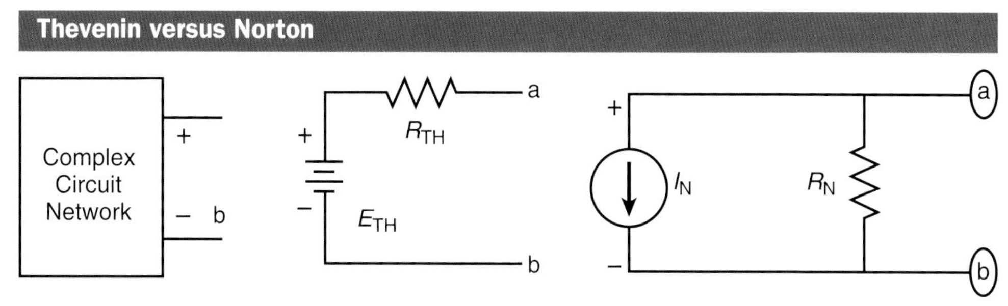

Figure 7–30 This illustration compares Thevenin's and Norton's circuit simplifications.

CREATING A NORTON EQUIVALENT CIRCUIT

Creating a Norton equivalent circuit for terminals A and B of Figure 7–25 means that any load attached to terminals A and B on the Norton circuit will result in the same voltage and current as if the same load were attached to the original circuit. The following explanation shows how to find the Norton equivalent values.

FINDING I_N

I_N is the current between terminals A and B when they are short-circuited. This was calculated earlier as:

$$I_{SC} = \left(\frac{36}{3}\right) = 12\text{ A} = I_N$$

This is the same as I_{TH}. The circuit used to develop this value is shown in Figure 7–26.

FINDING R_N

Calculate R_N as before or by turning the voltage source in Figure 7–25 to zero and calculating the impedance at terminals A and B. Figure 7–31 shows the method. R_N is now calculated as follows:

$$R_N = \left[\frac{(R_1 \times R_2)}{(R_1 + R_2)}\right] = \frac{18}{9} = 2\ \Omega$$

Note that R_N can also be calculated by turning the voltage sources to zero (short circuit) and then calculating the equivalent resistance between terminals A and B. The complete Norton's equivalent circuit is shown in Figure 7–32.

Resistance N

Shorted Voltage Source

$R_N = \frac{3 \times 6}{3 + 6} = 2\ \Omega$

R_1 3 Ω

R_2 6 Ω

a

b

Figure 7–31 The resistance *N* (Norton) is calculated by shorting the power source.

Norton's Circuit

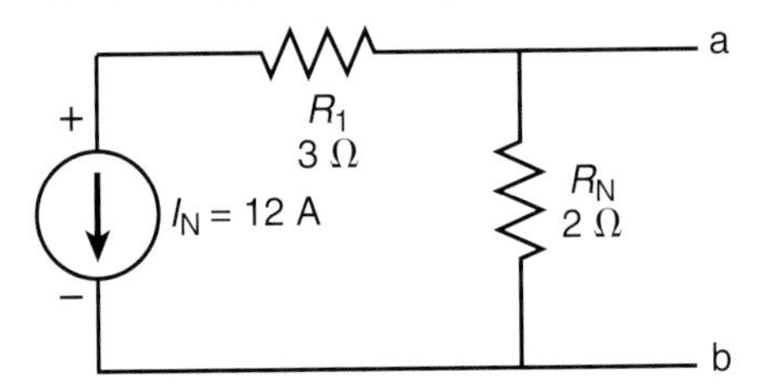

Figure 7–32 An equivalent Norton's circuit for Figure 7–25 is shown above.

CHECKING RESULTS

If your calculations have been accurate, any load placed at terminals A and B in Figure 7–25 should create the same load voltage (E_L) and load current (I_L). Earlier, you calculated the voltage drop with an 8-ohm load connected to the circuit (Figure 7–27). The voltage across the load can be calculated by realizing that R_L and R_2 are connected in parallel. The equivalent resistance of these two is:

$$R_{L,2} = \left[\frac{(R_L \times R_2)}{(R_L + R_2)}\right] = \frac{48}{14} = 3.43\ \Omega$$

Now the voltage drop across R_L can be calculated by using the voltage divider formula.

$$E_{L,2} = \frac{(36 \times R_{L,2})}{(R_{L,2} + R_1)}$$

$$= \frac{36 \times 3.43}{6.43} = 19.2\text{ V}$$

Because the voltage across $E_{L,2}$ is the same as the voltage across E_L, E_L = 19.2 V. Next, check the voltage drop across an 8-ohm load connected to terminals A and B of the Norton equivalent circuit (as shown in Figure 7–33). You can start by using a current divider formula to calculate the current through the load resistor.

$$I_T = 12\text{ A}$$

$$I_L = \left[12 \times \left(\frac{R_N}{(R_L + R_N)}\right)\right]$$

$$= 12 \times \left(\frac{2}{10}\right) = 12 \times 0.2 = 2.4\text{ A}$$

$$E_{RL} = 8 \times 2.4\text{ A} = 19.2\text{ V}$$

Clearly, the Norton equivalent circuit of Figure 7–32 performs exactly as the original circuit in Figure 7–25 and the Thevenin equivalent in Figure 7–32.

LIMITING CIRCUITS BY WATTAGE RATING

Often you will need to figure out the circuit values based on the wattage ratings of the load. This statement means that you will need to decide how much current or how much voltage a circuit could have without overloading the ratings of the load. For instance, if you have two 120 V rated lamps in series, what is the maximum voltage you could supply to the circuit without exceeding the voltage rating of either one?

You would of course answer that 240 V is the maximum, based on what you have already learned. Suppose you had many different rated devices in series. You might not be able to simply add them all together and get the same easy calculation.

SERIES CIRCUITS

For series-connected utilization equipment, you will need to calculate the maximum current a component can safely allow without exceeding its wattage rating. For instance, a 60 W 120 V lamp can safely carry 0.5 A. This means that the resistance of the lamp is 240 Ω when operating. A 25 W, 120 V lamp can safely allow 0.208 A, and the resistance is 577 Ω when operating. What would happen if you connected these two different loads in series to 240 V?

By calculating the total resistance of lamp 1 and lamp 2, you find the total resistance would be 817 ohms. Connecting to 240 V with the lamps in series would yield 0.294 A. The 25 W lamp has too much current. $I^2 \times R = 49$ W and the voltage is 0.294 A × 577 Ω = 169 V. The 25 W lamp burns very bright briefly before exceeding the wattage rating and burning open. The 60-W lamp is not in danger and in fact appears dim.

If you have several loads in series with different wattage and or voltage ratings, you will need to calculate the current they can allow without exceeding the wattage rating. Use Watt's laws of $I^2 \times R = \frac{P \times E^2}{R} = P$ and $E \times I = P$ to determine the current rating.

Norton's Simplified Circuit

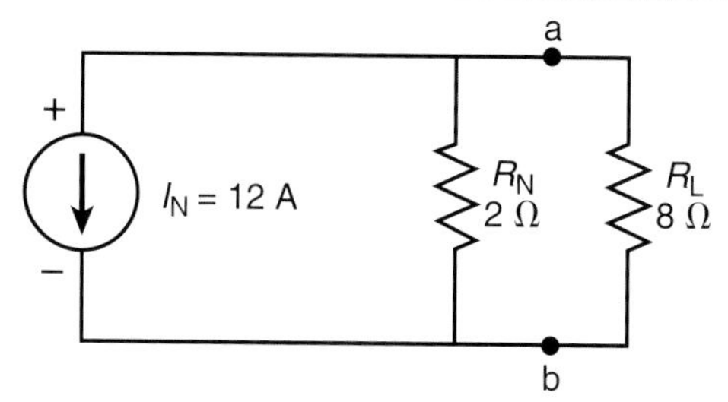

Figure 7–33 This Norton's simplified circuit has a load resistor at the output terminals a and b.

Three Resistors in Series

Figure 7–34 Of these three resistors in series, one is the limiting resistor.

Example

Three resistors in series are shown in Figure 7–34. The resistor R_1 is 500 Ω and 0.25 W, R_2 is 750 Ω and 0.50 W, and R_3 is 3 KΩ and 1 W. Which resistor is in danger of overheating?

Solution:

Find the current limit of each resistor, based on $I^2 \times R = P$ transposed to $I = \sqrt{\frac{P}{R}}$.

$$I_{R_1} = \sqrt{\frac{0.25 \text{ W}}{500\ \Omega}} = 0.0223 \text{ A}$$

$$I_{R_2} = \sqrt{\frac{0.5 \text{ W}}{750\ \Omega}} = 0.0258 \text{ A}$$

$$I_{R_3} = \sqrt{\frac{1 \text{ W}}{3000\ \Omega}} = 0.018 \text{ A}$$

Therefore, R_3 is the limiting resistor. To verify, place all three resistors in a series circuit. The total resistance would be 4,250 Ω. If we apply 76.5 V to this circuit, the total current will be 0.018 A and all resistors will be at or below their wattage rating. If we apply any higher voltage, the current will increase and the first resistor to overheat will be the 3,000 Ω, 1 W resistor because the rated current with that resistance will exceed the resistor's ability to dissipate enough wattage without damage.

COMPONENTS IN PARALLEL

Fixed resistive loads are rated in resistance and with power ratings. You may need to decide which resistors can be placed in parallel without exceeding the wattage ratings of the resistors. For example, with two resistors in parallel, which resistor will limit the amount of voltage that can be applied to the parallel branch? Use a 1 kΩ, 1 W resistor in parallel with a 5 kΩ, 2 W resistor.

If we use the wattage formula $\frac{E^2}{R} = P$ transposed to $E = \sqrt{P \times R}$, we can determine the maximum voltage to apply. The 1 kΩ resistor would be:

$$E = \sqrt{1 \times 1{,}000} = 31.62 \text{ V}$$

The 5 kΩ resistor has a maximum voltage of:

$$E = \sqrt{2 \times 5{,}000} = 100 \text{ V}$$

In this case, the maximum voltage that can be applied to the parallel branch of these two resistors is 31.62 V. The 1 kΩ, 1 W resistor is the limiting effect of this circuit.

Example

Imagine three resistors in parallel. Which is the limiting effect on the circuit and what is the maximum voltage that can be applied without danger to the components? R_1 is 1 kΩ –1 W, R_2 is 500 Ω –0.50 W, and R_3 is 1,500 Ω 0.25 W.

Solution:

$$E_{R_1} = \sqrt{1 \times 1{,}000} = 31.62 \text{ V}$$
$$E_{R_2} = \sqrt{0.5 \times 500} = 15.8 \text{ V}$$
$$E_{R_3} = \sqrt{0.25 \times 1{,}500} = 19.36 \text{ V}$$

R_2 is the limiting resistor, and the maximum voltage that can be applied to this parallel circuit is 15.8 V. Anything higher than 15.8 V will damage R_2.

SUMMARY

In this chapter, you learned the principles of the voltage divider circuit. Voltage dividers are used throughout electricity and electronics to provide multiple voltages. The law of proportionality of voltage allows you to calculate outputs from voltage dividers without knowing the current magnitude. Voltage dividers can be used to provide multiple voltages from single-voltage power supplies, provide selected voltages for biasing semiconductor circuits, set proportional voltage outputs for metering circuits, and perform a host of other uses. Voltage divider concepts are a very powerful tool that can be used in the analysis of complex circuits.

Current flows through parallel branches are inversely proportional to the amount of resistance in the branch. The higher the resistance, the lower the current. The amount of current can be calculated by using the law of proportionality. In circuits that have only two resistors or that are simplified to two resistors using equivalent circuits, the currents through each branch are proportional to the value of the resistance of the other branch divided by the sum of the branch resistances.

The superposition theorem is very useful in solving complex circuits with more than one voltage or current source. To use the theorem, the circuit has to have passive components that are linear and bilateral. Normally, resistors, capacitors, and air-core inductors meet these requirements. The steps in applying the theorem are as follows:

1. Reduce all but one voltage and/or current source to zero. Calculate the total resistance, voltage, and current for this redrawn circuit.
2. Repeat step 1 for each of the other sources in the circuit.
3. Combine the resulting voltages and currents through the different resistances. This is done algebraically. The results are the combined voltage and current effects on the various resistances within the circuits.

Kirchhoff's laws are very useful when analyzing complex circuits with more than one power source. Loop equations (Kirchhoff's voltage law) are written so that the algebraic sum of the I_R voltage drops equals zero. Care must be taken to observe the polarity and direction of current flow.

Node equations (Kirchhoff's current law) are written so that the algebraic sum of all currents entering and leaving a given circuit point must equal zero. Remember that the point where current enters a resistive element is considered negative, and you must always be consistent with your signs after you have assumed a polarity for any value.

Thevenin's theorem says that any network can be represented by a voltage source and series resistance, whereas Norton's theorem says that the same network can be represented by a current source and parallel resistance. In fact, the same example was used for both theorems, and with the same new load applied to both types of equivalent circuit, the resulting voltage drops were the same: 19.2 V. These two theorems will prove useful in circuit analysis for both power systems and electronic circuits.

Limiting components are critical factors as you design circuits to work under varying conditions or with variable components. A thorough understanding of the laws of series and parallel circuits and the power, current, and voltage limitations is essential.

CHAPTER GLOSSARY

Current divider A circuit of parallel circuit components that will split the total current according to the opposition of the branch, in inverse proportion to the total opposition.

Current source A source that keeps its output current constant regardless of the load applied. In the early days of electric power, street lighting circuits were often series circuits supplied with constant current transformers. In modern power systems, current sources are rarely if ever encountered. Current sources are more commonly used in electronics systems.

Voltage divider An electrical circuit (usually made up of resistive elements) that can be used to break down, or divide, a supply voltage into two or more smaller voltages.

Voltage source A source that keeps its output voltage constant regardless of the load applied. Such a device does not exist in reality. However, batteries and other such power sources approximate them.

REVIEW QUESTIONS

1. Voltage dividers in a series circuit may be calculated by using Ohm's law or what else?
2. State the formula for finding the total resistance in a series circuit.
3. State the formula for finding total voltage in a series circuit.
4. State the formula known as the law of proportionality.
5. In a parallel DC circuit, if you increase the resistance, what happens to the voltage drop?
6. In a parallel DC circuit, if you increase the number of resistive branches, what happens to the value of total current?
7. In a DC parallel circuit, current and resistance are __________ (directly/inversely) proportional.
8. In a DC parallel circuit, voltage and resistance are __________ (directly/inversely) proportional.
9. What is the constant in a series DC circuit?
10. What is the simple definition of the superposition theorem?
11. What are the steps that must be taken when using the superposition theorem to solve a problem with two voltage sources?
12. What are the steps that must be taken when using the superposition theorem when you have a voltage source and current source?
13. Write and explain the two Kirchhoff's laws.
14. Kirchhoff's law is useful for solving more complex circuits with more than one voltage source. Discuss the advantages and disadvantages of using Kirchhoff's law versus the superposition theorem.
15. When are Thevenin's and Norton's theorems generally used?
16. State and discuss Thevenin's theorem.
17. State and discuss Norton's theorem.
18. Explain what is meant by the limiting component in a series circuit.
19. Describe how to find the limiting component of a parallel circuit with various size resistance and wattage-rated components.

PRACTICE PROBLEMS

1. Using the equation for the law of proportionality, calculate the voltage drop across each resistor in Figure PP7–1.
2. Use the superposition theorem to calculate the voltage drops and current flows for all resistors in Figure PP7–1.
3. For the circuit shown in Figure PP7–2:
   a. Determine the Thevenin equivalent circuit at terminals a and b.
   b. Determine the Norton equivalent circuit at terminals a and b.
   c. Prove your answer by attaching a 10 ohm resistor to each circuit (at terminals a and b) and show that the same voltage and currents are present.

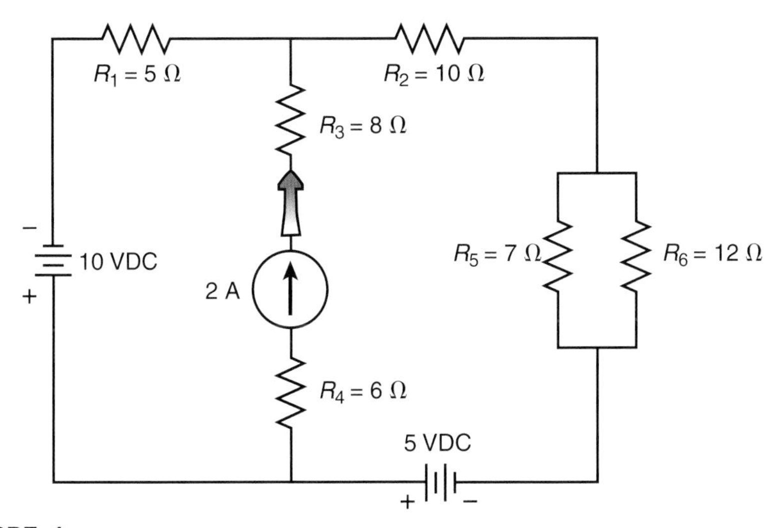

Figure PP7–1

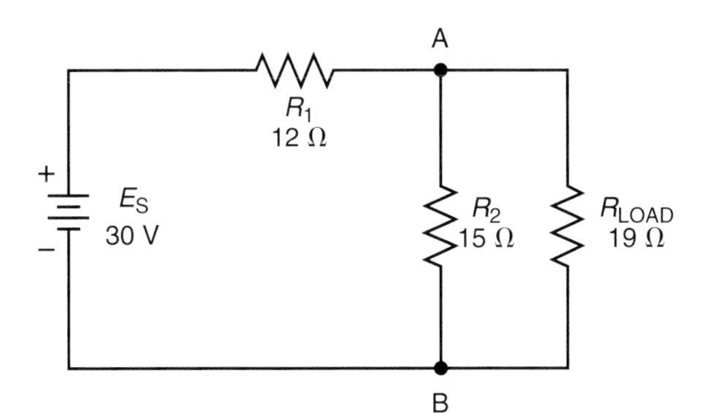

Figure PP7–2

8 Magnetism

TOPICS

- INTRODUCTION
- THE NATURE OF MAGNETISM
- THEORY OF MAGNETISM
- MAGNETIC REACTIONS
- MAGNETIC UNITS OF MEASURE
- MAGNETIC FLUX
- MAGNETIC SHIELDING

OVERVIEW

Magnetism is a force that occurs in nature. It is most often expressed as the ability of some materials to attract iron or other ferromagnetic materials. Most electricity is developed through the use of magnetism. No one knows exactly when the effects of magnets were first discovered, but the Greeks were studying magnetism and magnets more than 2,000 years ago.

The Chinese may have been studying magnetism even before then. However, most Western terminology is based on the Greek studies. They discovered that iron was attracted to a certain type of stone. This stone (magnetite) was originally discovered in a region of Asia called Magnesia. From *magnetite* comes our modern word *magnet*. Electrical systems are very often intertwined with the magnetic systems. If you have a thorough understanding of magnetic concepts, you will more easily understand the concepts of electrical interactions with magnetic fields.

OBJECTIVES

After completing this chapter, you will be able to:

- Identify the laws of magnetism
- Explain the theory of magnetism
- Predict the effects of permanent magnets in close proximity
- List some of the uses of permanent magnets
- Identify magnetic terms
- Calculate basic magnetic equations
- Describe how to shield a circuit from magnetic effects

INTRODUCTION

Magnets are used every day, whether you realize it or not. We use magnets to produce electricity, and electricity to produce magnets. Magnetic fields are used to operate electrical controls, and magnetic fields are used to make motors work. We use electromagnets to lift equipment. We use the theory of magnetism to operate transformers. Magnetism causes various reactions in DC circuits and other reactions in AC circuits. Knowledge of magnetism is essential to the understanding of electrical systems and electrical properties. If we can understand and apply rules for magnetism, we will be able to determine how magnetism and electricity are so closely related.

THE NATURE OF MAGNETISM

Magnets were first discovered as being useful when **magnetite** was observed to consistently align itself always in the same orientation when floating in water. Ancient Greeks placed magnetite on a piece of wood floating in water. Regardless of the magnetite's initial placement, it would eventually return to the same predicted position. The stones would align themselves with the Earth's magnetic field and were known as "leading stones" that altered into the term *lodestone.*

This phenomenon occurs because Earth is also a magnet. Modern scientists believe that the motion of Earth's molten iron core generates a huge magnetic field (Figure 8–1). There are many theories on how the Earth's huge magnetic field is produced. We do know that it is not a constant value or even that it stays in the same place. According to scientists, the Earth's magnetic field has, through time, reversed and will probably do so again.

When the ancient Greeks first learned that one end of a magnet would always point north, they decided to call that end of the magnet the North pole. Later, researchers learned that like poles repel and that unlike poles attract. This discovery clearly indicated that the magnet's North pole and Earth's geographic pole must be opposite. Because of this, a magnet pole that is attracted north is called the north-seeking pole. Usually, we just shorten this to the magnet's North pole.

As you look at Figure 8–1, it seems that the geographic North pole is offset from the magnetic pole by between 5 and 10 degrees (called the angle of declination). Actually, the actual North magnetic pole is always moving, although it has been in its same general location for thousands of years. This information is common knowledge as we set our compasses to compensate for the shifted magnetic pole.

What is not common knowledge is that the magnetic pole at the top of the Earth (geographic North) is actually the South magnetic pole.

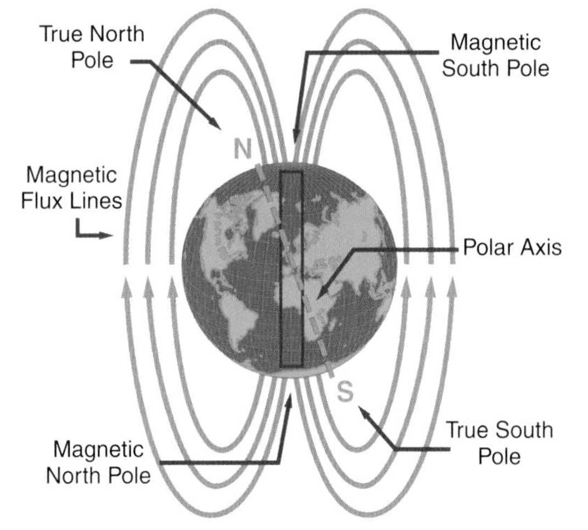

Figure 8–1 Earth's magnetic field has the South magnetic pole associated with the geographic North pole and offset by a varying angle of declination.

This becomes apparent as we use a compass to point toward the North geographic pole. As we stand on the outside of the Earth's core magnet, we are actually in the magnetic field set up by the Earth's core. As we will learn in the magnetic laws and principles, the magnetic field has an agreed-upon direction. The direction external to the magnet's core is from north to south.

We also know that a compass is a permanently magnetized piece of steel with the north end of the magnet marked North. This north end of the compass is attracted to the North geographic pole. In essence, the compass aligns itself in the direction of the lines of force (like the lodestone). Unlike magnetic poles attract, and thus the North geographic pole actually has a South magnetic polarity. This concept of unlike poles attracting and magnetic lines of force traveling from northern to southern magnetic polarities is an important point to remember as we study the effects of magnets and magnetic fields.

THEORY OF MAGNETISM

Through scientific study of magnets and the materials from which natural magnets are made, a theory of magnetism has been developed. This theory is based on the ability of a material to be magnetic because of the spin of the electrons in the material.

In Chapter 2, you learned that electrons orbit the nucleus of an atom. Being charged particles, electrons have an electrostatic field around them. The lines of force from this field travel toward the electron center (Figure 8–2). Electrons also spin on an axis, much like Earth. This makes each electron act like a small electromagnet. The combination of the electrostatic and magnetic fields creates a composite electromagnetic field somewhat like that shown in Figure 8–2. Note that the magnetic lines and electrostatic lines are perpendicular to each other in the electromagnetic field.

Electrons do not all spin in the same direction. Electrons that are in the same shell and spin in opposite directions usually form pairs. Because of their opposite spin, the magnetic properties of these paired electrons usually cancel each other (as we will again see in the rules for electromagnetic properties). This is why most materials do not exhibit natural magnetism.

The iron atom has a total of 26 electrons, with 22 of the electrons paired to become magnetically neutral. Iron has four electrons unpaired in its outer shell. If these electrons are spinning in the same direction, their individual magnetic fields will add and the iron exhibits the characteristics of a magnet. At very high temperature, the electrons change their spin characteristics, become neutral, and the iron becomes demagnetized. At very low temperatures, the magnetic field becomes more prevalent.

These magnetic fields develop in small packets called **magnetic domains**, or magnetic blocks, which react like small permanent magnets. This concept is referred to as the **domain theory**. Most materials that have magnetic domains are not magnetized because the domains are not aligned properly.

Spinning Electrons

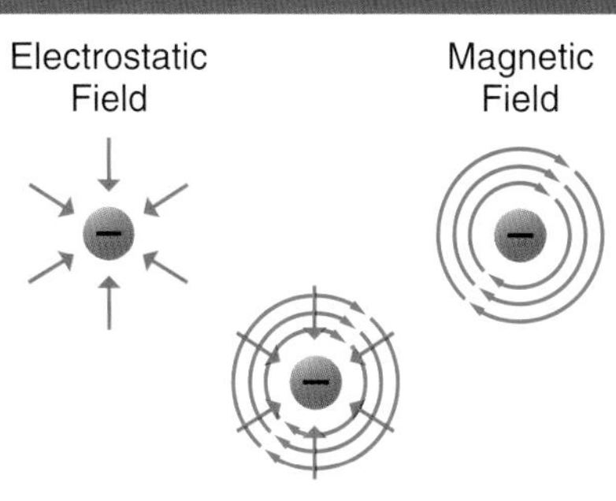

Figure 8–2 Spinning electrons have a magnetic and electrostatic field around them.

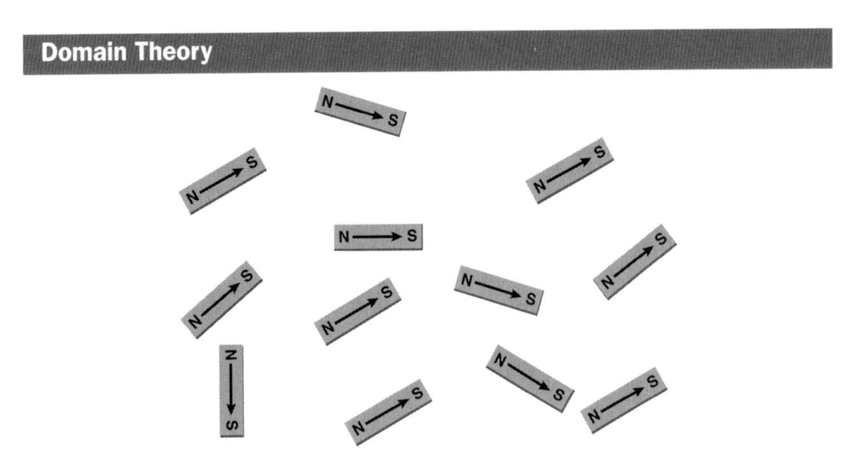

Figure 8–3 When the magnetic domains are not aligned, or are in random alignment, the material does not display the characteristics of a magnet.

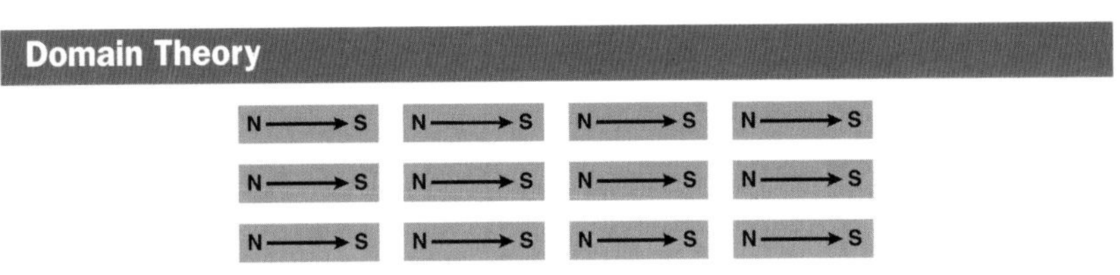

Figure 8–4 Aligned magnetic domains create a material that exhibits magnetic properties.

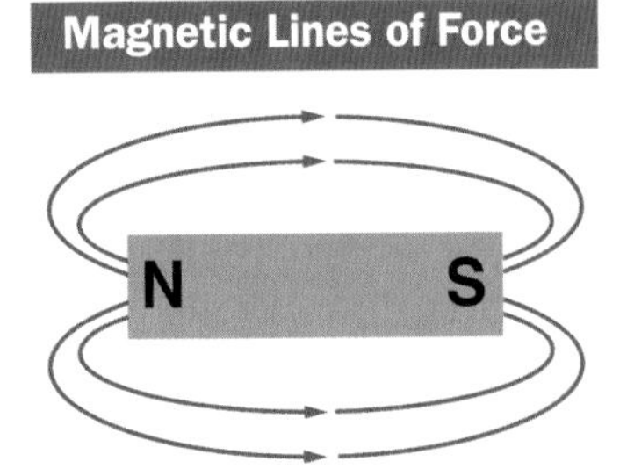

Figure 8–5 Magnetic lines of force travel from north to south.

This means that normally the domains are random and do not lie in the same direction (Figure 8–3). The process of aligning the domains into coordinated directions is the process we use to magnetize a material. We use external energy called magneto-motive force to align the domains.

In magnetic materials, many or most of the domains are aligned. This alignment allows the magnetic properties of the domains to be additive, making all of the small permanent magnets into one large permanent magnet (Figure 8–4). The more the domains line up, the stronger the magnet becomes. There is a point where no more domains can be aligned and the magnet is as strong as it will get. At this point, the magnet has reached **saturation**. The magnet cannot hold any more aligned domains and is magnetically saturated.

MAGNETIC REACTIONS

The following are laws of magnetism:

1. Energy is required to create a magnet (magnetic field), but no energy is required to maintain a magnet (magnetic field).
2. Like poles repel each other, and unlike poles attract each other.
3. The magnetic force between two poles is directly proportional to the pole strength and inversely proportional to the square of the distance between them.

The third law translates into mathematic formulas, such as the following:

$$F = M_1 \times M_2 \frac{\mu}{d^2}$$

Where:

F is the force of attraction or repulsion in dynes.

M_1 is the strength of one pole face, and M_2 is the other pole face measured in dynes.

1 dyne is 1/27,800 of an ounce of pull.

μ is permeability of the space between the poles. Air has a permeability of 1.

d is the distance in centimeters between them.

If two magnets are held close to each other, they will either attract or repel. The attraction or repulsion is caused by the **magnetic lines of force** that flow into and out of a magnet. Examine Figure 8–5. Note that the lines of force flow out of the north end of the magnet and flow into the south end of the magnet. This direction of magnetic force is the result of the early experiments with the lodestone being aligned with the Earth's magnetic field. We conclude that the lines travel from north to south.

A naturally occurring magnet always has a field around it, and thus we call it a permanent magnet. We can also create what we refer to as a permanent magnet, where the magnetized material retains its magnetic properties long after it is initially magnetized. Permanent magnets display one of the laws of magnetism. This law states that energy is required to build a magnetic field but that no energy is required to maintain one.

With modern technology, very strong permanent magnets can be manufactured. These magnets are stronger and last longer than naturally occurring magnets.

The magnetic laws indicate that it does take some energy to create a permanent magnet (typically electric power) but that the magnet will stay magnetized without any additional power. We call this lasting effect (to be able to retain magnetism) the **retentivity** of the magnet.

If you were to place a piece of paper on top of a permanent bar magnet that is on a piece of wood, isolating it from a metal surface, and sprinkle iron filings or iron bits on top of the paper, you would be able to see the results of lines of **magnetic flux**. With a single bar magnet, the lines of force (flux) travel in unbroken paths from one pole to the other. If you were to place a compass in the magnetic field (Figure 8–6), you would be able to determine which is the North pole and which is the South pole.

The compass points in the direction of the lines of force traveling from north to south. Remember that the field is three-dimensional even though you are viewing only two dimensions. The magnetic field you can see now is the external magnetic circuit of the magnet. Internally, the magnet has a magnetic field that travels south to north to complete the unbroken lines of flux.

This concept pertains to electrical circuits as well. In external circuits to a voltage source, the current flows from the negative terminal of the source through the circuit and back to the positive terminal. Internal to the power supply, the current moves from the positive toward the negative terminal to complete the path.

If you align two bar magnets (Figure 8–7) and again sprinkle iron filings on top of the paper, you can immediately see the lines of force travel from north to south and attract each other.

Likewise, when the same magnetic poles face each other, the flux lines repel each other and appear as in Figure 8–8. We cannot see magnetic lines of force but can see the effects as we use iron filings to trace the patterns created.

Compass and Magnetic Field

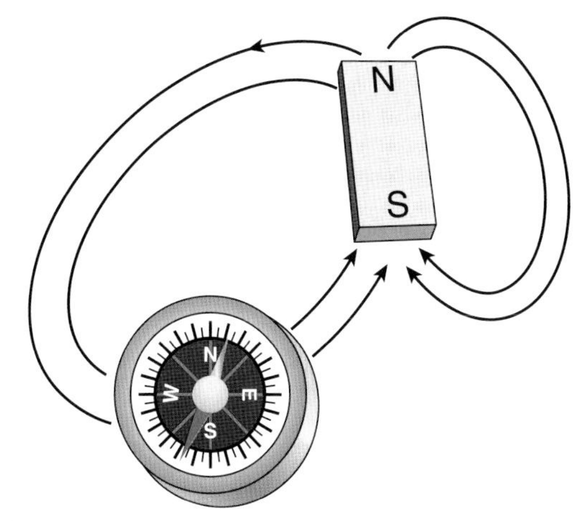

Figure 8–6 Placing a compass in a magnetic field will indicate the presence of lines of force and indicate direction.

Opposite Polarities

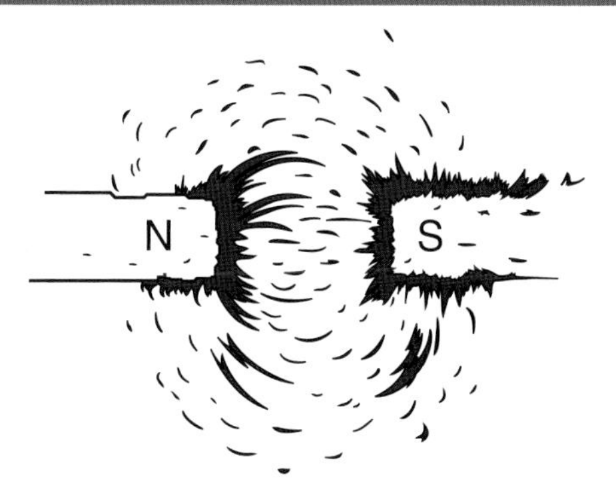

Figure 8–7 Magnetic lines of force attract between opposite polarities.

Magnetic Path

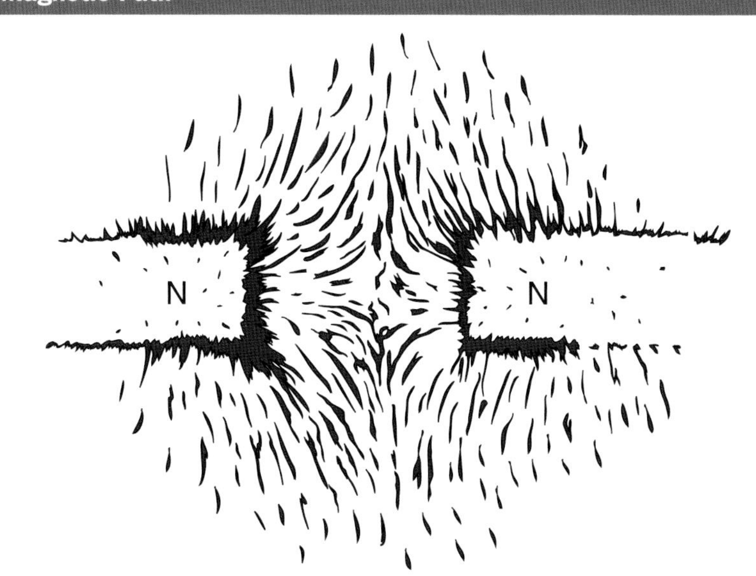

Figure 8–8 Magnetic lines of force repel between same magnetic polarities.

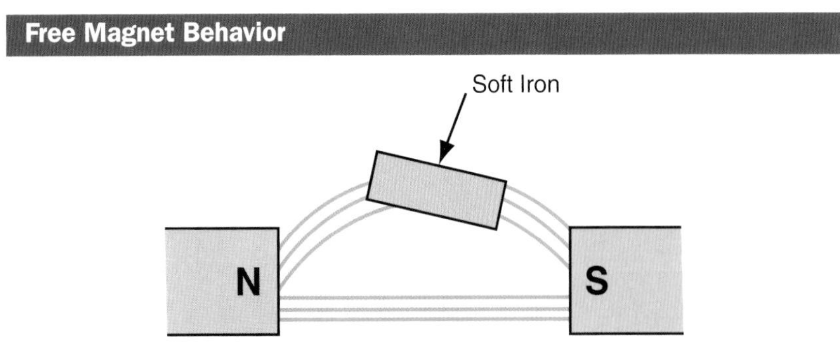

Figure 8–9 The magnetic path between pole faces can be altered by a more permeable path.

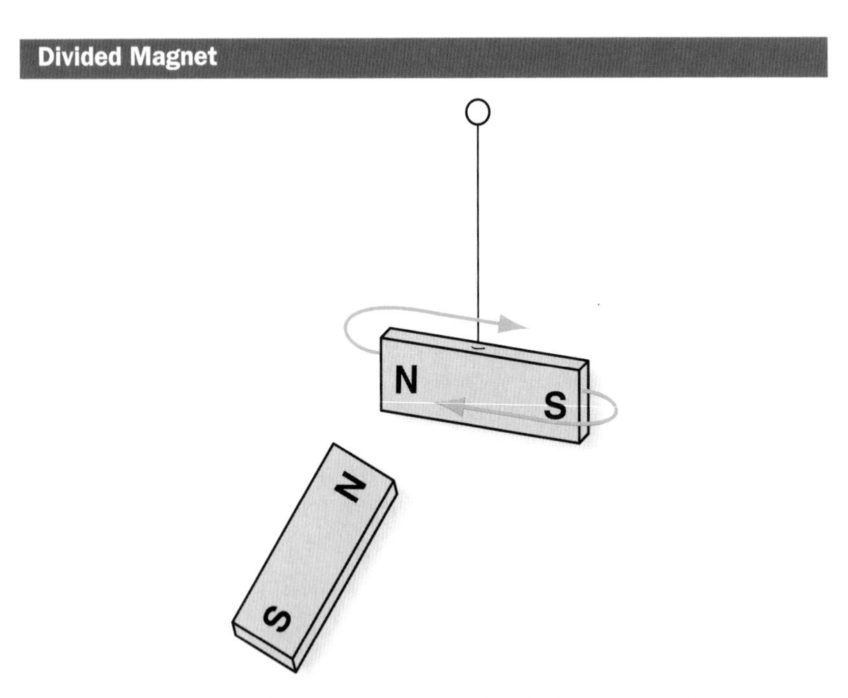

Figure 8–10 A magnet free to move is attracted or repelled by another magnet.

With two magnetic fields interacting and creating lines of magnetic force directly between them, if you insert a piece of soft iron into the path (Figure 8–9) at an angle to the original path, a change occurs. Note that the path is altered as the magnetic lines take an easier (more permeable) path between the magnets, thus illustrating the effects of a more permeable path than air. The soft iron has a higher **permeability** than the air.

Another demonstration of magnetic influence occurs when you suspend a magnet in the air so that it is free to move. As we approach the suspended magnet with another magnet (Figure 8–10), the suspended magnet moves. If the north end of the suspended magnet is approached with another north end, the magnet that is free to move moves away. If opposite poles are brought together, the suspended magnet moves toward the held magnet. You have seen the principle of an electric motor. One magnetic field is free to move (the rotor), and one is stationary (the stator). Exact operations of motors are explained in Chapter 10.

The effect observed is quantified in the third law of magnetism that states that the force between two magnets is proportional to the product of the magnetic strength of each and inversely proportional to the distance between them, assuming a consistent permeability. The formula ($F = M_1 \times M_2 \times \frac{\mu}{d^2}$) applied is the same whether it is attraction or repulsion.

Another principle of the magnetic force between the two magnetic poles is illustrated by the amount of force between the two poles. It is obvious that if the magnetic poles are typical of bar magnets and the two poles are feet apart, there is very little force of attraction or repulsion. However, if they are only inches or even a fraction of an inch apart, the forces are more apparent.

The rules of field strength and distance, as well as permeability, are in effect. The force between the magnets is inversely proportional to the distance squared between the magnets. As an example, if the distance is reduced to half the original distance, the force becomes the inverse of $(0.50)^2$, or four times the force.

If we were to break a bar magnet apart and into pieces, we would find that the domains stayed aligned in the small pieces. Now the patterns look like those shown in Figure 8–11, indicating that small magnetic fields are present in pieces and the total effect of the magnet is the sum of the individual small magnetic fields. These are all important concepts as we begin creating magnets and magnetic fields with electricity.

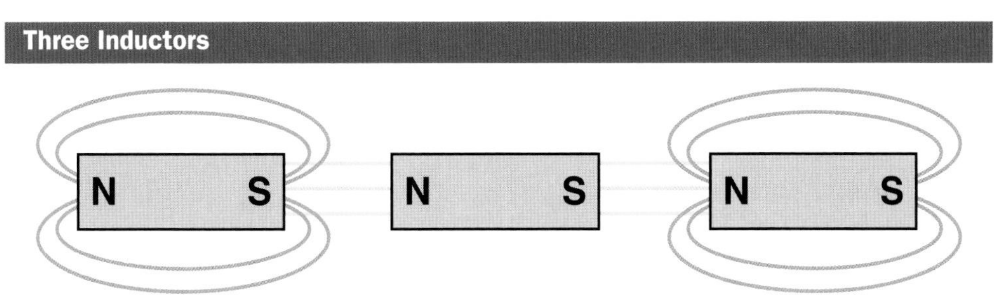

Figure 8–11 When a magnet is cut into pieces, each section displays the same magnetic effects as the original magnet.

MAGNETIC UNITS OF MEASURE

The magnetic system, like the electrical system, has developed from many different perspectives and with many different discoverers. Practitioners using the rules and laws of magnetics use different units of measure to measure the same phenomenon. There are three different systems for measuring magnetic effects.

The English system uses pounds of pull, inches for distance, and square inches for area. For instance, the units for measuring **flux density** are in magnetic lines per square inch. The flux density is a measure of the concentration of flux in a specific area. The flux density formula is $\beta = \frac{\phi}{A}$.

β (Greek letter beta) is the flux density at the face of a magnetic pole measured in Gauss. ϕ (Greek letter phi) is the total number of magnetic lines of force measured in **(Wb)**. A is the area of the pole face measured in square inches.

Example

What is the flux density at the pole face of a magnet that has 10,000 lines of force (Wb) in a 2-square-inch area?

Solution:

$$\beta = \frac{\phi}{A}$$

$$\beta = \frac{10{,}000 \text{ Wb}}{2 \text{ sq in}}$$

$$\beta = 5{,}000 \text{ Gauss}$$

FieldNote!

Permanent magnets are used for many common applications around the house. Refrigerators have magnetic door seals that keep the door closed and the seal pressed tight. Speaker systems use permanent magnets and electromagnetics to move the voice cone to produce sound waves you can hear. Many security systems use a magnet to pull a sensor switch to the closed position when a door is closed. When the door is opened, the magnet moves away and the switch opens, sounding an alarm.

Another very common application is the magnets that hold notes on the refrigerator door. Often exterior doors that are weatherproofed now use a magnetic seal like the refrigerator to seal against wind infiltration. A very common application is the use of magnets on the magnetic stripe on credit cards or bank cards. These magnetic marks are aligned to hold data about the account.

If the total flux lines are not given, we would have to find ϕ. We must use the algebraic equation (like Ohm's law) to calculate the number of lines of force. We use a magnetizing force called magneto-motive force (MMF) and a piece of ferromagnetic material. We can also find the resistance to magnetic influence, known as reluctance and represented by the symbol R (rel). The formula is $\phi = \frac{MMF}{R}$. As you can determine, the formula for the magnetic circuit is similar to the formula for the electric circuit. This is equivalent to ϕ equated to current, R is equated to resistance, and MMF is equated to voltage in an electrical circuit. Ohm's law states $I = \frac{E}{R}$. MMF is magneto-motive force measured in **ampere-turns**. R (rel) is the **reluctance** of the material measured in ampere-turns per weber; ϕ is the magnetic flux in webers.

Example

What is the number of magnetic lines of force (webers) with an MMF of 100 ampere-turns and a reluctance (rel) of 50 ampere-turns per weber?

Solution:

$$\phi = \frac{MMF}{\text{rel}}$$

$$\phi = \frac{100}{50}$$

$$\phi = 2 \text{ Wb}$$

Note that the number of magnetic lines is directly proportional to the magnetizing force and inversely proportional to the reluctance of the magnetic material.

Another calculation used in the English system is the force between two poles. The formula for pounds of attraction or repulsion is $F = \frac{(\beta \times A)}{72 \times 10^6}$.

Example

How many pounds of force are exerted between two magnets that have a flux density of 10,000 Gauss and pole face area of 3 square inches?

$$F = \frac{(\beta \times A)}{72 \times 10^6}$$

$$F = \frac{(10{,}000 \times 3)}{72 \times 10^6}$$

$$F = \frac{30{,}000}{72{,}000{,}000}$$

$$F = 0.000416 \text{ lb of pull}$$

The other forms of measurement conform to the CGS (centimeter-gram-second) system. One magnetic line of force is known as a maxwell. A Gauss represents 1 maxwell per square centimeter, and the MMF is referred to as gilberts. To convert ampere-turns (English) to gilberts (CGS), a conversion factor of 1.256 is used: 1 gilbert = 1.256 ampere-turns. A measure of a magnet's strength is measured with a standard pull at 1 centimeter distance.

If a magnet exerts a pull of 1 dyne at 1 cm, the magnet has a magnetic pull of 1. The MKS (meter-kilogram-second) system uses the force between magnets as 1 newton or 0.2248 pounds. The weber is the unit of flux and represents 10^8 lines of force, or 100,000,000 maxwells.

MAGNETIC FLUX

There are six rules or principles for magnetic lines of force, called flux lines.

1. The flux lines are continuous and form closed loops.
2. The lines never cross.
3. Lines of flux in the same direction tend to attract each other and strengthen the magnetic effect, and lines in opposite direction tend to repel each other and weaken the magnetic effect.
4. The flux lines try to contract, pulling unlike poles together.
5. Magnetic lines of force pass through nearly all materials, both magnetic and nonmagnetic.
6. The lines enter and leave the pole face at right angles and travel north to south around the magnet.

The lines of force that leave the north end of a magnet and enter the south end are called magnetic flux. We have seen the effects of magnetic flux by placing a magnet under a piece of paper and sprinkling iron shavings onto the paper. The filings will take the shape of the field lines, similar to those shown in Figure 8–7. Note that the lines of flux stay parallel to each other and never cross.

The same numbers of lines that leave the North pole enter the South pole. They are densest (closest together) near the poles and the least dense when farthest away from the poles. Each line of flux has the same energy. The closer the lines of flux are to each other, the stronger the resulting magnetic field. Therefore, the strongest field is near the center of the poles. The magnet pair attracts because the lines of flux move in the same direction and try to shorten.

Lines of flux leave the north end of the magnet and enter the south end. In Figure 8–8, the magnets have their North poles facing each other. They repel because the lines of flux are moving in opposite directions and pushing against each other. The amount of push or pull is the same force but in opposite directions.

FieldNote!

Flexible magnets are another form of permanent magnetic material. They are manufactured with the use of a combination of barium ferrite and strontium ferrite. These materials can be combined into a ceramic magnet in the form of granules or powder. The powder can be added to other materials such as ink to make magnetic ink (as in checking account printing) or with rubber or plastic to create flexible magnets.

TYPES OF MAGNETIC MATERIALS

As far as magnetism is concerned, materials are classified into three categories: ferromagnetic, paramagnetic, and diamagnetic. **Ferromagnetic** materials include cobalt, nickel, iron, and manganese and can be easily magnetized. These materials are attracted by magnets. These metals are more easily magnetized because their atoms have a slightly different configuration of electrons around the nucleus than nonmagnetic materials, and the magnetic domains align easier.

Iron is the most well-known material that is attracted to a magnet and becomes magnetized easily. A hardened version of iron is steel, which is more difficult to magnetize but maintains its magnetism longer after the magnetizing force is removed.

Paramagnetic materials include aluminum, copper, chromium, platinum, and titanium and can be magnetized but not as easily as ferromagnetic materials. These materials are weakly attracted to magnets. They can become magnets but exert very little magnetic effects. Because the force is so small, these materials are considered nonmagnetic.

Diamagnetic materials actually repel magnetic flux or tend to divert the magnetic flux, such as brass. When they are in a magnetic field, the material tends to create its own very weak magnetic field in the opposite direction of the magnetizing force, indicating a repelling magnetic field. Two of the strongest diamagnetic materials are graphite and bismuth.

Magnetic lines of flux will pass through all materials to one degree or another. Ferromagnetic materials allow the lines of magnetic flux to pass through easily and actually tend to focus the lines of flux. As previously noted, the ability of a material to concentrate lines of flux is called permeability. The opposition to magnetic lines of flux is called the reluctance.

Most permanent magnets are made of steel, which is an alloy of iron and other metals. One alloy (Alnico 5) is a combination of aluminum, nickel, cobalt, and iron. This alloy, once magnetized, retains its magnetism indefinitely. A magnet made solely of soft iron does not maintain its magnetism long. For this reason, equipment that requires a strong permanent magnet will have its magnet made of a steel alloy. Those pieces of equipment that need to be magnetized but should not retain their magnetism (such as motors and generators) will have their structures made of iron.

To demagnetize materials, we can heat the material, which allows the domains to go back to their original random pattern. The temperature at which a magnet loses it magnetic properties is referred to as the Curie temperature. We can strike the material, which jars or scrambles the domains to rearrange them into a random pattern. We can also deliberately use an AC magnetic field to "mix up" the domains and leave them in a scrambled condition. This process is referred to as de-gaussing the magnet.

Driving under power lines, you have no doubt noticed that your car radio becomes scrambled and often different stations or no stations are heard temporarily. The problem is in the magnetic fields. High-voltage power lines exert a magnetic field whenever current flows. These magnetic fields interfere with the radio transmissions and the signals become scrambled and mixed up. As you drive out of the proximity of the power lines, the magnetic field returns to normal and the radio signals return to normal.

MAGNETIC SHIELDING

Magnetic fields are present everywhere, whether produced by the Earth's magnetic field, by permanent magnets, or intentionally or accidentally by electrical circuits.

As you have studied, magnetic fields pass through all materials but some pass more easily than others. The idea of reluctance, or the inverse of permeance or permeability, affects how magnetic lines of force travel. As we have seen, when a piece of soft iron is placed in a magnetic field, it tends to alter the path of flux.

We will use this knowledge to shield equipment from magnetic flux by altering the normal path of magnetic flux and causing the flux to be diverted around a component. We will shield a system from outside magnetic flux. As mentioned earlier, you may hear a buzz or a distortion of radio (magnetic) waves in your radio or even broadcast TV signals. This can be caused by other sources of magnetic influences nearby.

Shielded Cable

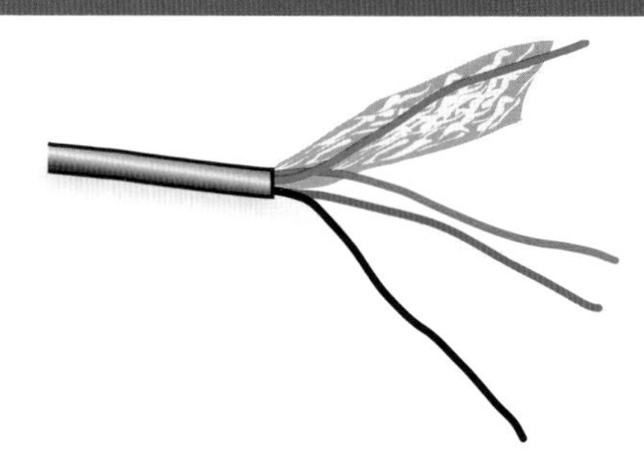

Figure 8–12 Wire that carries signals, especially sound reproduction signals, needs magnetic field shielding (such as aluminum wrapped around circuit conductors).

You may hear a buzz or hum in a speaker system when the speaker wire is placed too close to a magnetic field and the frequency is induced into the speaker lines. The term often used is *electromagnetic interference* (EMI) or *radio frequency interference* (RFI). If we do not want these magnetic fields to interfere with our circuit functions, we will need to shield our circuit or our lead wire.

One common type of shielding is in shielded circuit wire (Figure 8–12). The shield consists of an aluminum foil or a mesh of aluminum or copper wire that surrounds the circuit conductors. Remember that aluminum and copper are paramagnetic and act like a magnetic shield for magnetic lines of force that would otherwise induce voltages into the wire. The magnetic shields are usually grounded at one end of the cable to take any induced voltage and current that are developed in the shield to ground potential.

SUMMARY

Magnetism is the ability of a material to attract iron or other ferromagnetic materials, such as nickel, cobalt, and manganese. Materials that have this ability are referred to as magnets. Magnets have two magnetic poles: a North pole and a South pole. If a magnet is suspended and free to spin, by definition the North pole of the magnet is the pole that points toward the geographic North pole of Earth. The geographic North pole has a magnetic South pole polarity. Like poles of magnets repel each other, and unlike poles attract each other.

Magnets are generally divided into two groups: permanent magnets and temporary magnets. Permanent magnets, once they become magnetized, retain their magnetism indefinitely. They usually have cores made of steel or alloys. Temporary magnets lose their magnetic properties when their magnetizing force is removed. Their cores are usually made of iron.

Magnets have lines of force called flux lines. Flux lines are invisible to the human eye, but their presence can be detected by sprinkling iron filings on a piece of paper placed on top of a magnet. Flux lines have some specific rules and characteristics that must be understood in order to study magnetic-electrical principles.

- They flow from the North pole of a magnet to the South pole.
- They always form closed loops. The number of flux lines leaving the North pole of a magnet must eventually return to its South pole.
- They never cross each other.
- They tend to cancel each other when they meet flowing in opposite directions.
- They repel each other as they leave the magnet's North pole.
- They take the path of least reluctance.
- They pass through all materials.
- They are denser at the poles of the magnet. As the flux lines move away from the North pole of the magnet, they repel each other and become less dense.
- They always try to find the shortest and lowest reluctance high-permeability path from the North pole to the South pole.

The laws of magnetism tell us that the force between two magnets is proportional to the magnetic strength of each magnet and inversely proportional to the distance squared. We also know that the permeability of the medium between the two magnetic materials affects the magnetic pull. If the space between the two magnets is highly permeable, the magnetic pull is increased.

Materials are classified as ferromagnetic, paramagnetic, and diamagnetic, referring to their ability to be magnetized. Paramagnetic materials are often used for shielding circuits from magnetic influence.

CHAPTER GLOSSARY

Ampere-turns The strength of an electromagnetic field calculated by multiplying the current flow times the number of turns for an electromagnet used to produce a magnetic field.

Diamagnetic Materials diametrically opposite to magnetic materials. When exposed to magnetizing force, they produce a weak field that opposes the magnetizing force.

Domain theory This theory relates to the effects of magnetic domains. If enough domains are aligned so that the small magnetic fields have the same magnetic polarity, the entire material will exhibit magnetic effects.

Ferromagnetic Properties displayed by certain substances (such as iron, nickel, cobalt, and various alloys) that exhibit extremely high magnetic capability.

Flux density Flux density is a measure of the pole strength of a magnet. It indicates how many lines of flux are either leaving or entering the pole face per unit area.

Magnetic domains As the electrons in the atoms of a material spin in the same direction to create small magnetic fields, the fields of many atoms add together to create atomic level magnets called magnetic domains.

Magnetic flux Flux is the name given to the invisible magnetic lines that encircle the magnet.

Magnetic lines of force The force of a magnetic field expressed as lines, also called magnetic flux.

Magnetite The mineral form of black iron oxide. Magnetite is a naturally occurring magnet and was possibly the first type of magnet studied by early scientists.

Paramagnetic Materials that have weak magnetic abilities. They can become magnets with a large magnetizing force, but do not generally exhibit magnetic capability.

Permeability The degree to which a material focuses magnetic lines of flux, or the ease with which magnetic lines of force distribute themselves through a material. Permeability is a factor compared to air (permeability factor of 1). There is no unit of measure, just a factor compared to air. The symbol used for magnetic permeability in a formula is the lowercase Greek letter mu (μ).

Reluctance The opposition that materials present to the flow of magnetic lines of flux (similar to electrical resistance). Permanent magnets have a high reluctance and temporary magnets have a low reluctance. The symbol is the Greek letter rel (R) and the force is measured in ampere-turns per weber.

Retentivity The ability of a material to hold magnetism after the magnetizing force is removed.

Saturation The point at which the magnetic domains are all lined up. Any further magneto-motive force will not produce a stronger magnet.

Weber (Wb) Field strength equal to 100,000,000 lines.

REVIEW QUESTIONS

1. Define magnetism.
2. List some ferromagnetic materials and paramagnetic materials.
3. What are the laws of like poles and unlike poles?
4. Magnets are divided into two groups. What are they?
5. What is meant by the reluctance of a magnetic material?
6. Give two examples of where you use permanent magnets.
7. Which material is better for a permanent magnet: steel or soft iron?
8. Explain how to determine magnetic field strength.
9. Explain the domain theory.
10. Name five properties of flux lines.
11. Within a magnet, which direction do flux lines travel? Around a magnet, which direction do they travel?
12. Why would you need to magnetically shield an electric circuit? How could you do it?

9 Electromagnetism

TOPICS

- INTRODUCTION
- ELECTRICITY AND MAGNETISM
- MAGNETIC DIRECTION
- FIELD STRENGTH
- ELECTROMAGNETIC WAVES
- MAGNETIC POLES OF ELECTROMAGNETS
- INDUCING VOLTAGE WITH MAGNETIC FIELDS
- FACTORS AFFECTING VOLTAGE PRODUCTION

OVERVIEW

The basis for our electrical and magnetic principles originates with the core of the earth. The current theories are that the earth has a solid iron core about the size of our moon. It is called the solid inner core. Around this core is a molten liquid outer core that is constantly in motion.

The theory is that this motion of the molten outer core actually generates a magnetic field. This action of moving magnetic materials is a concept that helps explain how magnetism and electricity are linked. We will see how moving electricity creates magnetic fields and how moving magnetic fields create electricity.

OBJECTIVES

After completing this chapter, you will be able to:

- Explain how to determine the magnetic field direction around a conductor
- Determine if a conductor's magnetic fields aid or reduce magnetic influence
- Explain why electromagnets work with a coiled conductor
- Give a brief explanation of how electromagnetic waves can be used to transmit information
- Identify the points on a hysteresis loop regarding the magnetization of a material
- Determine the polarity of magnetic poles, or electrical polarity, of an electromagnet
- Identify the requirements of generation and determine circuit polarity

INTRODUCTION

You use electromagnets every day, whether you realize it or not. Electromagnets are used to produce the electricity you need everyday. They are used in generators, no matter what the source of the mechanical power. Electromagnets are used in the motors of refrigerators, freezers, and air movement fans, as well as in vehicles.

Small electromagnets are used in your doorbells, in relays for heating and cooling controls, and in writing data to credit card magnetic strips. Magnetic waves are produced so that you can hear radios, TVs, cell phones, and many other broadcast media. We will investigate how and why these processes are used to combine electricity and magnetics through electromagnetics.

ELECTRICITY AND MAGNETISM

There are some applications in which a magnet is useful but only part of the time. For example, when you press a doorbell button, a small temporary magnet pulls on a piece of ferrous metal, causing it to move and strike the chime, creating the bell sound. This magnet does its work only when the button is pressed. Permanent magnets cannot do this. Permanent magnets cannot be turned off and on, and thus a different type of magnet (one that can be turned off and on) must be used: an electromagnet.

Direction of Current Flow

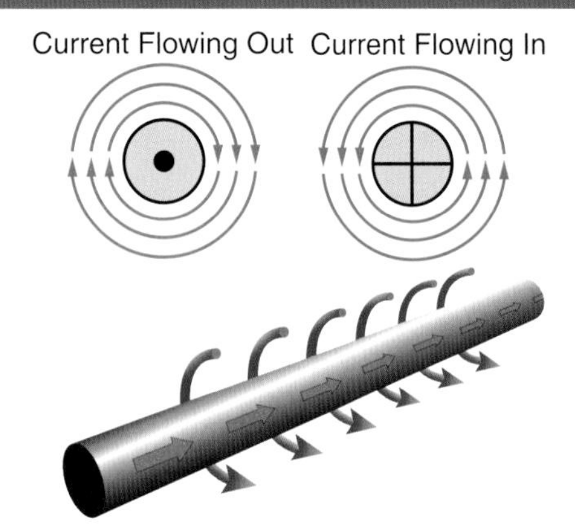

Figure 9–1 The magnetic field around the conductor has a direction determined by the direction of current flow.

MAGNETIC DIRECTION

When current flows through a conductor, a magnetic field is produced in a concentric circular form around the conductor (Figure 9–1). The two diagrams are cross-sectional views of conductors, and both have current flowing through them. The one on the left has current flowing out of the page. What you see is a dot representing the head of an arrow coming toward you. The one on the right has current flowing into the page.

This is represented by the crosshairs or the tail feathers of an arrow moving away from you. The arrows on the concentric circles show the direction the magnetic fields move. The fields around the conductor on the left move clockwise, and the fields around the conductor on the right move counterclockwise. The diagram of the conductor at the bottom of the figure shows the direction of the magnetic field, with the electron current flowing from left to right. You can verify that the magnetic field has a direction much as we did with the earth's magnetic field. If you hold a compass perpendicular to the conductor, as shown in Figure 9–2, you will see that the compass points in the direction of the magnetic field.

Direction of Magnetic Flow

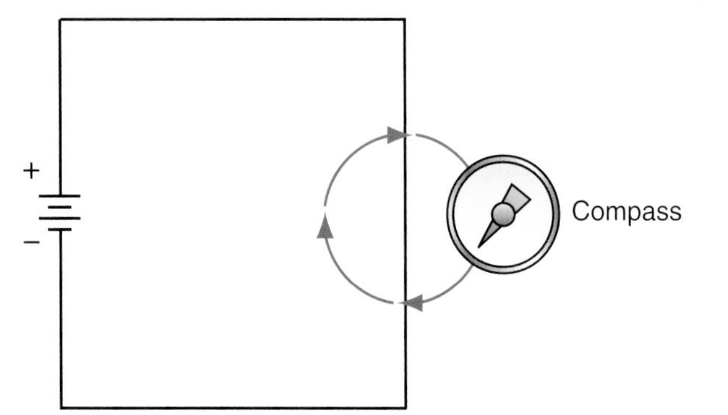

Figure 9–2 A compass held near a conductor with current flow will indicate the direction of the magnetic field.

The direction in which the field encircles the conductor can also be determined by using a simple memory device called the **left-hand rule for a conductor** (Figure 9–3). This rule states that if you grasp the conductor with your left hand, with your thumb pointing in the direction of the electron flow, your fingers will encircle the conductor in the direction of the magnetic field surrounding that conductor.

The magnetic field is a concentric field, meaning that it originates at the center of the conductor and moves out in unbroken circles perpendicular to the direction of the current flow. The size of the magnetic field is proportional to the amount of current. More moving current means a larger magnetic field, and conversely, a lower current means a smaller magnetic field.

ATTRACTION AND REPULSION

There is a physical force between conductors carrying current. This force comes from the interaction of the magnetic fields coming from the wires while current flows through them. Figure 9–4 shows three sets of parallel wires. The top two sets show current flowing in the same direction. In these two cases, the magnetic fields aid each other because at their points of interaction (between the conductors) the fields are moving in the same direction.

This causes the wires to be repelled from each other. This is one of the rules for magnetic flux, as explained in the previous chapter. The interactions of the fields in the bottom pair of wires oppose each other. This opposition causes the conductors to attract each other. The amount of current determines the amount of interaction between the magnetic fields.

The magnetic forces around a wire are also addressed in the way wiring is permitted in the *National Electric Code*® (*NEC*®). As you need to add more current to a load, the conductors need to be larger and larger. As the wires become larger they are more difficult to pull into conduits and bend into boxes or fittings. Electricians overcome this problem by taking several conductors and paralleling them, connecting them together at each end.

Often it is not feasible to pull all parallel conductors into the same conduit, and they must be split up. It is important to split them correctly. We actually must take one conductor of each circuit into a metal conduit so that the total magnetic field of the bundle of conductors is neutral. If you were to take the supply conductors all in one metal pipe and all the return conductors in another metal pipe, the magnetic fields would actually magnetize the metal conduits and cause them to heat.

Magnetic fields also play a part as you pull current-carrying conductors into metal enclosures. If you need to pull separate conductors into metal cabinets, be sure not to pull only the feed or return line through an individual knockout. The magnetic field around a single conductor can induce voltage and current into the surrounding metal and heat the metal to red hot. Care must be used to always be aware of the magnetic influence of current-carrying conductors and how the magnetic fields add to or cancel the magnetic effects.

As you study safety in the electrical industry, you will need to understand the OSHA requirements and in turn the requirement of NFPA document 70E. These safety requirements are designed to keep electrical personnel safe while working with live circuits.

André Ampère and Hans Oersted determined that a current flow would produce a magnetic field. They found that flowing current produced a magnetic field around the conductor that had a magnitude. This discovery, in 1820, led to many other discoveries regarding the links between magnetism and electrical phenomenon. Michael Faraday believed that if electric current created magnetism then magnetism should be able to create electrical current. Faraday demonstrated this principle of creating electrical current by magnetic induction in 1831.

Left-Hand Rule

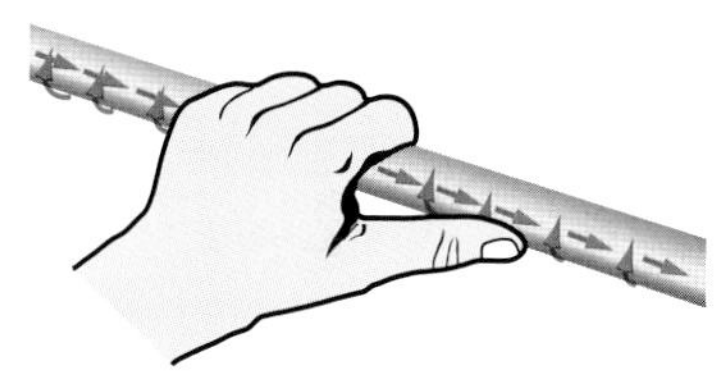

Figure 9–3 The left-hand rule for a conductor indicates current flow and magnetic field.

Interacting Magnetic Fields

Aiding Fields

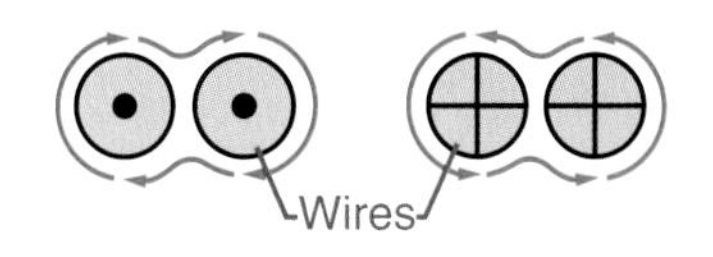

Opposing Fields

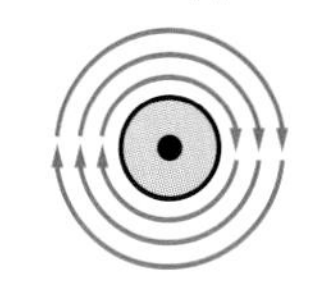

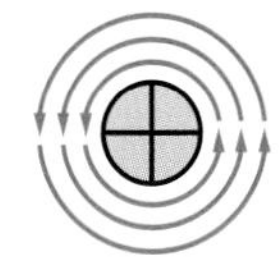

Figure 9–4 Conductors placed near each other have magnetic fields that interact.

You will study the effects of arc flash and arc blast as they relate to high currents that occur in short-circuit conditions. One of the reasons for arc blast or explosions during a short circuit is the extreme magnetic effects of the metals involved. The magnetic reaction between conductive parts often causes metal parts to be physically blown out of the enclosures when a fault occurs.

When there is a short circuit in a set of conductors in a conduit, you can often hear the conductors moving and slapping the side of the pipe. This is caused by a large amount of current (creating a large magnetic field around the conductor) and the magnetic fields reacting. Under normal conditions, the two conductors have the current flow in opposite directions at any point in time. This has the effect of creating magnetic concentric fields that are in opposite directions, thereby canceling the total magnetic effect, and the circuit wiring remains magnetically neutral.

Electromagnetic Field

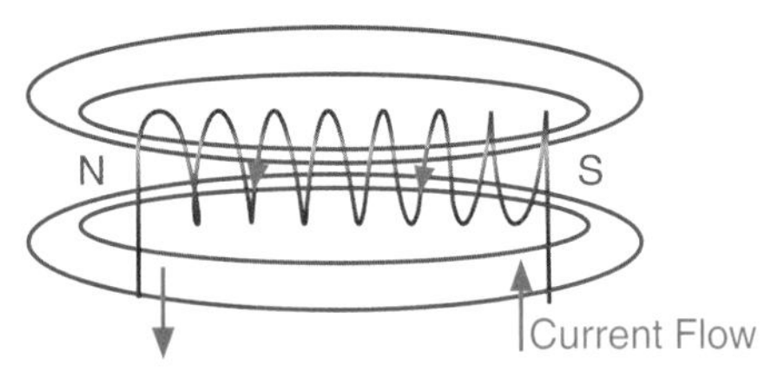

Figure 9–5 Small conductor magnetic fields are concentrated in a coiled conductor to create a larger electromagnetic field.

Ampere Turns

Original coil with an electromagnet strength of 10 ampere-turns

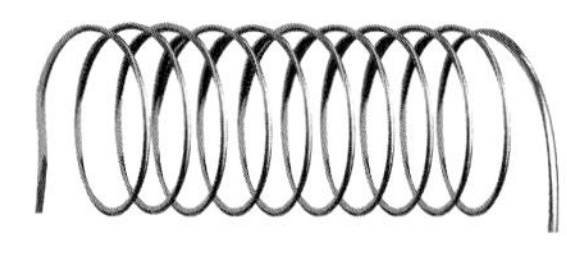

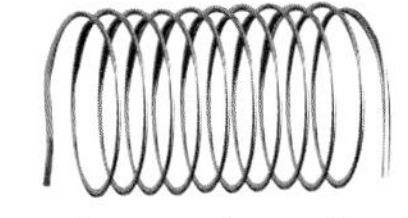

Closer (more dense) coil (same amount of turns) with electromagnet strength of 30 ampere-turns

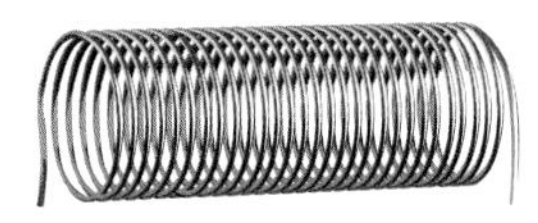

Closer (more dense) coil + more turns with electromagnet strength of 300 ampere-turns

Figure 9–6 More turns of wire in a specific distance will increase the MMF (ampere-turns) of a coil.

FIELD STRENGTH

The strength of an electromagnetic field can be increased by coiling conductors so that the magnetic field from each conductor adds to the one next to it (Figure 9–5). This aiding effect creates one large magnetic field, with a north and south pole at the ends of the iron core (like a permanent magnet). These coils of conductor used to produce magnetism are called **electromagnets**.

The strength of the magnetic field depends on several factors. These factors include the number of turns of conductor in the coil, the current measured in amperes, and the spacing between the coils. If the coil has more turns of conductor, there are more small magnetic fields around the conductor that will add together to create a stronger magnetic field. If there is a larger current flowing in the conductor, the individual magnetic fields around the conductor are larger, and the effect is to create a stronger electromagnet.

Multiplying the number of turns times the current flow results in a measurement called **ampere-turns**. As we saw in the last chapter, the magneto-motive force (MMF) is measured in ampere-turns. The measurement in ampere-turns gives us an indication of the relative field strength of the coil. Wrapping the coils closer together, by reducing the space between coils, further increases the strength of the field, because the amount of flux that is linked or combined with the next flux line is not lost or decreased (Figure 9–6). So far, we have changed the amount of magnetic flux produced by changing the ampere-turns and have changed the flux linkage to other magnetic flux by changing the spacing.

We can also change the overall magnetic effect of the coil by changing the shape, size, and material composition of the core.

The permeability of the core will affect the flux density of the magnetic field. If the core has a higher permeability, the flux produced by the coil is more easily transferred to the pole faces and in turn creates more flux density at the pole face. This is verified when an iron core replaces a brass core. Iron is much more permeable than brass and produces a stronger magnetic field. Another factor is the diameter of the core or the pole piece.

If the diameter of the core is greater, the cross-sectional area of the core is larger, allowing more flux line to be developed. In other words, the larger the cross-sectional area, the greater the magnetic field strength. The last factor is the length of the core material. If we assume the same number of turns as a reference, by lengthening the core the spacing of the coil becomes greater (fewer turns per length). This is the same effect as stretching the coil out, whereby the total magnetic effect becomes less with a longer core. All of these factors are involved when designing a magnetic component for a particular application.

ELECTROMAGNETIC WAVES

As part of the theory of electromagnetism, you need to be aware of the sources of magnetic energy. You have already studied the effects of the earth's magnetic field as sources of magnetic flux. Permanent magnets are sources of magnetic flux, as are many other man-made sources of magnetic energy. These sources do not ordinarily affect our working with electrical systems but occasionally can cause some reactions. The transmission of information by electromagnetic waves, particularly radio frequency waves, is part of the electrical/magnetic interaction.

The transmitter of a radio signal is simply a magnetic flux produced by moving current. A simple radio transmitter can be produced as shown in Figure 9–7. By simply closing the switch connecting the wire to the battery, a magnetic wave is produced. As you open the switch, you break the current and the magnetic field around the wire collapses. You can detect (receive) the magnetic field by placing another wire in the magnetic field and allowing current to flow through induction.

We generally think of the sending magnetic field as a transmitter. The signal is received by the antenna on the receiver. You have just created a magnetic pulse. This can be replicated at various frequencies by turning the magnetic field on and off. The result is a pulse modulation signal, such as the type of signal that can drive radio-synchronized clocks all around the country.

Another variation in the magnetic waves for information transmission is the amplitude modulation (AM) radio signal.

Pulse Modulation

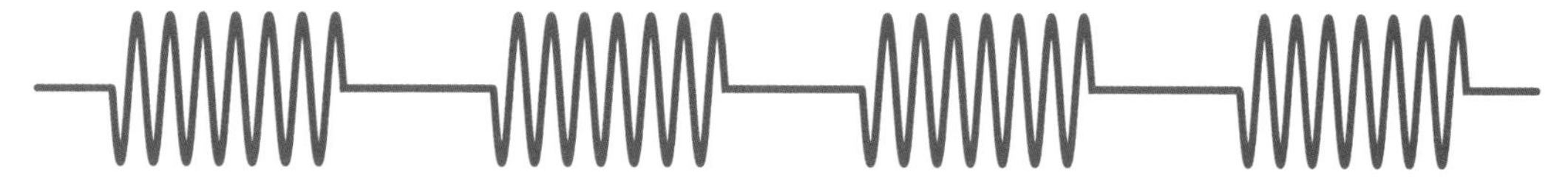

Figure 9–7 A form of electromagnetic transmission uses pulse modulation.

Figure 9–8 shows the effect of changing the size or strength of the signal as it is added to the base frequency that will change the amplitude of the received signal. You can now decode the received magnetic wave into an AM radio station.

Frequency modulation (FM) is another variation that allows us to transmit information. In this type of magnetic wave, the frequency of the magnetic wave is produced and adjusted by the frequency of the carrier or base wave. By superimposing the information signal on top of the base frequency (Figure 9–9), the output magnetics are altered slightly. The receiver senses these slight alterations and reproduces the FM information signal as anything from FM radio to cell phones, cordless phones, TV audio signals, wireless Internet, and so on.

All of these radio signal transmissions create a tiny voltage and current in the receiver, which is amplified many times and typically sent to another magnetic device (speaker) so that you can hear the signal as sound waves. If you work in a building that has a great deal of steel structure, the magnetic signals may be shielded from your receiver and you cannot "hear" the information transmission.

MAGNETIC CORES AND HYSTERESIS

One way to create a stronger electromagnet is to increase the number of flux lines that pass through the coil's center. You learned earlier that certain types of materials (ferromagnetic and paramagnetic) cause the lines of flux to concentrate. These lines of magnetic flux focus because they tend to pass through the ferromagnetic and paramagnetic materials better than through air.

If a coil were wrapped around a core of ferromagnetic material (iron, for instance), the number of flux lines passing through the center of the coil would be increased and so would the strength of the electromagnet. This happens because the individual domains in the core material become polarized (aligned) in relation to the magnetic field produced by the coil, as studied in magnetic domain theory.

Amplitude Modulation

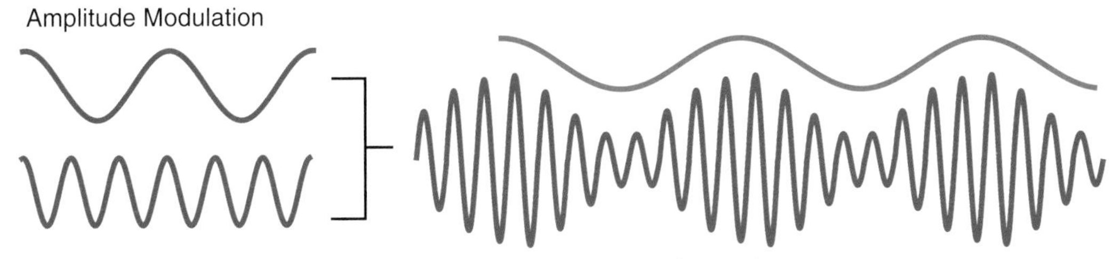

Figure 9–8 AM radio uses amplitude modulation for transmission of signal.

Frequency Modulation

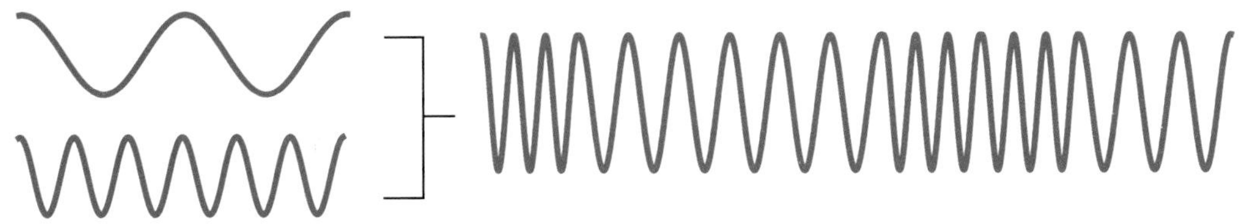

Figure 9–9 FM radio waves use frequency modulation for transmission of information.

We state that the core is permeable and allows flux lines to pass easily and concentrate in the iron. A material that is highly permeable has low reluctance. The magnetic formula $\phi = \frac{MMF}{Rel}$ indicates that there are more flux lines when the reluctance is low.

As the MMF in ampere-turns becomes stronger (either with more current or more turns of the coil), the magnetic core may saturate. The saturation point is the point at which increasing the amount of MMF has no more effect on the magnetic field strength. This occurs because all magnetic domains that can be aligned magnetically have been aligned. If the electromagnet needs to be stronger, a different core material has to be chosen (Figure 9–10). Therefore, different cores of ferromagnetic materials or paramagnetic materials will create electromagnets of different strengths.

A good material for a temporary magnet core, in which we do not want the magnetic effect to remain after the MMF is removed, is soft steel or iron. Soft iron has a high permeability. It easily passes magnetic lines of flux and makes the electromagnet stronger. Furthermore, soft iron is a good core because it does not become permanently magnetized. See the previous chapter on magnetic properties for more details on permanent magnets.

As you study electromagnets, keep in mind the concept that it takes energy to make a magnet but does not take energy to maintain a magnet. You can now study the amount of energy needed to create a magnet as MMF and relate it to the ability to remain a magnet. The amount of effort needed to turn all domains the same direction depends on the **hysteresis** of the magnetic material. As we try to magnetize a material we will need to overcome the hysteresis (i.e., the difficulty encountered in aligning the domains) of the material.

Some materials have high hysteresis and some low. If we need a material that becomes magnetized quickly and demagnetized quickly, we want a material with low hysteresis. Manufacturers create this easy movement of domains by utilizing low-hysteresis silicon steel.

Figure 9–11 shows a graph of MMF versus magnetic flux density. This graph (points 1 to 2) is referred to as a B-H curve. The curve is different for different types of magnetic materials because they create different amounts of flux density per MMF and saturate at different points. The B on the vertical axis is β (beta), representing the Gauss (flux density) of the electromagnet.

Concentrated Magnetic Flux

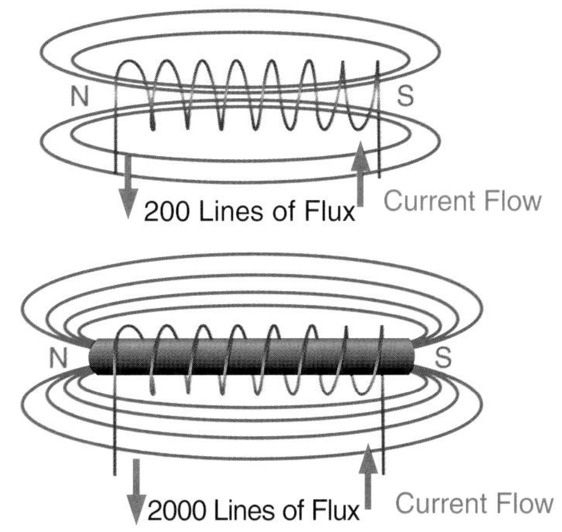

Figure 9–10 The amount of magnetic effect a coil has is partly dependent on the core material's ability to concentrate magnetic flux.

B-H Curve

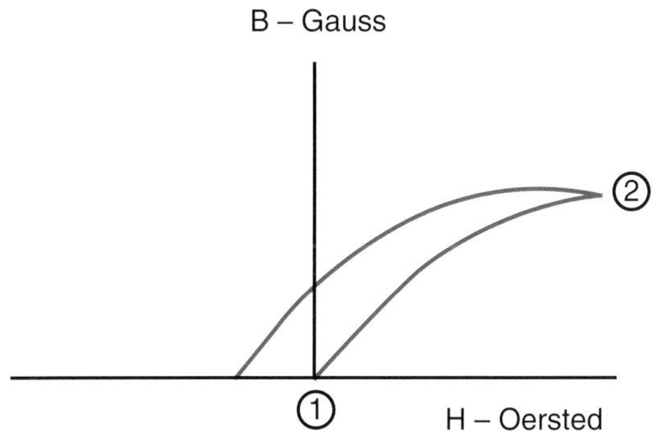

Figure 9–11 The B-H curve is a graph of flux density versus MMF applied.

The H is the magnetizing force in Oersted's, or the MMF represented on the horizontal axis. These units are in the CGS system. Note that the curve starts at the axis 0-0 (point 1) and increases in flux density as the MMF increases. It then begins to level off as the material begins to saturate magnetically (point 2).

Refer to Figure 9–12 for the following explanation of the **hysteresis loop**.

Hysteresis Loop

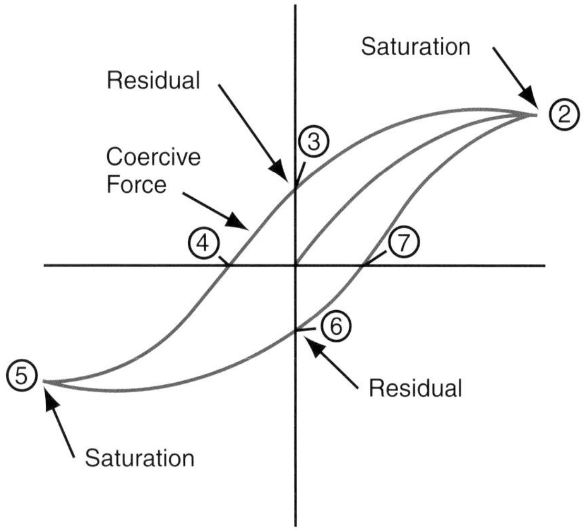

Figure 9–12 The hysteresis loop completes the B-H curve and identifies the magnetic characteristic of the magnetic material.

For a particular metal, you can extend the B-H curve to see the effects on the magnetic material as we decrease and reverse the MMF force. The B-H curve includes the additional results. From point 2, the Gauss decreases again as the MMF is decreased (back to 0) and there is no MMF applied to the coil (point 3). However, not all of the magnetic effect (B) is back to zero. The amount of magnetic effect left is called **residual magnetism**.

Now the current flow in the coil is reversed and the electromagnetic field (MMF) is reversed. It takes some energy to reduce the residual magnetism to zero flux at point 4. The amount of energy to bring the magnet back to neutral is referred to as the **coercive force**. Finally, the magnetic field of the material reverses and the magnet now has a field in the opposite direction compared to the original direction. Again, the magnetic field saturates at point 5, and any further increase in MMF will not change the magnetic influence.

As the MMF is reduced to zero, a small amount of residual magnetism remains at point 6. A coercive is needed to coerce the magnetic field back to zero at point 7. As the magnetic field builds back again in the original direction, the magnetic flux density again increases until saturation at point 2. We now have a B-H curve for a particular core material.

Comparing Figures 9–12 to 9–13, you will note the same basic shape but that Figure 9–13 is much slimmer. This means that the residual magnetism is less, and there is less coercive force needed to return the magnet to neutral. The MMF and the coercive force are both forms of energy. The effect of high hysteresis is heating of the material, as it takes energy to move the domains. The heated iron is caused by a wattage loss and is known as **hysteresis loss**. The second B-H curve is for a low-hysteresis steel. This low-hysteresis steel is a type of steel that is best to use when we want the magnetic field to change quickly with little lost energy.

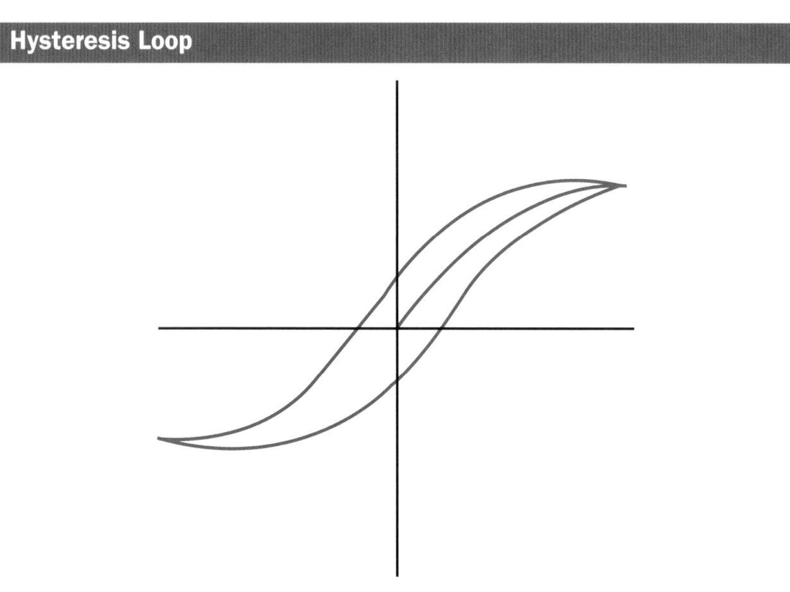

Figure 9–13 The shape of the hysteresis loop or the same MMF can be used to evaluate comparative materials used for magnetic cores.

MAGNETIC POLES OF ELECTROMAGNETS

The magnetic polarity of the electromagnet is determined by the way the individual conductor magnetic fields interact. As you look at the coiled conductor in Figure 9–14, note that the concentric magnetic fields are in close proximity to the neighboring conductor's magnetic field. Because all of the magnetic fields on the left side of the coil in the diagram are moving upward, the effect of the magnetic fields is cumulative, creating a large magnetic field on the left traveling upward.

Note the far right side of the coil. The individual magnetic fields are again accumulating and traveling from bottom to top. This is the external circuit of the magnet. On the inside of each of the coiled conductors, the magnetic fields are accumulating and traveling downward. This is the internal circuit of the magnet. In the external circuit, magnetic flux travels north to south. Internally, lines of force travel south to north. This is exactly the same as you learned in the preceding chapter. Therefore, the North pole of this magnet with the current flow is at the bottom of the magnet and the South pole at the top.

A quicker way to determine the orientation of the North pole of an electromagnet is to use the left-hand rule for a coil (Figure 9–15). To use the left-hand rule, wrap the fingers around the coil, pointing in the direction of electron flow. In this position, the thumb points in the direction of the North pole of the electromagnet. The left-hand rule for a coil is used to quickly determine magnetic polarity based on the electrical polarity and the electron flow principles. To determine current flow on the basis of magnetic polarity, use a compass to determine the magnetic polarity and your left-hand rule for a coil to verify that the current flow direction is correct (Figure 9–16).

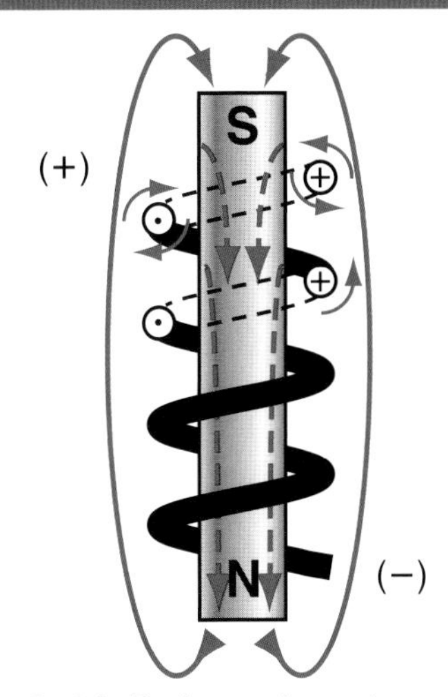

Figure 9–14 Each conductor's magnetic influence is added together to create a large magnetic field.

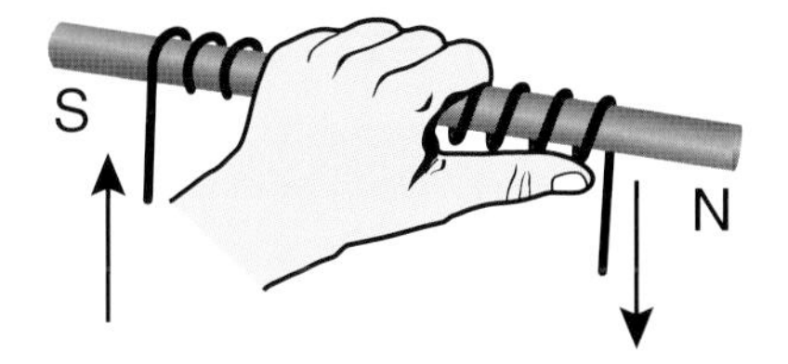

Figure 9–15 In the left-hand rule for a coil, fingers point in the direction of electron flow and the thumb points to the north pole of the magnet.

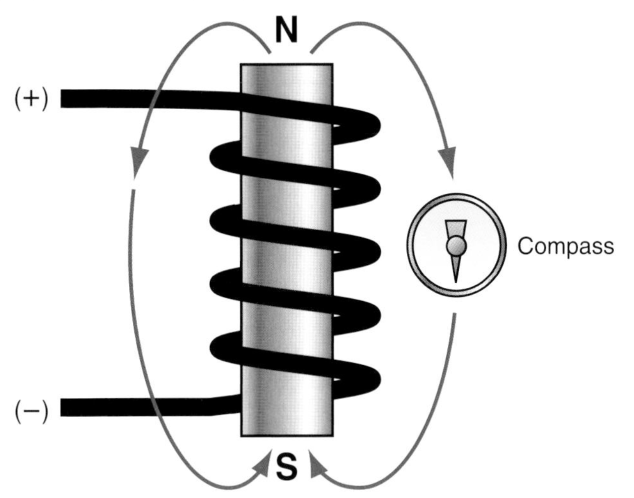

Figure 9–16 Using a compass near an electromagnet will determine direction of current flow.

In 1831, 11 years after it was discovered that a current flow produced a magnetic field, Michael Faraday discovered that a magnetic field created an induced voltage and a resultant current. The process of creating electricity from a magnetic field is called induction.

Inducing Voltage

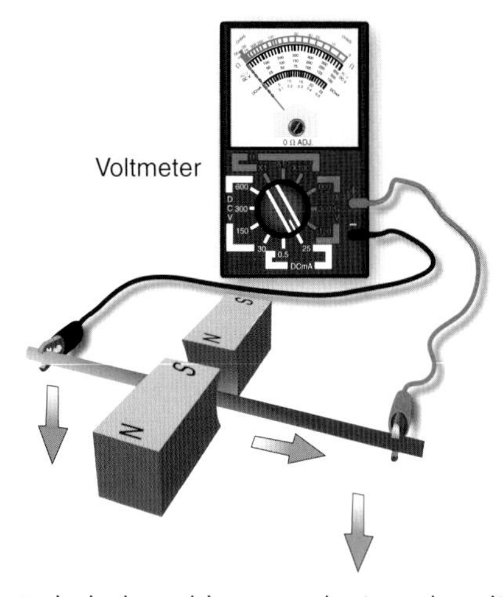

Figure 9–17 Voltage is induced in a conductor when it moves through a magnetic field.

Reversing Induced EMF

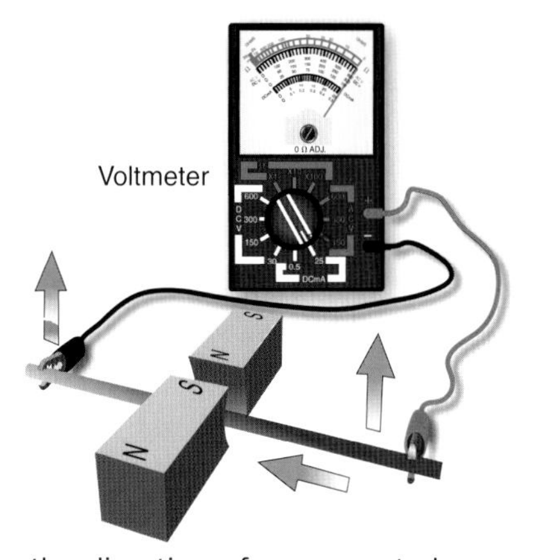

Figure 9–18 When the direction of movement changes, the direction of the induced EMF is reversed.

INDUCING VOLTAGE WITH MAGNETIC FIELDS

All of the previous discussion has centered on magnetism and the magnetic field. A magnetic field is generated around a conductor that is carrying current. This section discusses the reverse principle, which says that a voltage will be induced into a conductor that is passed through a magnetic field. If this conductor is connected to a complete electrical circuit, current will flow. Note that in Figure 9–17 a complete circuit is not shown. There must be a complete circuit for current to flow, even though the voltage is still induced.

To better understand this phenomenon, examine Figure 9–17. The conductor is being moved downward between the poles of the two magnets. The poles of the magnets create a field that has a direction, from North pole to South pole. This downward motion of the conductor causes a voltage to be induced into the conductor, the polarity of which causes the current to flow in the direction shown if a complete circuit exists. The electrons move because of the interactions between the magnetic field of the stationary magnets and the electromagnetic fields of the electrons in the conductor.

The energy to make the electrons in the conductor move is provided by the physical movement of the conductor through the magnetic field. Just placing the conductor in the magnetic field does not add energy, but the relative motion does add the needed energy.

If the physical motion is upward, the magnetic interactions cause the voltage polarity to reverse the flow of current (Figure 9–18). This is because the direction of the field interactions is opposite. The same would be true if the poles of the magnets were reversed and the motion were downward. The field interactions would be reversed, and current would flow in the opposite direction.

A rule that can be used to determine the direction of current flow based on magnetic field direction and physical motion is the **left-hand rule for generators** (Figure 9–19). This rule says that with the thumb, index, and middle fingers of the left hand at right angles to one another and the thumb pointing in the direction of motion, the index finger points in the direction of the magnetic field flux (north to south) and the middle finger points in the direction of current flow.

An important point is that for the current to flow there must be relative motion between the conductor and the magnetic field. Either the conductor must move through the field or the field must move past the conductor. Note also that the conductor and the stationary field flux are perpendicular to each other (Figures 9–17 and 9–18). This is necessary for the conductor to move through the magnetic field or to cut through the lines of flux. If no lines of flux are cut, no voltage will be induced. If the conductor were parallel to the lines of flux and the motion were in the same plane (north to south), the conductor would not cut the flux lines and no voltage would be induced.

Left-Hand Rule

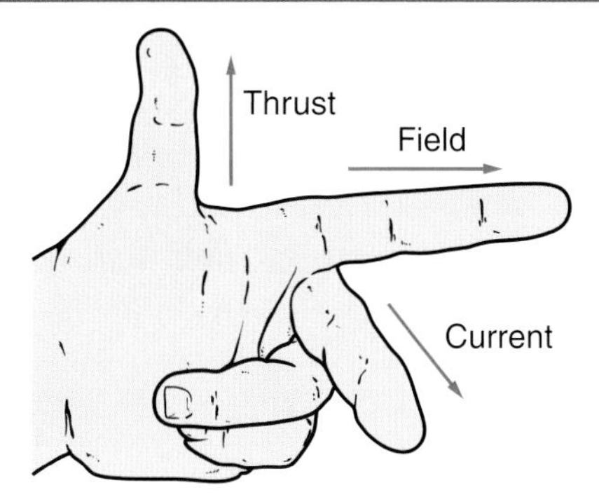

Figure 9–19 The left-hand rule for generators can be used with electron flow theory.

FACTORS AFFECTING VOLTAGE PRODUCTION

The amount of voltage produced depends on different factors. The effects on the electrons in the conductor are dependent on the strength of the magnetic field of the stationary magnetic field. The stronger the magnetic field, the more the interaction. The more displacement of the electrons, the more voltage is produced.

As you see in the diagrams, the current that would flow in a completed circuit flows out of the generator conductors at the negative terminal. As you have noticed in previous chapters, the current leaves the power source at the negative terminal and returns to the power source at the positive terminal. However, the current inside the generation conductor flows from positive to negative. The internal circuit is positive (+) to negative (–) and the external circuit is negative (–) to positive (+). This is the same concept used in the magnetic principles of magnetic directions.

Another factor that determines the amount of voltage—and therefore the current—is the rate at which the conductor cuts through the magnetic lines of force of the stationary magnets. One volt is produced when 10^8 lines of force are cut in 1 second. If we move faster (more imparted energy), the voltage produced is greater. Now the speed of the movement which determines the amount of effort per time is added to validate that we need to add energy to the conductor and magnet generator to create electrical energy. To generate more voltage, you can increase the number of flux lines in the magnetic field (more flux cut per second yields more voltage) or increase the speed of the cutting action.

SUMMARY

When current flows through a straight conductor, a magnetic field is created. This field can be easily detected by passing a compass over a current-carrying conductor and noting how the needle is deflected. When the current is turned off, the conductor's magnetic field collapses. We use this reaction every day, just as electricity is used to create magnetic fields for many applications and magnetism is used to create electricity.

The magnetic field around a conductor is a series of concentric flux lines situated perpendicular to the conductor. They do not form a magnetic polarity with definite North and South poles. The direction in which the field encircles the conductor can be determined by using the left-hand rule for conductors. The left-hand rule states that if you wrap the left hand around the conductor, with the thumb pointing in the direction of the current flow, the fingers will encircle the conductor in the direction of the magnetic field surrounding that conductor (Figure 9–3). Magnetic action is present in all electrical conductors, and thus the placement of conductors is critical when installing conductors in ferrous metal conduits and boxes.

If the conductor is wound in the shape of a coil, a magnetic field (with polarity orientation) will be formed when current flows through it. The polarity of the magnetic field can be determined by using the left-hand rule for coils. If you grasp the coil with your left hand, placing your fingers so that they point in the direction of the current flow, your thumb will point toward the north magnetic pole (Figure 9–15). The strength of the magnetic field of a coil depends on several factors:

- The amount of current
- Number of turns
- Core permeability
- Cross-sectional area of core material
- Length of core and spacing between turns

Although electromagnetic waves are not an integral part of the electrician's daily tasks, a basic understanding of the electromagnetic world includes the understanding of electromagnetic radio waves. Such knowledge may affect how you go about troubleshooting problems with radio-controlled equipment or signals from remote monitors and the like.

As you expand your understanding of equipment used in the electrical industry, you will be able to answer more questions on design problems and suggest solutions. Some of the problems and solutions depend on your understanding of the core materials and how they react magnetically under various conditions. The B-H curve and the hysteresis loop of the materials offer insight into the reasons some equipment is more or less efficient than other equipment, works better or worse than other equipment, or causes more electrical problems as a result of magnetic problems.

Whenever a conductor cuts through a magnetic field, a voltage will be induced into that conductor. If the conductor is part of a complete circuit, current (electrons) will flow. Reversing the direction the conductor travels through the magnetic field will result in change in the direction of the electron (current) flow. The direction of the current can be determined by using the left-hand rule for generators (Figure 9–19).

CHAPTER GLOSSARY

Ampere-turns The strength of an electromagnetic field calculated by multiplying the current flow times the number of turns for an electromagnet used to produce a magnetic field.

Coercive force The amount of reverse MMF that must be applied to a magnetized material to reduce residual magnetism to zero.

Electromagnet An electromagnetic temporary magnet created when current flows through a coiled conductor. The magnet is usually an iron core that is easily magnetized and easily demagnetized when the electric current stops.

Hysteresis The lagging effect of moving the magnetic domains when magnetizing or demagnetizing a material. The difficulty encountered in aligning the domains is due to the hysteresis of the material, and likewise, the difficulty in allowing the domains to go back to random alignment is due to hysteresis.

Hysteresis loop The graph of the flux density versus magneto-motive force as applied to a particular material. The graph forms a loop showing magnetic saturation, residual magnetism, and coercive force as the MMF changes in strength and direction.

Hysteresis loss Losses in the core material of magnetic materials are partly caused by the wattage loss in overcoming hysteresis. The amount of energy lost is related to the movement of domains from one direction to another that causes heat.

Left-hand rule for a conductor Using your left hand, grasp a conductor with your thumb in the direction of the electron flow. Your fingers will wrap around the conductor in the direction of the magnetic field.

Left-hand rule for generators Use your left hand, with thumb, first finger, and center finger at right angles. The thumb indicates the direction of thrust. The first finger indicates the direction of flux (north to south), and the middle finger indicates the direction of current flow.

Residual magnetism The magnetic effect left in a magnetic material after the MMF has been removed. The remaining magnetic effect is due to residual magnetism because the domains do not go back to random arrangement.

REVIEW QUESTIONS

1. Give some examples of electromagnets used regularly.
2. Describe the left-hand rule used to easily determine magnetic field direction when a conductor has current flow.
3. Explain how you could verify the direction of the concentric magnetic field around a conductor.
4. What problems may arise if you place only one supply conductor of a circuit in a metal conduit and the return conductor in another conduit?
5. What type of core is typically used in an electromagnet?
6. Explain what hysteresis is and why it is important.
7. What is the name of the force used to reduce residual magnetism to zero?
8. Name some of the factors that determine the strength of the magnetic field of a coil.
9. Briefly describe how AM and FM radio transmissions work.
10. Explain how a coiled conductor with current flow, on an iron core, produces a north and south magnetic pole.
11. Describe how the left-hand rule for a coil is used.
12. Explain how the amount of voltage can be increased in a generator that moves a conductor through a magnetic field.
13. Describe the left-hand rule for a generator.

10

DC Generators and Motors

TOPICS

- INTRODUCTION
- CALCULATING THE LEVEL OF INDUCED VOLTAGE
- DC GENERATORS
- PARTS OF A GENERATOR
- DC GENERATOR CONNECTIONS
- SELF-EXCITED SHUNT-CONNECTED GENERATOR
- SELF-EXCITED SERIES GENERATOR
- SELF-EXCITED COMPOUND GENERATORS
- DC MOTOR THEORY
- DC SHUNT MOTOR
- SERIES MOTOR
- COMPOUND MOTOR
- PERMANENT MAGNET MOTOR

OVERVIEW

English physicist Michael Faraday discovered the basic theory of electric generation in 1831. He discovered that a magnetic field could be used to produce an electrical current. Practically all commercial power is produced according to the principle of magnetism. Mechanical energy delivered from water, wind, and fossil fuels (such as diesel or gasoline) directly turns an armature in a magnetic field to generate electricity.

Coal, oil, and nuclear energy are used to create heat to boil water, creating steam to turn the steam turbines connected to the rotors of a generator. The topics in this chapter are based on the fundamentals of electrical power generation and are intended to show the relationship between magnetism (which you have previously studied) and the production of electrical power. These are the basic theories of generation and thus do not cover all aspects of the generation and use of power.

OBJECTIVES

After completing this chapter, you will be able to:

- Explain the basic principles involved in generating DC power
- Describe the basic construction and operation of simple DC generators
- Demonstrate the left-hand rule for generators to show relative motion, magnetic flux, and electron flow
- Describe the variables that determine the level of voltage generated by a generator
- Identify the components of a DC machine
- List the advantages of different connections for DC generators
- Explain how a DC motor turns and determine direction of motion
- List the advantages of different DC motor connections
- Apply the formulas for voltage regulation and speed regulation
- Explain the differences between self-excited and separately excited generators

INTRODUCTION

GENERATION OF ELECTRICITY THROUGH INDUCTION

Earlier you studied the concept of **electromagnetic induction**. You learned that a voltage is induced into a conductor that is passed through a magnetic field. If the conductor is connected to a complete electrical circuit, or closed loop, then current will flow. This concept is used whether you are generating DC or AC power through electromagnetic induction.

Producing EMF

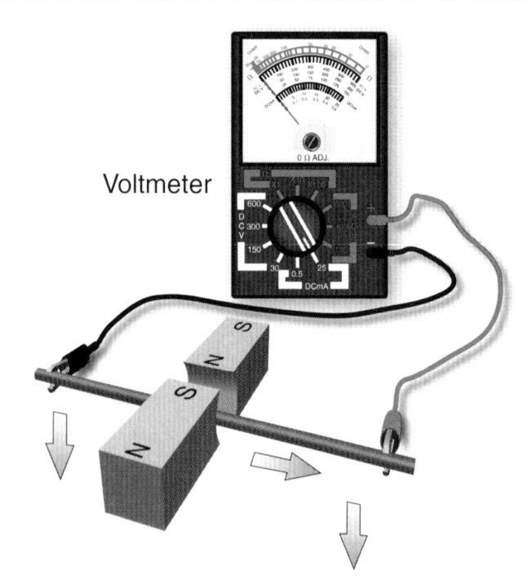

Figure 10–1 A conductor in motion, cutting through a magnetic field, produces an EMF.

Reversing EMF Polarity

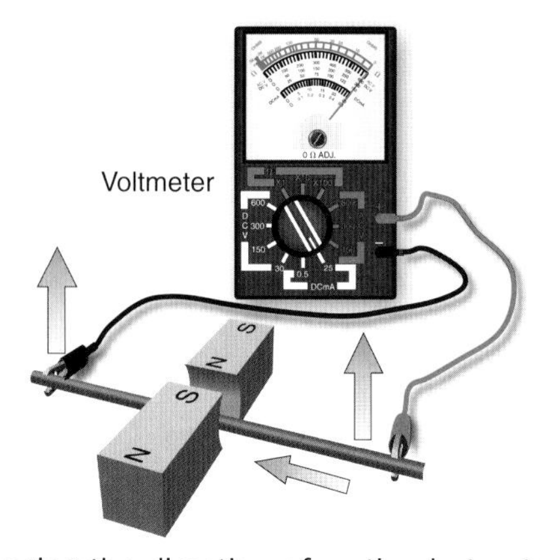

Figure 10–2 Changing the direction of motion but not the field will reverse the EMF polarity.

Figure 10–1 shows a conductor being moved downward between the poles of the two magnets. The poles of the magnets create a field that has a direction, from north to south pole. The downward motion of the conductor causes a voltage to be induced into the conductor. The polarity of the induced voltage would cause the current to flow in the direction shown. The electrons move because of the interactions between the magnetic field of the permanent magnets and the electromagnetic fields of the electrons themselves. Note that without movement there is no induced voltage or current flow.

If the motion is upward, the magnetic interactions cause the voltage polarity to reverse the flow of current (Figure 10–2). This is because the direction of the field interactions is opposite. The same would be true if the poles of the magnets were reversed and the motion were downward. The field interactions would be reversed, and current would flow in the opposite direction.

One of the key factors to remember is that relative motion between the magnetic field and the conductor must exist before a voltage will be generated. This motion can be the motion of the magnetic field instead of the conductor. That is, the magnets can move instead of the conductor. In DC generation, the conductor is moved through the magnetic field.

The left-hand rule for generators (Figure 10–3) indicates that with the thumb, index, and middle fingers of the left hand at right angles to one another and the thumb pointing in the direction of thrust or motion (and first or index finger pointing in the direction of the magnetic field flux), the center finger points in the direction of current flow.

This is used for determining direction of current by using the electron flow theory. When following the conventional flow theory, the right-hand rule for generators is used.

This point is mentioned only to avoid confusion when discussing theory with persons who are using conventional flow theory.

CALCULATING THE LEVEL OF INDUCED VOLTAGE

The magnitude of the voltage being produced by electromagnetic induction is determined by how many lines of magnetic flux are being cut by the conductor per second. For the simple system such as that shown in Figure 10–1, three main factors determine this magnitude: (1) the strength of the magnetic field, (2) the speed of relative motion cutting the magnetic field, and (3) the number of loops in the conductor. To produce 1 V, the conductor must cut through 100,000,000 (10^8) lines of flux in 1 second. Field strength equal to 100,000,000 lines is called a weber (Wb).

If a conductor moved through twice the number of flux lines (2 Wb) in 1 second, 2 V would be produced. If the same wire cut through 1 Wb in half the time, 2 V would be produced. The formula:

$$V = \left(\frac{Wb}{\text{sec}}\right) \times \text{Number of conductors}$$

represents the amount of voltage based on field strength and speed of cutting for one conductor. If the number of loops in the wire is increased, it effectively increases the number of flux lines being cut per second so that a larger voltage is produced (Figure 10–4).

This creates a higher amplitude of the waveform and consequently a higher voltage. The waveform you see is called a full-wave DC waveform. To calculate the value of the DC, you would measure with a DC voltmeter and calculate **voltage peak × 0.636.** This factor is applied to the sinusoidal wave produced by the rotating conductors and compensates for the peak and zero points to determine an average DC voltage.

Example

In the full-wave DC waveform shown in Figure 10–5, if the peak amplitude of the wave is 100 V, what is the average DC value read by a DC voltmeter?

100 V Peak × 0.636 = 63.6 V

Left-Hand Rule

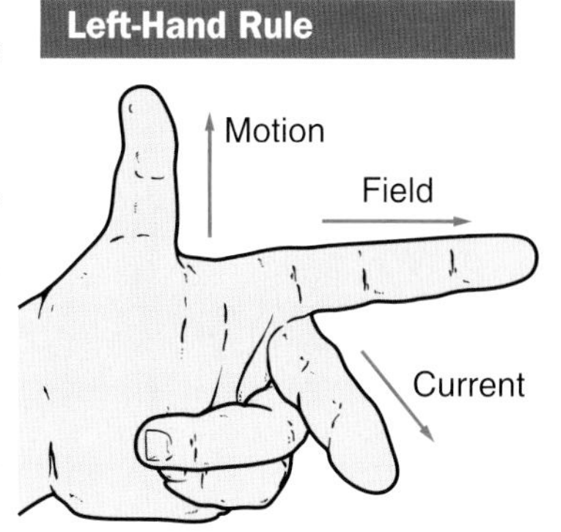

Figure 10–3 The left-hand rule for generators is used with electron flow theory to indicate current flow direction.

Increasing Output Voltage

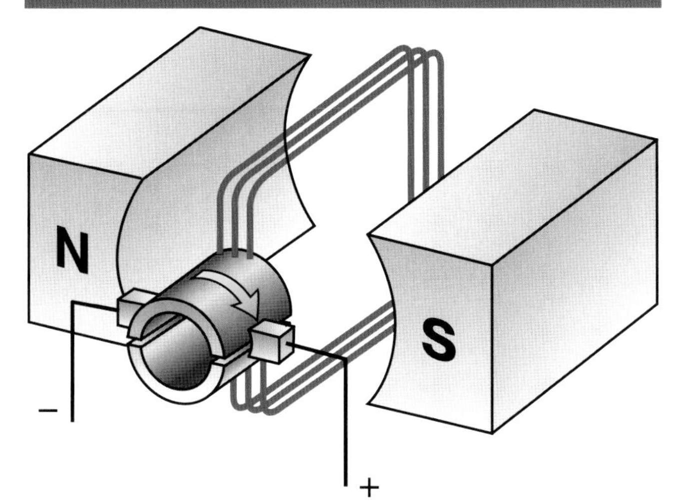

Figure 10–4 Increasing the number of conductors that cut the field will increase the output voltage.

Full-Wave DC Waveform

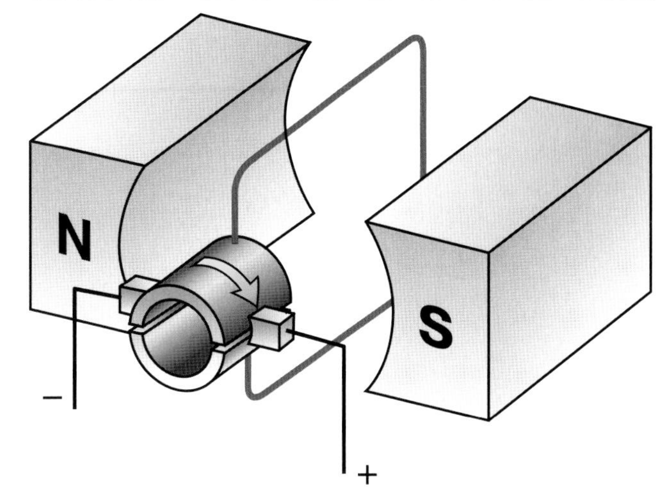

Figure 10–5 The output of the generator with a commutator is a full-wave DC waveform.

Simple AC Generator

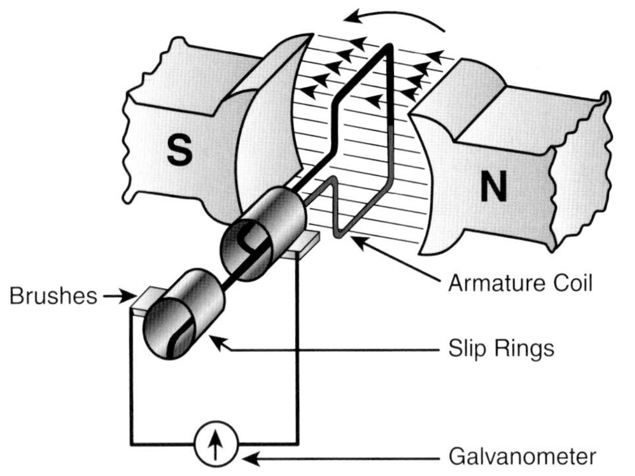

S
N
Brushes
Armature Coil
Slip Rings
Galvanometer

Figure 10–6 A simple AC generator rotates through a magnetic field, but the output is connected to slip rings.

AC Waveform

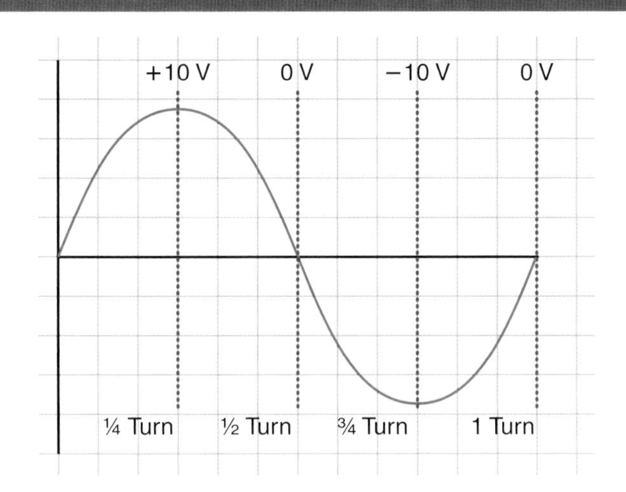

Figure 10–7 The output of the generator without a commutator is an AC waveform.

AC versus DC

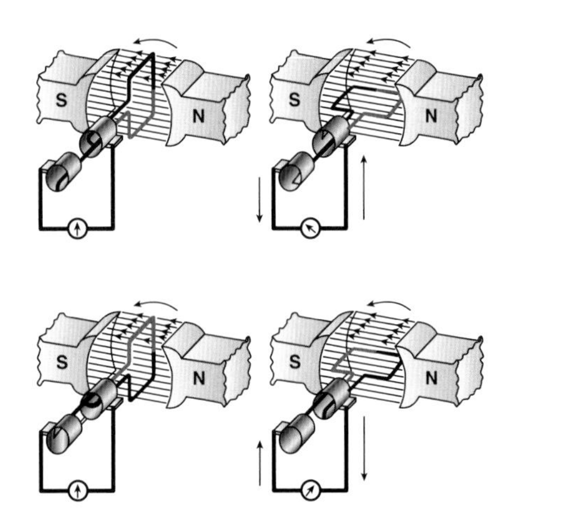

Figure 10–8 AC is taken from a generator through slip rings; DC is taken through a commutator.

Increasing Average DC Voltage

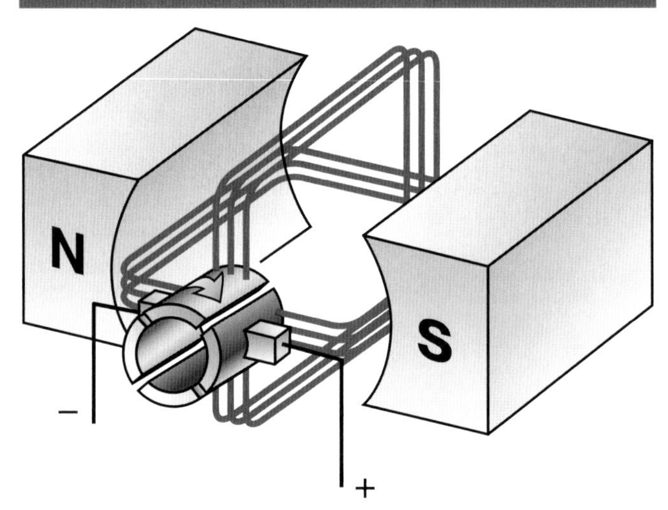

Figure 10–9 With the addition of coils, the waveform begins to fill in and the average DC voltage increases.

As the loop wire shown in Figure 10–6 rotates through one complete turn, it will generate voltage. Figure 10–7 shows the waveform of the voltage being generated. Note that the darker portion of the spinning coil actually has a current flow that changes direction based on which magnetic pole it is moving across. The current flow in a particular conductor is an alternating current. If we had the coil ends connected to slip rings, as in Figure 10–8, the output of the generator would be an AC voltage and current. However, we are concerned with DC voltage and current in this chapter.

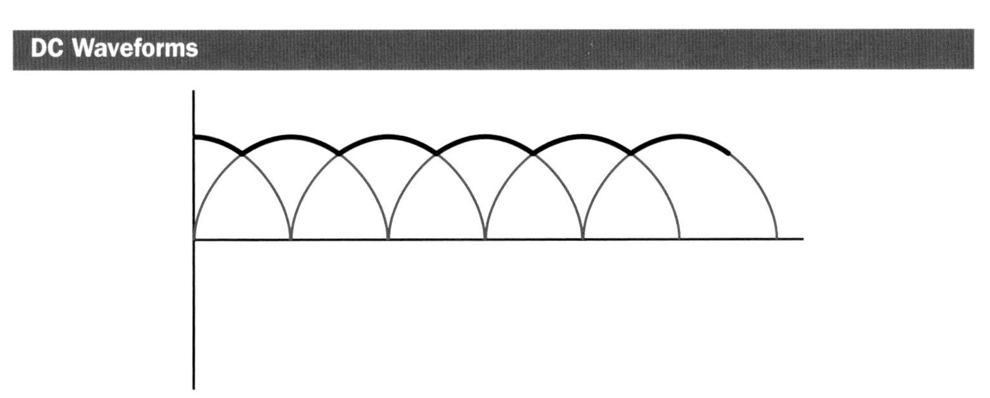

Figure 10–10 The output waveforms of a DC generator with multiple coils are shown above.

If you wanted to create a higher value of DC without increasing the number of loops in each coil, you would add coils (Figure 10–9). If you add coils so that there are always conductors cutting magnetic flux, some induced voltage will always be connected to an output brush and the output voltage waveform will begin to fill in (Figure 10–10).

The waveform is a smoother DC, and the DC output voltage average is higher than the original 0.636 × peak. Note in the figure that the commutator has been cut into four pieces so that the coil ends can be connected to the appropriate brush. The four-segment connection to the moving coils is an example of how separating the commutator segments is used to increase the DC output and create smoother DC waveforms from a DC generator.

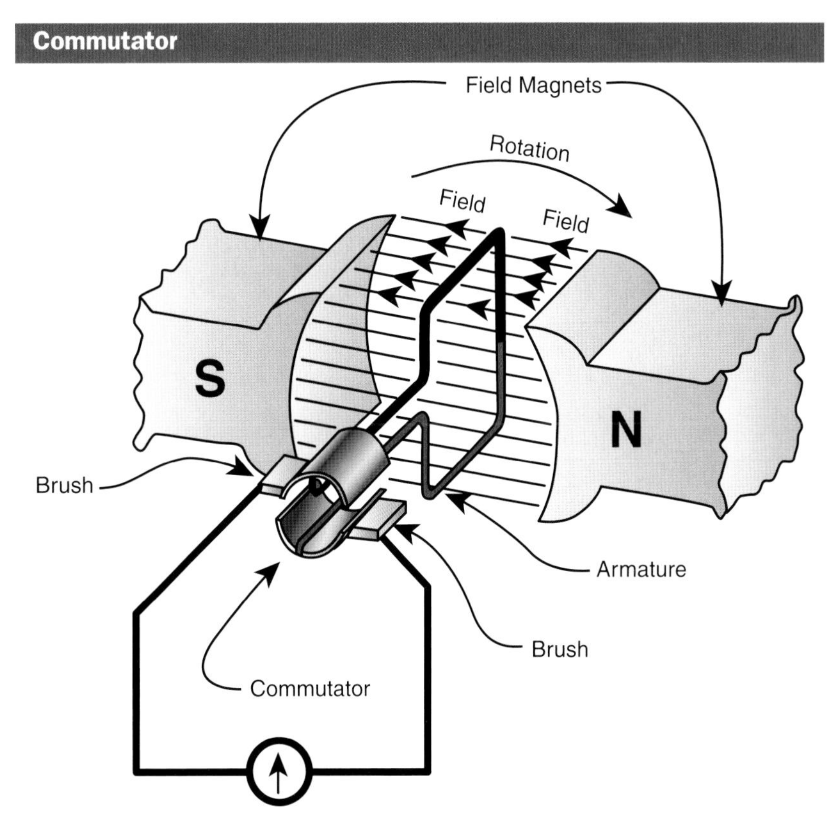

Figure 10–11 The commutator is segmented so that each end of the coil is connected to a segment.

DC GENERATORS

DC generators are similar to AC generators except that the voltage is removed from the armature by a device called a commutator, instead of by slip rings. A commutator can be as few as two segments and as many as are required to make connections to multiple coils (Figure 10–11). Each commutator segment is attached to the end of one side of each loop in the generator.

DC generators have largely been replaced by AC generators for the following reasons:

- DC generators are less efficient than AC generators (or alternators).
- Output produced by DC generators is cost-prohibitive to move any distance at low voltage and cannot efficiently be converted from high voltage to low voltage. DC cannot be converted to AC without suffering significant energy loss.
- DC generators require more frequent servicing than AC generators.

Figure 10–12 shows the output voltage at the brushes. For the first half a rotation, the output of the DC generator is a sine waveform with a zero/peak/back-to-zero wave. However, when the armature loop reaches the 180-degree point in its rotation, the commutator segments connect to a different connection brush so that the output to each brush maintains the same electrical polarity.

The effect is that a particular brush (such as the darker brush) is always electrically connected to the armature wire that is traveling across the same magnetic pole. That brush will always have the same direction of current flow through it and the same voltage polarity. The resulting waveform is not AC but a varying DC voltage. This voltage waveform always creates a current in the same direction in the external circuit but varies in value as it goes from zero to maximum and back to zero.

PARTS OF A GENERATOR

The names of the parts of generators and motors are among the most often misunderstood of all electrical terms. Understanding these terms will help you as you progress in your understanding of DC machines. There are many types of DC generators and motors, known as DC machines, but all have similar components.

STATOR AND ROTOR

The stator, or stationary frame, makes up all nonrotating (stationary) electrical parts of a generator or motor, and the rotor makes up all rotating parts. This is true whether the machine is a DC generator or motor.

FIELD AND ARMATURE

The **field** of a machine is the winding placed on the stator that creates the magnetic field. The current in the field does not alternate but is a DC

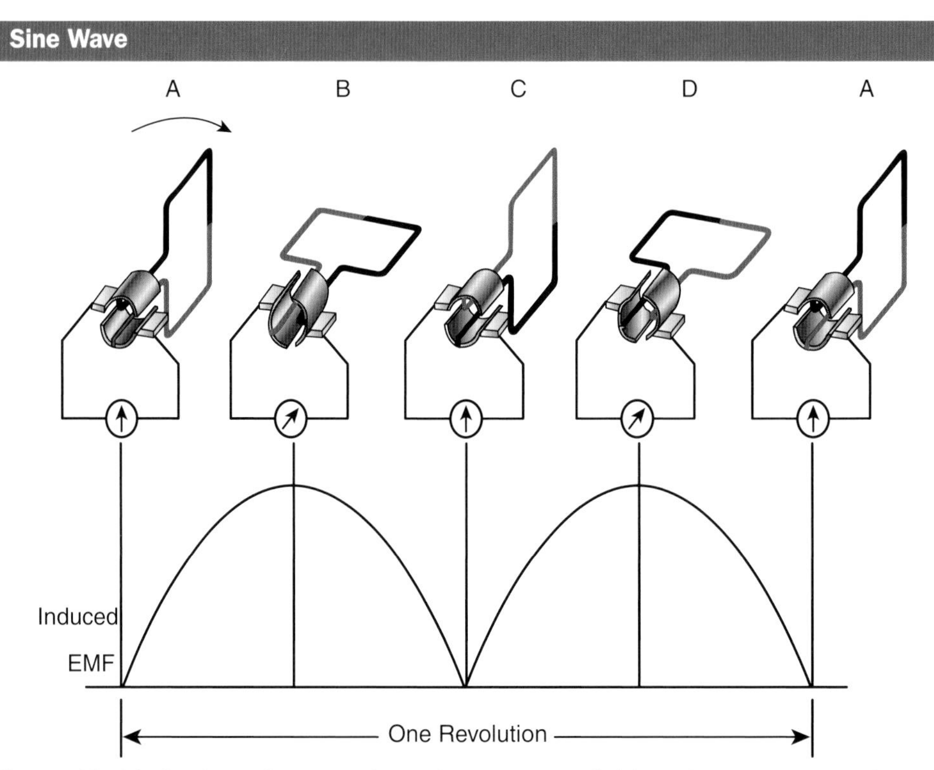

Figure 10–12 As the coil moves through a magnetic field, a sine wave output is developed.

current that creates a specific magnetic field strength and direction. In Figures 10–6 and 10–11, for example, the field is the permanent magnet and its associated lines of magnetic flux. The **armature** winding is that which generates or has an induced voltage created within it. Usually, the terms *armature* and *field* are applied to DC motors and DC generators.

DC MOTORS AND GENERATORS

In DC motors and generators, the armature is the rotor, and the field is held on the stator. Because the armature is always the rotor on DC machines, many electricians and engineers mistakenly believe that the armature is the rotor on all motors and generators. This is not true, because the armature is often the stator on AC generators. This is explained in more detail in *AC Theory*.

COMMUTATOR

In DC machines, a commutator is used instead of slip rings to "collect" the output voltage from the armature. The commutator has segments, with one pair of segments for each coil in the armature. Each of the commutator segments is insulated from the others by an insulator material, such as mica. As discussed previously, the commutator converts the AC generated in the armature winding to a DC output from the generator (Figure 10–12).

Sometimes this is referred to as a mechanical rectifier. In a generator, the commutator segments reconnect the loop connections to the brushes every half cycle to maintain constant polarity of voltage. Earlier, you saw a two-segment commutator. Figure 10–13 shows a multisegment commutator that is typical of those used in commercial machines.

BRUSHES

Brushes are usually made of carbon and graphite and are spring loaded so that the brush is held firmly against the spinning rings on the rotor or commutator. Brush leads are used to connect the armature to the external circuit. Figure 10–14 shows a single-brush assembly.

Commutator for DC Generator

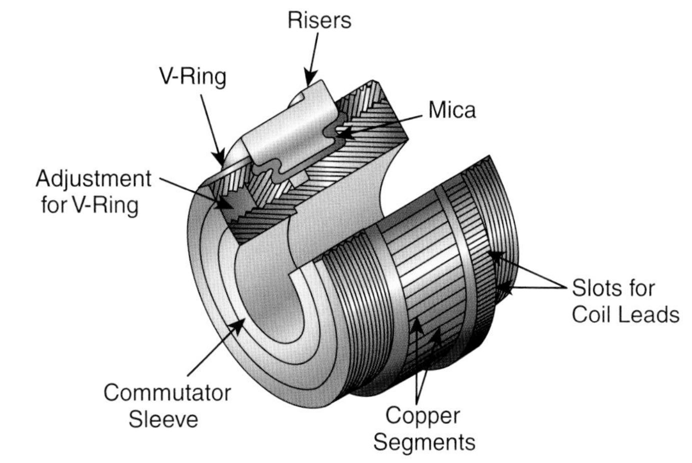

Figure 10–13 The construction of a commutator with many coil connections, typically used for DC generators, is shown.

Carbon Brush

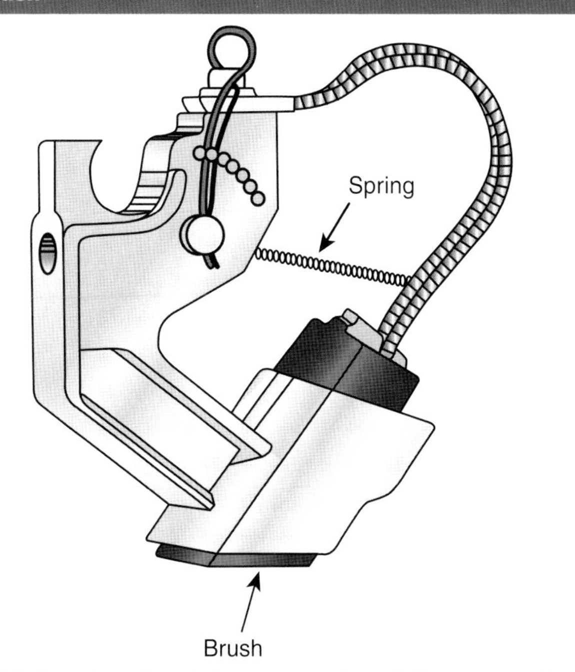

Figure 10–14 A carbon brush is used as the sliding connection to draw-connect the external electrical circuit to the armature output voltage.

POLE PIECES

The magnetic field in most DC generators and motors is created with a DC electric current rather than the permanent magnets discussed previously. If the field assembly is a permanent magnet, it will be made of a material (such as steel) that retains its magnetic field. Electromagnetic field poles are made of laminated soft iron.

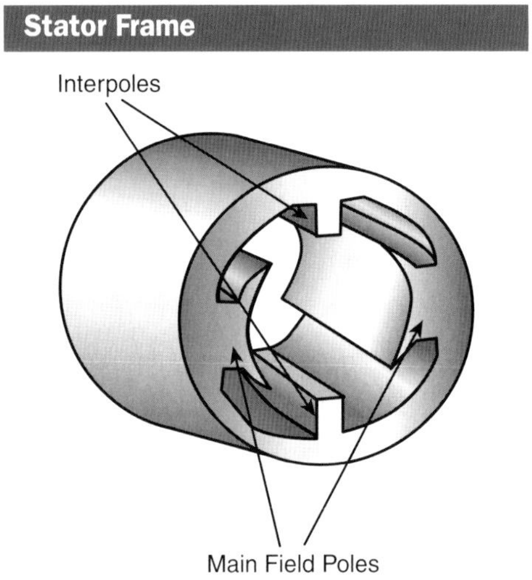

Pole pieces are found inside the generator's stator housing. Figure 10–15 shows the frame of a DC motor with the two main pole cores and two smaller ones called interpoles. Figure 10–16 shows a pole piece with the wiring. Interpoles are explained in more detail in references dealing with generators and motors.

Stator Frame

Interpoles

Main Field Poles

Figure 10–15 The frame of the stator holds the field windings for the DC generator.

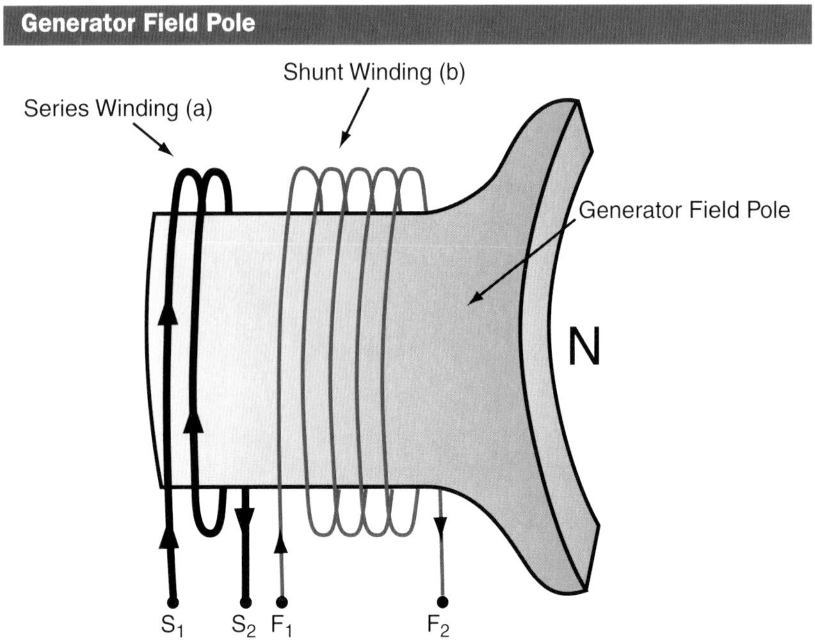

Generator Field Pole

Series Winding (a)

Shunt Winding (b)

Generator Field Pole

N

S_1 S_2 F_1 F_2

Figure 10–16 The field windings are wound on the stator field pole pieces. Shunt and series fields are each wound separately.

DC GENERATOR CONNECTIONS

Generator types vary as the needs for the output power vary. DC generators are becoming less common because the electronic methods of producing DC have greatly improved over the years. However, DC machines still provide power for various specialized applications. DC generators are connected in different configurations to give the desired output voltages and currents to satisfy various needs. The generators can be self-excited, meaning that the current for creating the magnetic field comes from the generator itself.

The generators can be separately excited, meaning that the field current is provided by a separate source. The other methods of connection refer to the way the field is connected to the generator when self-excitation is used. The connections are related to the pattern of connection of the various magnetic field coils. The fields can be connected as shunted (parallel to the armature) or as series, or both fields can be connected to compound the effects of the two types of field coils. This is explained in detail as you learn each style of generation.

SEPARATELY EXCITED GENERATORS

Separately excited generators are the least common. The magnetic field coil is wound on the stator of the generator (Figure 10–17). As previously discussed, this coil produces the magnetic flux field the armature conductors cut through. This coil and armature rela-

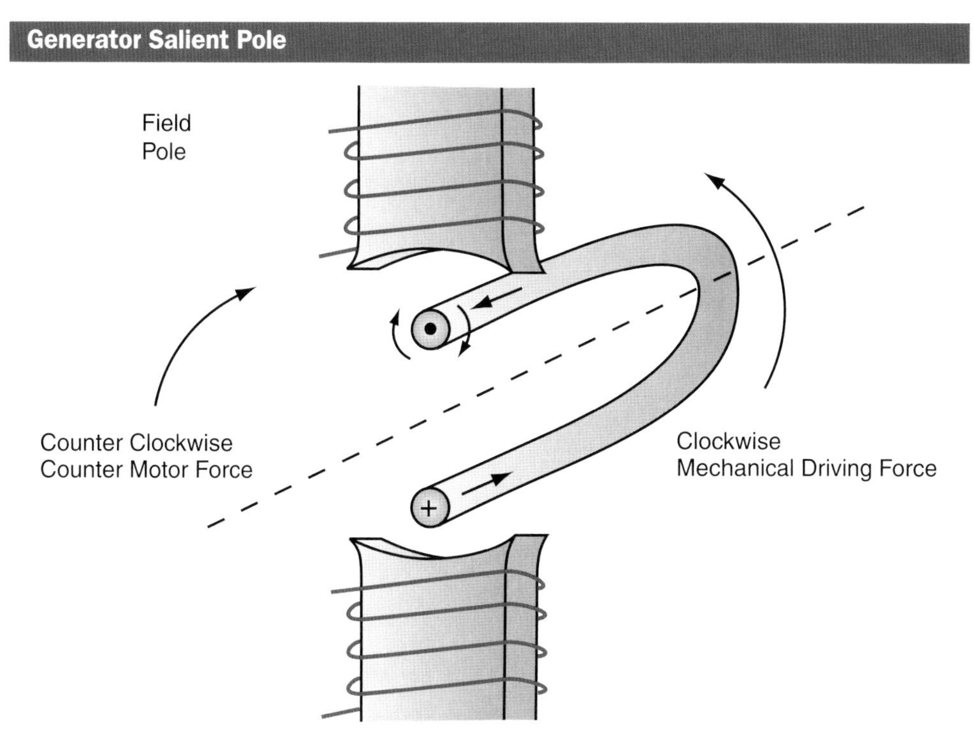

Figure 10–17 The generator magnetic field coil is wound on the generator salient pole piece.

tionship is depicted schematically in Figure 10–18. The field winding F_1–F_2 is supplied by a DC source of power.

The level of field current is controlled by a variable resistance known as a field rheostat. By controlling the amount of field current, the amount of field flux is controlled, and therefore the amount of output voltage can be adjusted. The more flux, the more output voltage—within limits. Normally, more field current yields more magnetic flux. As you learned in magnetic basics, this occurs until the iron core that is holding the coil becomes magnetically saturated.

The armature is shown as a circle in Figure 10–18, denoting the rotating cylinder where the armature conductors are installed. The armature is labeled A_1–A_2. The brushes riding on the commutator are the output points of the generator and are shown connected directly to the load.

In the previous induction theory information, you learned that the speed of cutting action (density of the flux) and the number of turns in the coil determine the level of induced voltage. The size of the armature conductor and the brush capacity will determine the current output the generator can supply. Too much current load will cause the conductors and the brushes to overheat and become damaged.

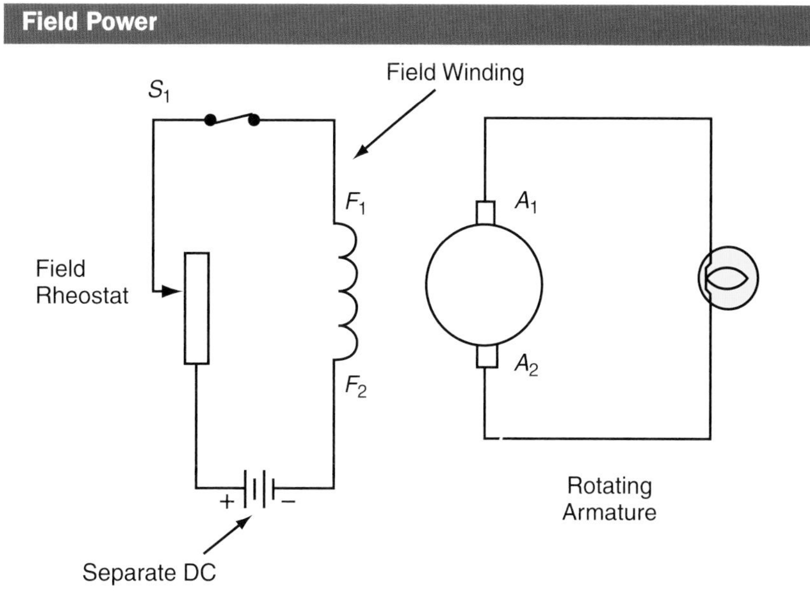

Figure 10–18 A separately excited DC generator uses an external source of DC for the field power.

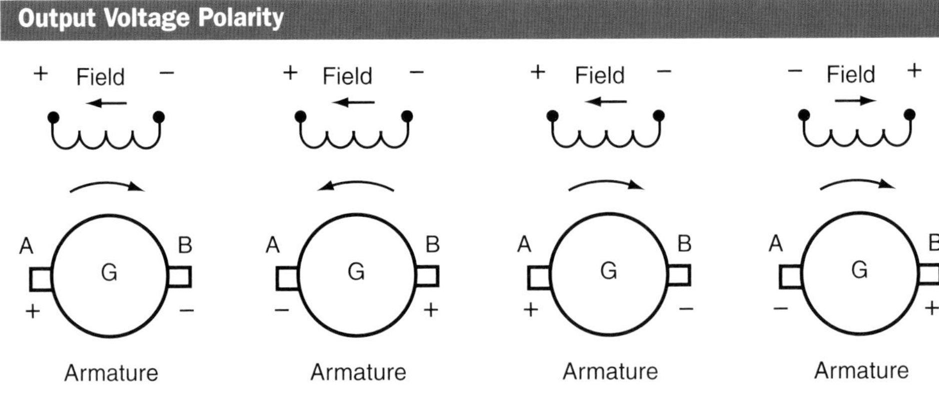

Figure 10–19 Magnetic field direction and direction of rotation determine output voltage polarity (G = generator).

The direction of the DC current or the electrical polarity will be determined by the magnetic field polarity and the direction of rotation. If the magnetic polarity of the field is constant but the armature direction of motion is reversed, the polarity of the DC output is reversed (Figure 10–19). If the direction of rotation is the same but the magnetic field is reversed (Figure 10–19), the electrical output polarity is reversed.

The separately excited generator is not as common because it requires a separate source of DC to excite the magnetic field. This separate source requirement is a disadvantage. The advantage of this type of generation is that the field can be maintained at a set value and the output voltage remains fairly constant. This is not always the case with self-excited machines.

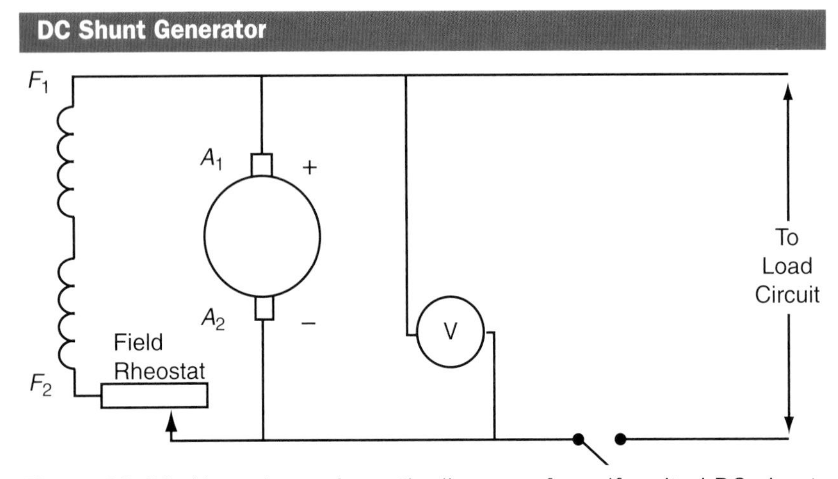

Figure 10–20 Above is a schematic diagram of a self-excited DC shunt generator.

SELF-EXCITED SHUNT-CONNECTED GENERATOR

The self-excited generator connected as a shunt generator is shown in Figure 10–20. As you can see, the magnetic field winding F_1–F_2 is shown connected to the armature brush connections of A_1–A_2. There is a field rheostat (variable resistor) to control the amount of field current, just as there was in the separately excited generator. The field is in parallel with, or shunted across, the armature. The output of the generator is also connected to the armature connection. As you can determine, the output voltage will be the same as the voltage supplied to the field.

The shunt field winding is a coil made up of relatively small wire and many turns. This gives the winding a high resistance, and therefore it does not have a large current. Most of the armature current goes to the load, but a portion of the current is "shunted" to the field to create a magnetic field.

See Figure 10–16 for an idea of how a coil is wound onto a pole piece of the generator's stator. The question you may ask is: How does the field produce flux to generate the voltage in the spinning armature that is needed to create a self-excited shunt field current to get the process started?

The answer is again in the basics of magnetism. The iron of the stator does not lose all of its magnetic effect even after the magnetizing effect of the current carrying coil is removed. There is a small amount of residual magnetism left in the iron from the last time the field was used. As the armature first begins to spin, the armature conductors spin through the residual magnetism and produce a residual voltage.

This is a very small voltage to start with. As the residual voltage appears at the brushes of the armature, the effects are felt on the shunt field, strengthening the magnetic field. The voltage starts to build as the armature spins through a stronger field. The voltage increases until the field rheostat limits the field current and the output voltage stabilizes at the desired value.

If the speed stays constant and the load stays constant (current output is steady), the output voltage will remain constant. If, however, the load current demand changes, the output voltage will fluctuate. This is the result of several factors. As the current load increases, the motor effect of a generator causes the rotor to slow down. Think of the rotor doing more work to produce more output power and therefore slowing slightly.

As the armature slows, it cuts fewer lines of force per second and the output voltage drops. This drop in output voltage is felt by the shunt field and it produces less flux, which also reduces the output voltage. The effect is that the voltage output drops as load (current) increases. This affects the voltage regulation of the machine.

Voltage regulation is calculated by this formula:

$$\% \text{ Regulation} = \left[\frac{(E\,@\,NL - E\,@\,FL)}{E\,@\,FL}\right] \times 100$$

NL stands for No Load and *FL* stands for Full Load. We compare the voltage at no load to the designed full-load voltage to see how well the generator can hold the voltage constant. A small percentage of regulation is good, or the generator has "good" voltage regulation. A graph of the voltage output would look like that shown in Figure 10–21.

Example

What is the percentage voltage regulation of a DC shunt generator that has an output voltage of 100 V with no electrical load connected and of 90 V with the full-current load connected?

$$\% \text{ Regulation} = \left[\frac{(E\,@\,NL - E\,@\,FL)}{E\,@\,FL}\right] \times 100$$

$$\% \text{ Regulation} = \left[\frac{(100\text{ V} - 90\text{ V})}{90\text{ V}}\right] \times 100$$

$$\% \text{ Regulation} = \left[\frac{10}{90}\right] \times 100$$

$$\% \text{ Regulation} = 11.1\%$$

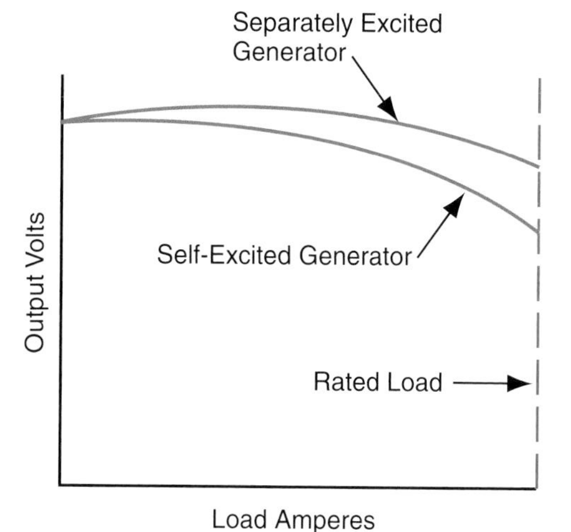

Figure 10–21 Output voltage is compared with load current curves for self- and separately excited shunt generators.

SELF-EXCITED SERIES GENERATOR

A series generator has a magnetic field connected in series with the load. A DC series generator must be self-excited because the field is connected in series with the armature. The schematic view of this connection is shown in Figure 10–22. The series field is constructed of coils of relatively large wire and fewer turns than are used in a shunt field winding.

The pictorial view is represented by Figure 10–23. With this type of connection, there must be a load connection made to allow current to flow

Series DC Generator

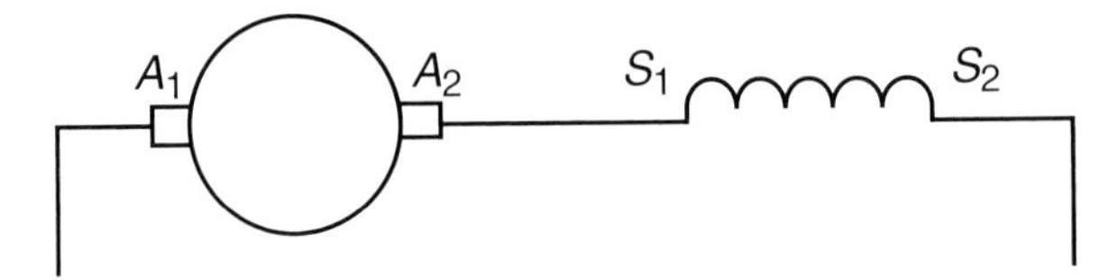

Figure 10–22 This schematic shows a series DC generator.

Series Field Connection

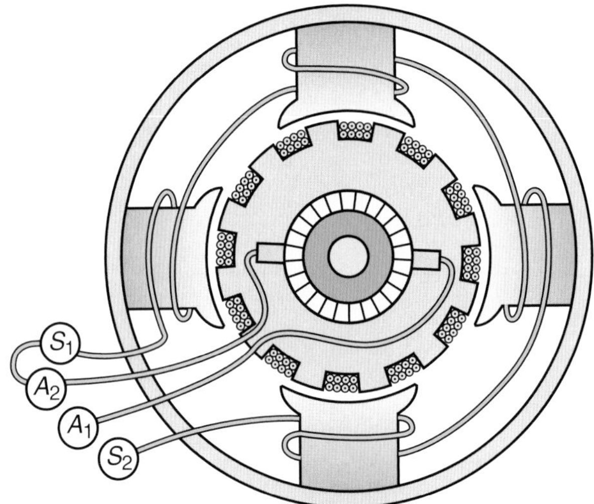

Figure 10–23 This pictorial view illustrates how the series field is connected to the armature.

Diverter Variable Resistor

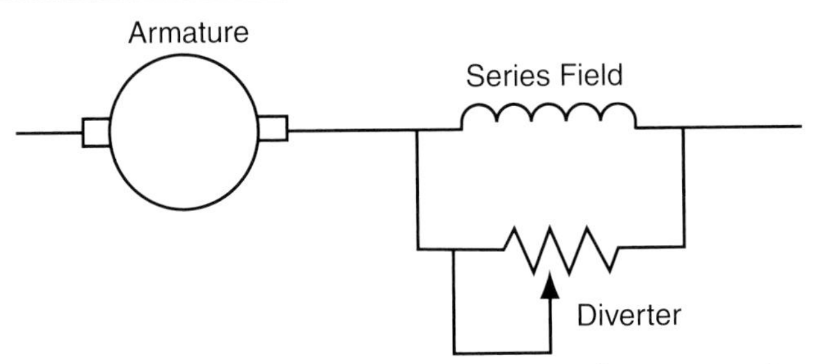

Figure 10–24 A "diverter" variable resistor is used to divert current around the series field to control the flux of the series field.

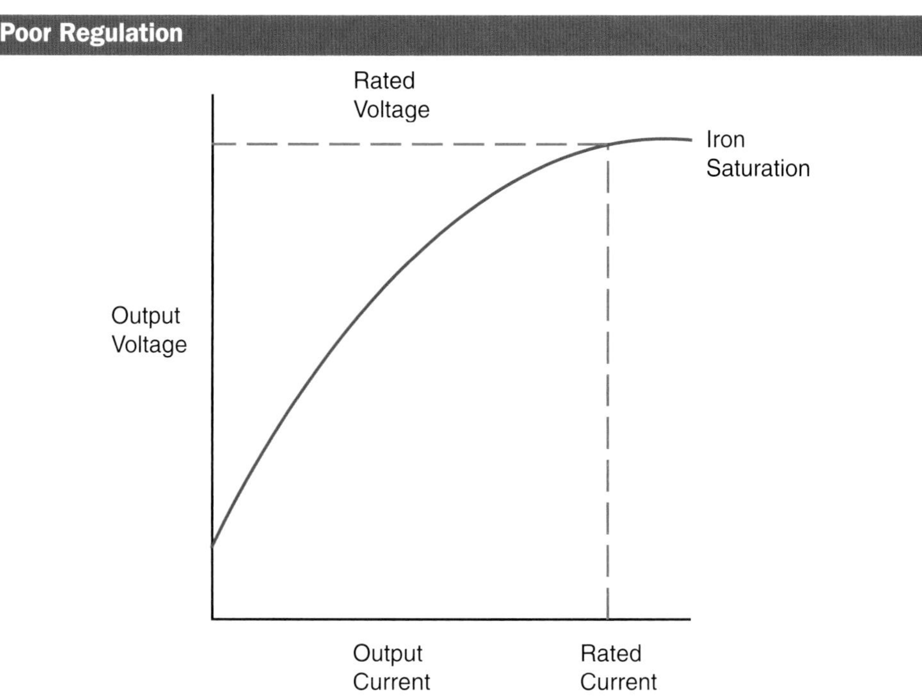

Figure 10–25 The voltage characteristic of a series generator shows very poor regulation.

through the series field. After a small residual voltage is developed from the residual magnetism, the remaining increase must come from a stronger series field. As the load current flows, it builds the magnetic field flux in the stator (labeled S_1-S_2) and the voltage builds up. Current can be diverted around the series field to control the actual output voltage connected from A_1 to S_2.

The diverter resistor is shown in Figure 10–24. It diverts current around the series field without affecting the current to the load. As you can surmise, the output voltage will vary greatly as the load current changes. Every time the load has more current flow, the magnetic field of the stator increases, and therefore the output voltage increases. The regulation of the generator voltage is very poor, as is indicated by the graph of the output versus current shown in Figure 10–25. This generator is not used very often because of the extreme changes in voltage as current changes from no load (NL) to full load (FL).

Example

What is the voltage regulation of a series-connected DC generator with a no-load voltage of 50 V and a full-load voltage of 100 V?

$$\% \text{ Regulation} = \left[\frac{(E @ NL - E @ FL)}{E @ FL}\right] \times 100$$

$$\% \text{ Regulation} = \left[\frac{(50 \text{ V} - 100 \text{ V})}{100 \text{ V}}\right] \times 100$$

$$\% \text{ Regulation} = \left[\frac{50}{100}\right] \times 100$$

$$\% \text{ Regulation} = 50\%$$

The automobile alternator is an excellent illustration of a technique for generating a highly regulated DC output voltage. The alternator, a three-phase AC generator, delivers output to a circuit mounted on the alternator body. This circuit, called a rectifier, converts AC to DC. A portion of the DC, through a small electronic regulator circuit, provides exactly the right amount of field-winding current to produce the specified DC output voltage. Typical output voltage ranges from 13.8 to 14.25 volts, regardless of engine speed.

SELF-EXCITED COMPOUND GENERATORS

The compound generator is a combination of the shunt and the series generators. The schematic diagram (Figure 10–26) shows a short-shunt connection compared to an alternate connection called the long-shunt connection (Figure 10–27). The short-shunt connection is the more common connection for DC compound generators. The term *short-shunt* refers to the method of connecting the shunt field to the armature.

The shunt field is essentially connected as a short across the A_1–A_2 brush connection, along with the field

Short-Shunt Compound DC Generator

Field Rheostat

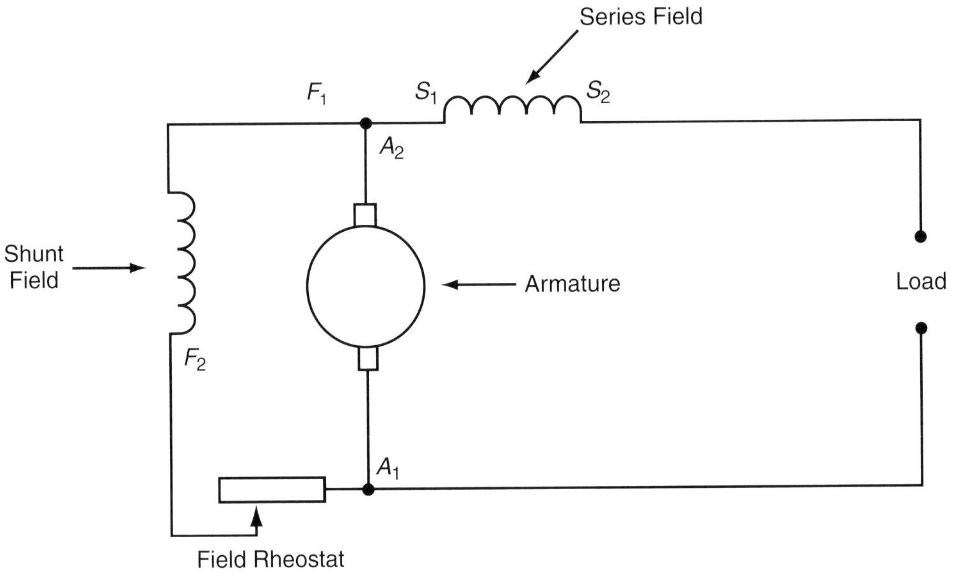

Figure 10–26 This schematic diagram depicts a short-shunt compound DC generator.

Long-Shunt Compound DC Generator

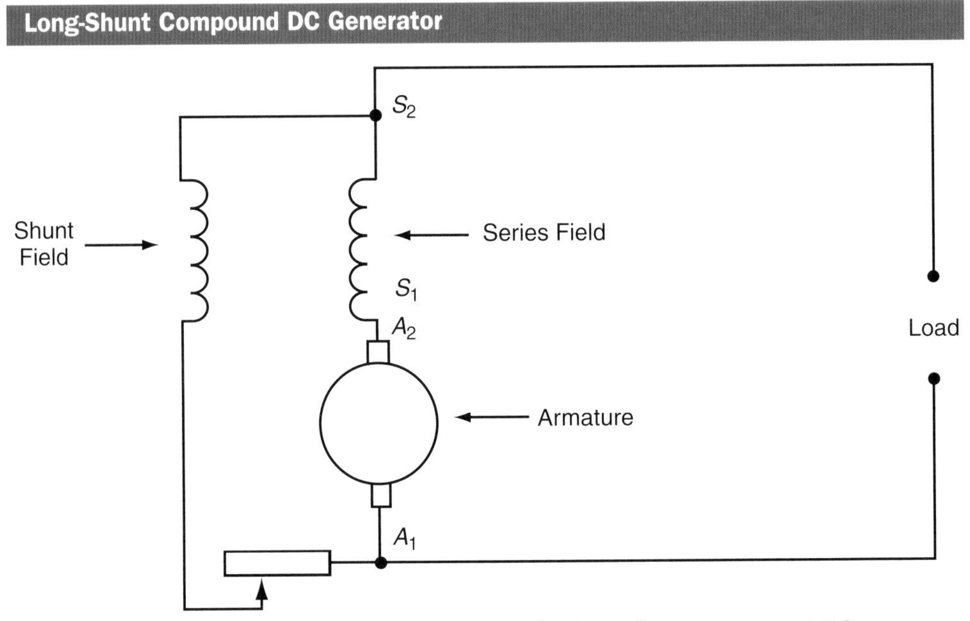

Figure 10–27 Above is a schematic diagram of a long-shunt compound DC generator.

rheostat. The series field is still connected in series with the load. Figure 10–16 shows how the two field windings might be wound on the stator pole pieces. Again, the shunt field has many turns of small wire, and the series field has fewer turns of larger wire.

The principle of generation is the same as described in the shunt or series generator. However, the voltage effect is a combination of each of the other generators. If the two fields aid each other and produce flux in the same direction, the fields are said to be cumulatively compounded. If they oppose each other, the fields are differentially compounded. When the rotor has mechanical energy applied and begins to spin, the residual magnetism allows a small residual voltage to be developed.

This residual voltage is applied to the shunt field and the shunt field winding produces flux. The voltage can build up even with no electrical load applied. The current to the field is controlled to produce a certain output at no electrical load. Remember that with just a shunt field the voltage tends to drop as electrical load is added, as discussed previously. To combat this "drooping" effect, the series field can add flux to the field based on the amount of load.

With a higher load current, the series field is able to add flux and the output voltage tends to stay more constant. As load varies, the series field responds because it is directly affected by load current. The load voltage can be regulated, and the voltage regulation can be almost zero. This means that the voltage at no load can be the same as the voltage at full load. This particular situation is referred to as flat compounding and is graphed as shown in Figure 10–28.

If you allow the series field to add more flux at a greater rate as current load is added, you can create more flux at full load than you had at no load, and the output voltage will rise with added load. This effect is referred to as overcompounding and is illustrated in Figure 10–28. If the series field does not add enough flux to keep the output voltage constant, then the output voltage drops with added load (not as much as a self-excited shunt generator). However, the voltage curve appears as the undercompounded curve shown in Figure 10–28.

When the shunt and the series field are connected to oppose each other, the shunt field establishes the no-load voltage. As current load is increased, the series field actually counteracts the shunt field flux, and the voltage drops quickly (as the voltage curve shows in Figure 10–28). The differentially compounded generator may be used for DC arc welding applications. The voltage starts high and is used to strike an arc. As the current flows, the voltage drops to reduce the arcing action at the electrode but supplies large amounts of current at the arc.

LONG-SHUNT COMPOUND GENERATORS

The long-shunt compound wound generator can be used for some purposes. Note in Figure 10–27 that the shunt field is connected the "long" way across the armature and the series field.

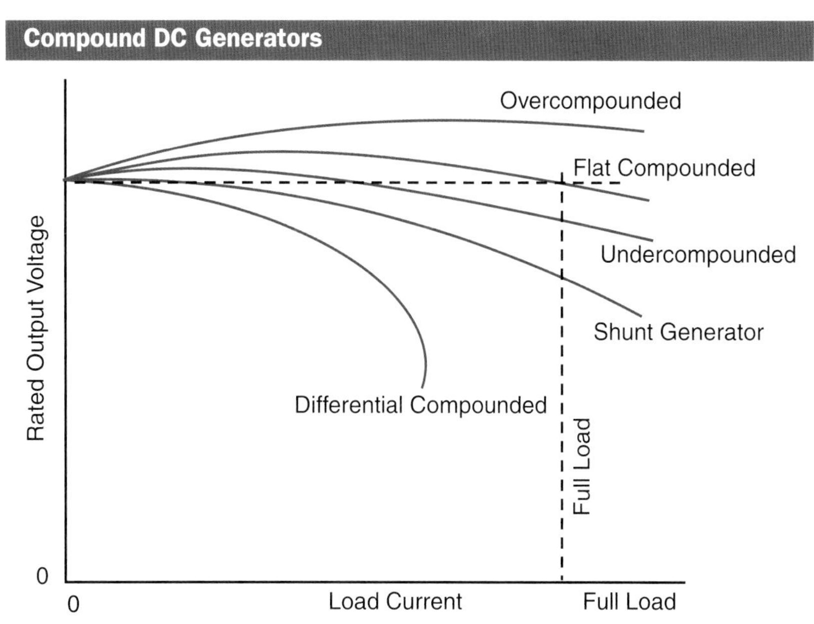

Figure 10–28 Voltage is compared with load current characteristics for compound DC generators.

This connection can be used when the load is expected to be constant and the series and shunt field controls to adjust field current are left constant.

Any change in the load would affect the series field, which in turn would have a direct impact on the shunt field available voltage and current. The voltage would tend to change more when the generator is run from no-load current to full-load current. Figure 10–29 shows the various voltage characteristic graphs for the different generator connections.

The generator connections produce effects on the output voltage. As discussed earlier, the series connected generator increases in output voltage as the load current increases until the iron core saturates and the increase in field current cannot increase the magnetic field any further. The graph shows that if the voltage increases with an increase in load current, to full load and slightly more, then the voltage does not increase further.

The shunt generator has a definite no-load voltage, and as the load current increases, the output voltage has a tendency to drop off for reasons as described in the section on self-excited shunt generators. The graph shows the voltage drops as load current increases.

Compound generators have different degrees of compounding and are either cumulatively compounded or differentially compounded. In the

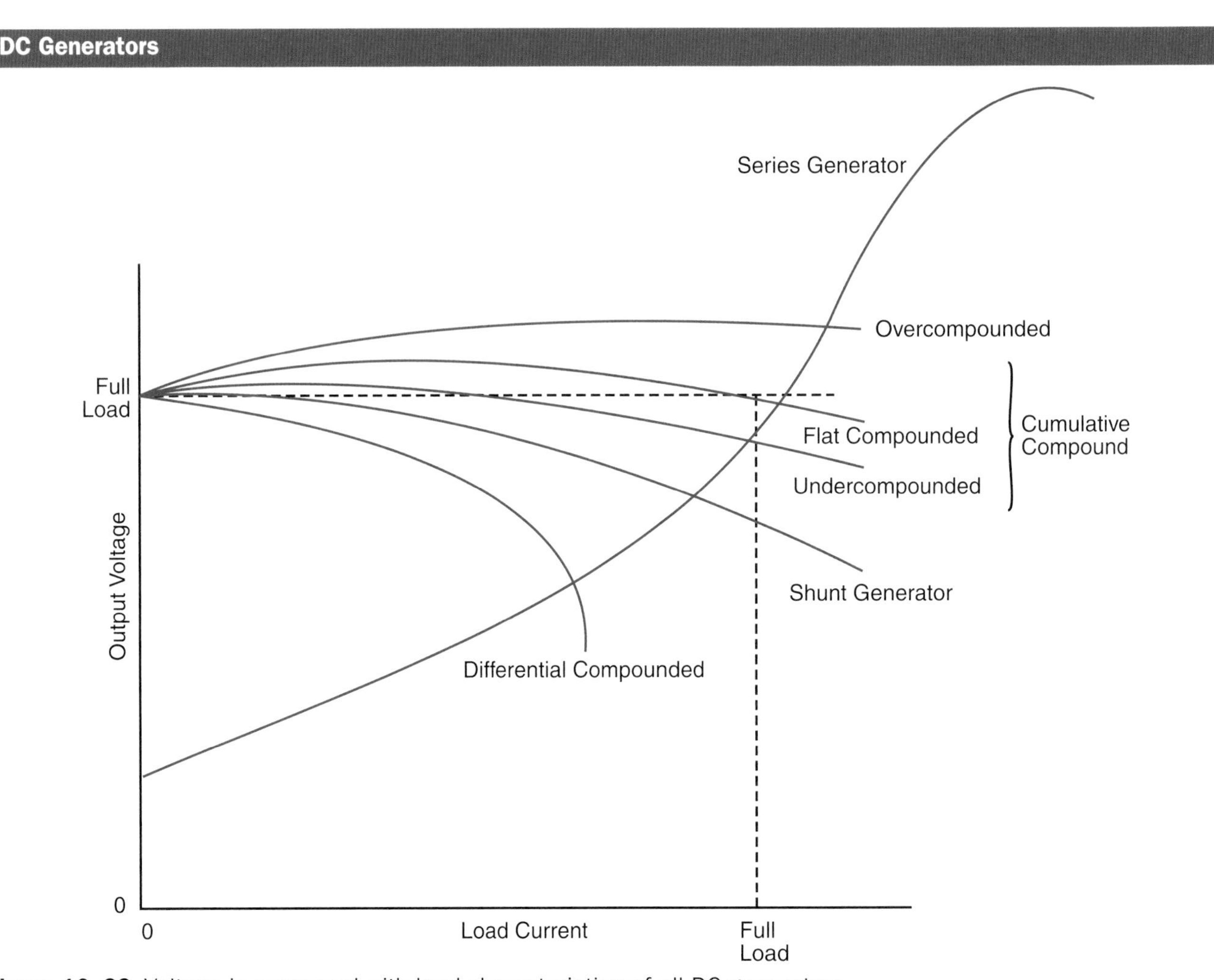

Figure 10–29 Voltage is compared with load characteristics of all DC generators.

differential compound connection the series field flux opposes the shunt field flux, so as load current flows through the series field, the output voltage decreases. The other method of connecting the series field is to have the series field flux aid the shunt field flux. The degree to which the aiding flux helps will determine the degree of compounding. As you can see by the graph, a flat compounded–cumulative connection allows the series field to add just enough flux to compensate for the shunt field's weakening flux. Now the no-load voltage is equal to the full-load voltage. Another variation is the overcompounded generator where the series field more than compensates for the weakening shunt field. Now the compounding of the series field creates a voltage at full-load current that is over the no-load voltage. The last variation is an undercompounded cumulative connected generator. As shown in the graph, the output voltage at full load is under the voltage at no load but not as severely reduced as with a simple shunt generator. Each generator connection has possible uses for specific purposes.

DC MOTOR THEORY

DC motors are in use today in many different capacities because the special characteristics of the motor connections satisfy specific requirements for speed and torque that are not common with AC motors. As you learned with DC generators, many of the components of the motors are the same as generators. The connection patterns are similar to DC generator connections. The difference between **generators** and **motors** is that generators convert mechanical energy to electrical energy, and motors convert electrical energy back to mechanical energy.

The motor action relies on the effect of magnetism that tells us that if a current-carrying conductor is placed in a magnetic field, the conductor will try to move. Refer to Figure 10–30 to see which way the conductor will move. The direction can also be determined by the right-hand rule for motors (Figure 10–31). The center finger is used to indicate the current flow in the conductor, the first finger to indicate the direction of the magnetic flux, and the thumb to indicate the direction of the thrust or motion of the conductor.

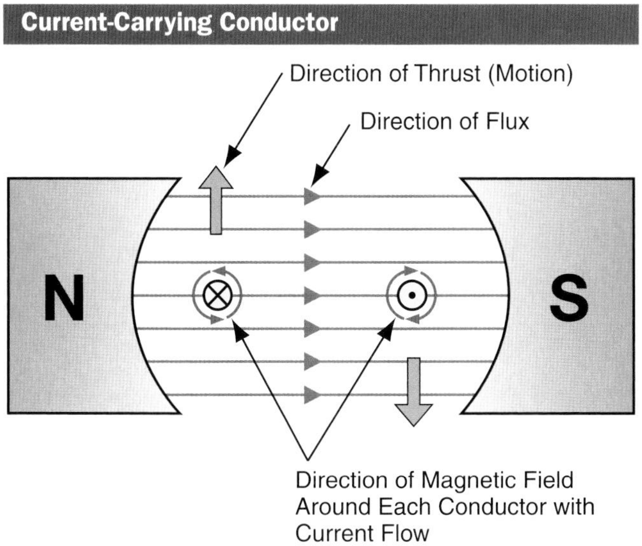

Current-Carrying Conductor

Direction of Thrust (Motion)
Direction of Flux
N
S
Direction of Magnetic Field Around Each Conductor with Current Flow

Figure 10–30 A current-carrying conductor will try to move when placed in a magnetic field.

Right-Hand Rule

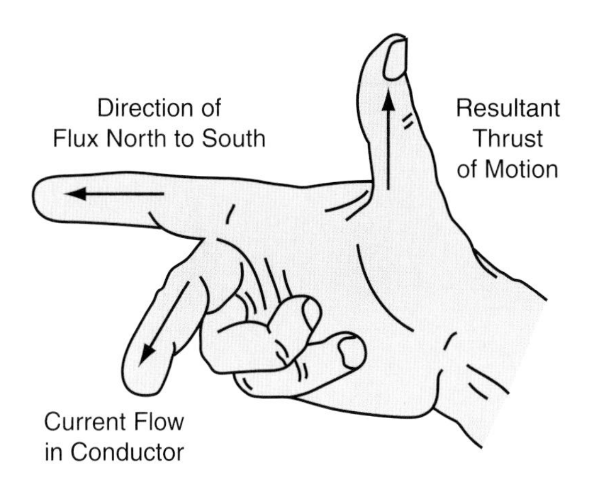

Figure 10–31 The right-hand rule for motors is used with electron flow theory.

DC Motor Connections

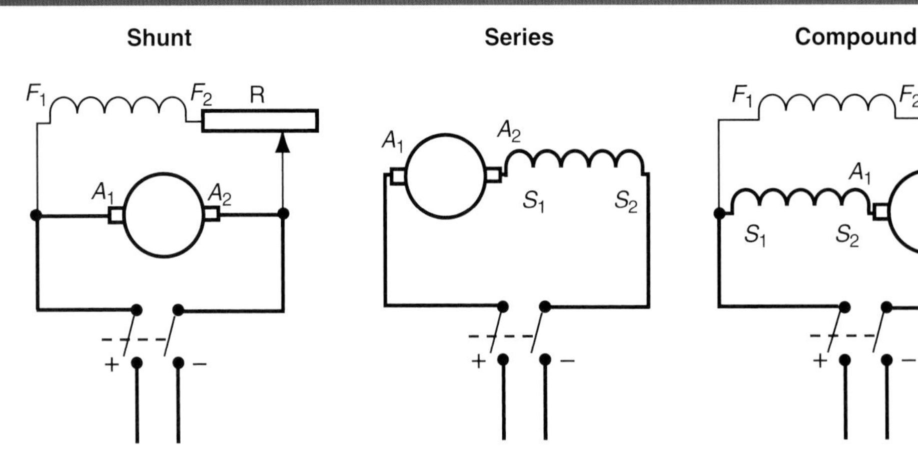

Figure 10–32 These schematic views depict DC motor connections.

Commutator

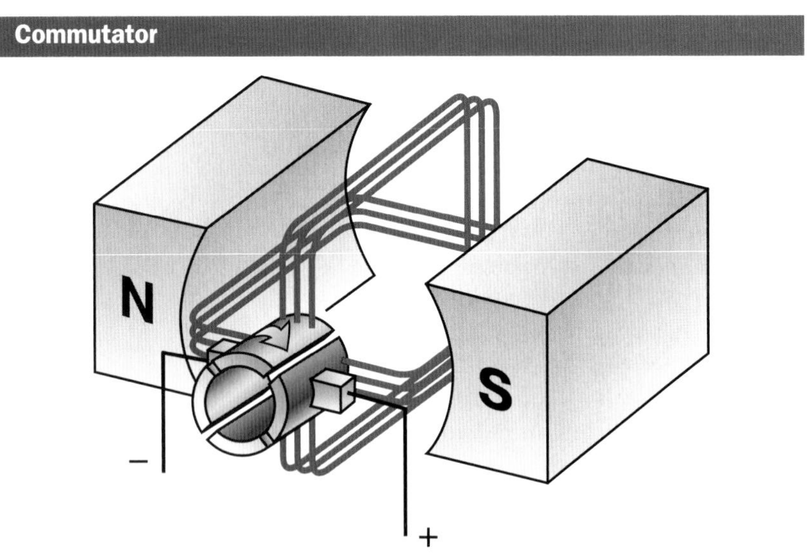

Figure 10–33 A commutator connects the supply voltage to the proper coil under the pole piece.

The rest of the motor information is needed to determine how the current gets into the conductor and how the magnetic field is produced and controlled. The connections are very simply shown in Figure 10–32. A fourth general type of motor uses permanent magnets instead of winding on the rotor.

DC SHUNT MOTOR

The shunt-connected motor is essentially the same as a shunt-connected generator. The stator field is developed by a winding on the pole pieces consisting of many turns or by relatively small wire comprising the shunt field. As indicated by the schematic diagram, the shunt field is connected across the supply conductors and is shunted across the rotor. The rotor is wound with coils, much the same as a generator, and the coils' ends are connected to commutator segments.

Carbon brushes are used to conduct current to the rotor windings from the source of power. Normally, the control of this motor is accomplished by adjusting the current or voltage to the rotor and leaving the shunt field at full strength. The motor rotates because of the motor effect of the current-carrying conductors, but the current-carrying conductors must be in the proper proximity to the stator's magnetic field.

Referring to Figure 10–33, you can see that the brushes are stationary and are used to make connection to the coil that is aligned with the middle of the magnetic pole. For instance, the positive brush is connected to the ends of the coil that are in the center of the South pole, whereas the negative brush is connected to the ends of the coil that are directly under the North pole. Using the right-hand rule for a motor on the coil under the North pole and the current flowing into the coil, you can determine that the thrust will be upward.

As this happens, the coil moves and the rotor spins, so that the negative brush now makes contact with the next coil (which is again under the north pole). The DC motor is spinning as long as current flows to the rotor. If we reduce the voltage and consequently the current to the rotor, known as the armature, we will have less magnetic interaction, and the twisting effort will diminish, and the motor will slow.

The torque of the motor is controlled by the strength of the magnetic field poles and the strength of the magnetic field of the rotor magnetic coils. The speed is generally controlled by adjusting the voltage and current to the rotor.

The speed characteristic of the shunt motor is fairly constant, from no mechanical load to full mechanical load. This is because of several factors. The shunt field is unaffected by the mechanical load because it is essentially connected to the line power. The armature reacts to the mechanical load by drawing more current as it does more work. This reaction of the rotor to draw more current is a result of the generator action of a spinning rotor in a magnetic field.

As the armature spins in the magnetic field, it produces its own voltage. This voltage is a countervoltage to the applied voltage and is known as counter-electromotive force (CEMF). The voltage that sends the current through the rotor is known as the differential voltage, or the difference between the applied voltage and the CEMF. As the generator does more work, the speed slows, the CEMF decreases, the differential voltage increases, and more armature current results.

Therefore, as the mechanical load increases (requiring more torque), the speed slows slightly, allowing more current to flow in the armature and increasing the armature magnetism to keep the speed nearly steady. There is some reduction in speed to cause the increase in torque. The speed curves shown in Figure 10–34 indicate how rpm reacts to increased load. The relatively straight line of the shunt motor indicates that the speed regulation is very good.

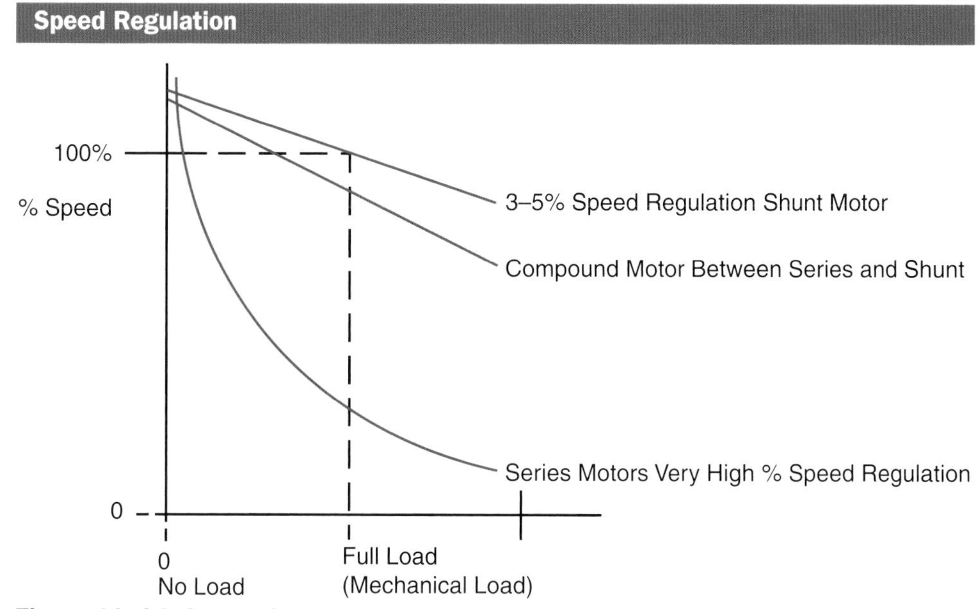

Figure 10–34 Curves for speed versus load indicate the speed regulation of DC motors.

The formula for this is:

% Speed regulation =

$$\left[\frac{(N@NL - N@FL)}{N@FL}\right] \times 100$$

Example

What is the speed regulation of a motor that has a no-load speed of 1,500 rpm and a full-load speed of 1,400 rpm?

% Speed regulation =

$$\left[\frac{(N@NL - N@FL)}{N@FL}\right] \times 100$$

% Speed regulation =

$$\left[\frac{(1{,}500 - 1{,}400)}{1{,}400}\right] \times 100$$

% Speed regulation = 7.14%

In many parts of the world, including the United States, there are electrically driven rail systems involving an overhead electrical system or a third rail system in subway systems. The locomotives that do not have external power sources are actually diesel engines that drive electric generators.

In either case, the power used to move the drive wheels comes from electricity and is controlled and then connected to the drive motors, called traction motors. Typically, these are DC series motors, although there are variations to that process now. The high-speed trains in Europe and Asia use electrically driven motors that can take the trains from standstill at the stations to more than 150 mph in a short amount of time.

SERIES MOTOR

The DC series-connected motor is a very common DC connection. As the name implies, the motor has an armature and just a series field (Figure 10–35). The standard connection pattern for clockwise and counterclockwise rotation is noted by reversing the connection to the armature brushes. The DC series motor has the unique characteristic of producing maximum torque, or twisting effort, at very low speed.

A typical series motor may produce up to 500% of full-load torque at starting rpm. This type of motor is used for traction motors (e.g., electric fork lifts and electric drives for rail cars). As the starting speed is slow and torque is high, they can get very heavy loads moving. As the speed increases, the torque goes down and allows the load to move with less required torque.

The effect on the motor fields is fairly simple. As discussed previously, if the armature is not spinning at rated speed during starting, it is slower than normal. The effect is that the spinning rotor does not produce much CEMF compared to the line voltage. The differential voltage is high and the current to the rotor (armature) is high.

The series field is directly in series with the armature and has high initial current. Now the stator field and the rotor field are very strong, with high current flow creating strong magnetic fields. The reaction between the magnetic fields is strong, and the attraction and repulsion create a very strong torque.

As the rotor speed increases, the CEMF increases, the differential voltage

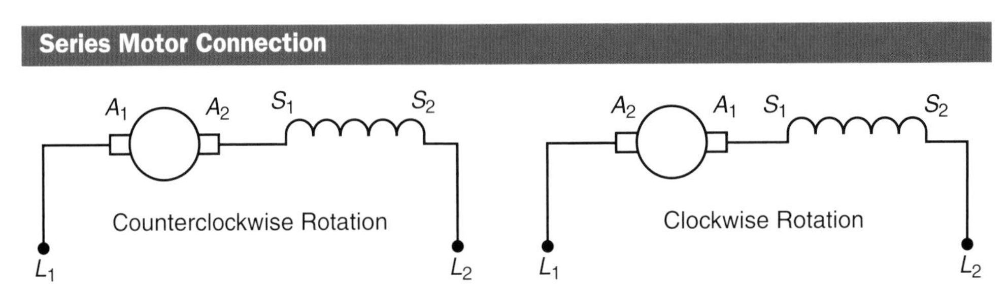

Figure 10–35 This schematic diagram depicts a series motor connection with standard terminal connections for rotation.

decreases, and the torque falls. Figure 10–36 shows the relationship between torque and speed in a series motor. The speed regulation is poor for exactly the reasons just outlined. The speed changes drastically for heavy loads to light loads because the field and the armature are affected simultaneously.

The rotor is designed to carry large currents, and the series field will carry high currents. The effect of disconnection of the mechanical load on the series motor is that the small current in the armature and the small current in the field can cause the motor to run very fast, or "overspeed." In fact, with no mechanical load, the rotor can spin fast enough to cause the centrifugal forces on the rotor to throw the commutator's bars out of the motor. Series motor controls have safeguards to prevent motor damage on overspeed situations.

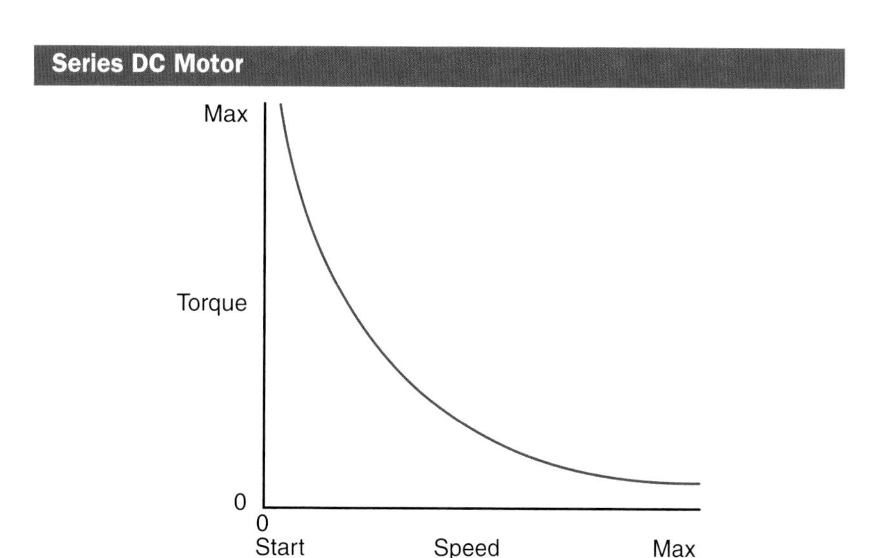

Figure 10–36 The speed-versus-torque curve for a series DC motor is shown.

COMPOUND MOTOR

You have learned that compound generators are a combination of a shunt generator and a series generator. The compound motor is also a combination of the shunt motor and the series motor. Just as in the generator, the two fields can be connected in different patterns, such as the long-shunt compound motor and the short-shunt compound motor. The two connections are drawn schematically in Figure 10–37.

Electric drills are series motors. The motor field windings are connected to the armature through brushes. You can sometimes see the sparking of the brushes as the connection to the armature is made and broken. The motors used with cordless drills are designed for the specific DC voltage of the battery pack. When using a cord-connected drill, the motor is still a series motor, referred to as a universal motor. The windings are slightly different on AC line voltage than a battery pack. However, both are series motors with high speed at light mechanical load and high torque at low speed.

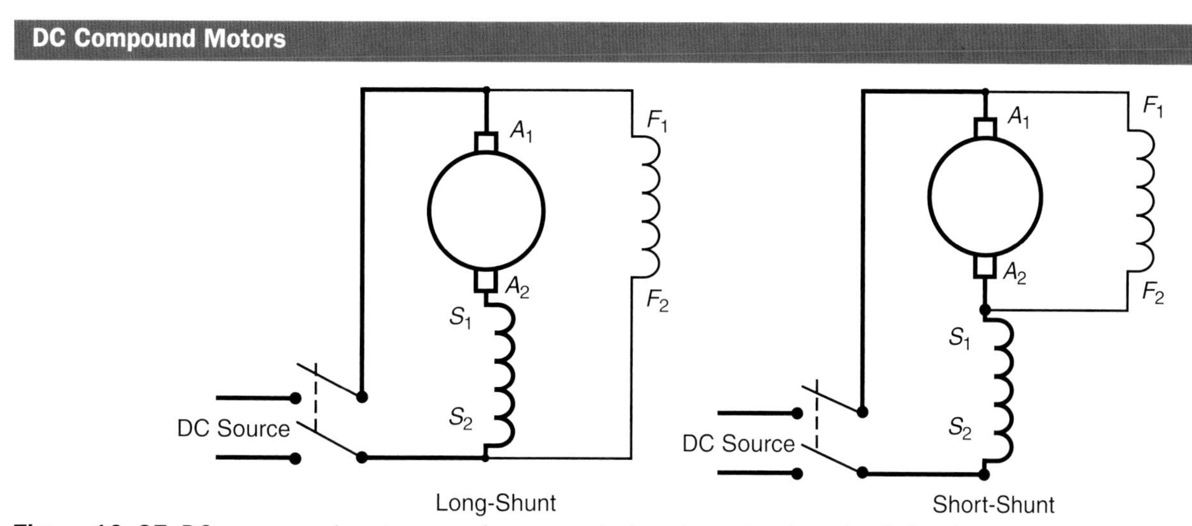

Figure 10–37 DC compound motors can be connected as long-shunt or short-shunt.

Compound connected motors have characteristics that are a compound of the series and shunt effects.

The higher starting torque of the series motor is combined with the better speed regulation of the shunt motor. For motors, the long-shunt connection is more typical. In the long-shunt connection, the shunt field is connected to the line power, thereby keeping the shunt field fairly constant. The series field, connected in series with the armature, allows the motor to respond to changes in mechanical load and add or reduce flux in the field according to the current flowing to the armature.

The cumulative compound connection is almost always used for a typical motor load. However, the motor can be connected with a differential compounded connection when the series field opposes the flux of the shunt field. It is important when connecting the compound motor to check the field connections. If the cumulative compound connection is desired, connect the motor first with only the shunt field in the circuit and note the direction of rotation. Then connect the motor with just the series field and note the direction of rotation. They should both cause the same direction of rotation.

With both fields connected, you can see that the speed regulation falls somewhere between the shunt motor and the series motor (Figure 10–34). Likewise, the torque curves for the motors in relation to armature current (Figure 10–38) show the torque characteristics of the shunt versus series motors.

Motor Torque versus Armature Current

Max
Shunt Motor
Compound Motor
Series Motor
Armature Current
0
0
Torque
Max

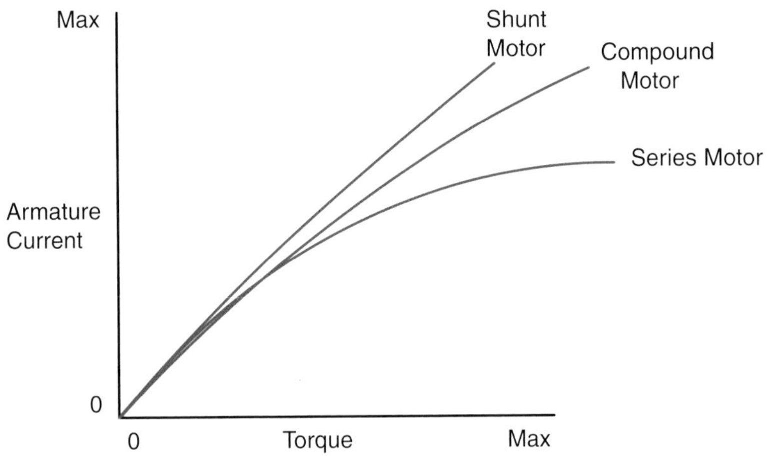

Torque Curves for DC Motors

Figure 10–38 This graph represents motor torque versus armature current for DC motors.

PERMANENT MAGNET MOTOR

Permanent magnet motors are a variation on the shunt type of motor. Instead of a wound field coil on the stator, the field is provided by ceramic permanent magnets. Figure 10–39 shows an example of a DC permanent magnet stator. The permanent magnet provides a constant magnetic field, and the rotor is connected with brushes and a commutator, just like a shunt motor. The advantage is that there are no field windings to use power or produce heat. The motor can be smaller and is more efficient than a comparable shunt motor.

Permanent Magnet Motor Stator

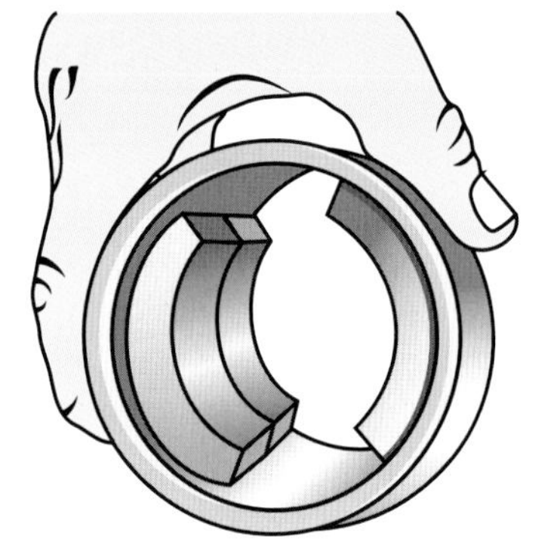

Figure 10–39 This photograph shows a permanent magnet motor stator.

SUMMARY

Most modern commercial electric power is produced by electromagnetic induction. In all DC generators, a rotating winding (the armature) passes through a magnetic field to induce voltage into itself. This is called a rotating armature. The methods of connecting the armature and the magnetic field coils depend on the desired application. The generators can be shunt generators, series generators, or compound generators.

The generators can be separately excited or self-excited. When they are self-excited, the buildup of voltage depends on the residual magnetism left in the field iron. DC motors are very similar in appearance and construction to DC generators. Motors can be shunt connected, series connected, or compound connected to meet the needs of torque and speed of the intended mechanical load.

DC generators use segmented commutators and brushes to collect the output voltage from the coils of wire generating the most EMF, directly under the magnetic poles pieces. Likewise, motors use commutators and brushes to supply current to the correct armature coils, directly centered under the desired pole pieces.

Every motor is a generator when you consider the CEMF generated that helps control current. Every generator is a motor when you consider that as more current is drawn by the load, the generator acts like a motor trying to turn the other way, thus slowing the generator.

A conductor loop must pass through 1 Wb of flux in 1 second to produce 1 V. The direction of current flow through the armature can be deduced by using the left-hand rule for generators. The direction of a motor conductor placed in a magnetic field can be determined by the right-hand rule for motors.

CHAPTER GLOSSARY

Armature The winding in a generator where an induced voltage is created.

Electromagnetic induction The generation of electricity by passing a conductor through a magnetic field. The same thing can be accomplished by moving a magnetic field across a conductor.

Field The winding placed on a stator that creates the magnetic field.

Generator A machine that converts mechanical energy into electrical energy.

Motor A machine that converts electrical energy into mechanical energy.

REVIEW QUESTIONS

1. Explain how to determine current flow direction with the left-hand rule for generators.
2. How is DC output voltage created when the coil current in the armature is actually AC?
3. Where are the field coil windings actually located?
4. How is the DC output voltage calculated on the basis of the physical components of the generator?
5. Create a schematic drawing of a shunt-connected self-excited generator. Label all components and leads.
6. Explain what is meant by a compound generator being overcompounded when connected in the cumulative connection.
7. What is meant by residual voltage in a self-excited generator, and why is it important?
8. How is the right-hand rule for motors used to determine direction of rotation for a motor?
9. How is the direction of rotation typically reversed when connecting a DC compound motor?
10. Explain why a series motor has highest torque at slowest speed.
11. Write the formula for speed regulation for DC motors.
12. What connection of a compound motor is most common for DC motors and for long-shunt or short-shunt motors, and why?
13. How would you be able to tell the difference between the series and shunt field winding by visual inspection?
14. How is the speed of a DC motor usually controlled?
15. Give an example of where a DC series motor would be used.

G L O S S A R Y

Ampacity The amp-carrying capacity of an electrical conductor. The contraction of amp and capacity create an electrical term for how much current a conductor can carry safely.

Ampere The unit of electrical current flow, often abbreviated as amp or just A. It is a measure of the movement of electrons as 1 coulomb per second past a fixed point.

Ampere-turns The strength of an electromagnetic field calculated by multiplying the current flow times the number of turns for an electromagnet used to produce a magnetic field.

Armature The winding in a generator where an induced voltage is created.

Assumed voltage method In a parallel circuit, the total resistance can be calculated by using an assumed voltage with the known resistive branches. This assumed voltage produces values of current, but will calculate actual resistance. The formula is:

$$\frac{E_{T\ assumed}}{I_{T\ assumed}} = R_{T\ actual}$$

Atom The smallest particle still characterizing a chemical element.

Battery Series and/or parallel combination of cells. A group of cells connected in series (more voltage) or parallel (more current).

Cell A single chemical structure composed of an electrolytic solution (sulfuric acid) and two different metallic electrodes (lead and lead peroxide).

Circuit A completed path for current to flow from a source of current through a load and back to the source of current.

Coercive force The amount of reverse MMF that must be applied to a magnetized material to reduce residual magnetism to zero.

Combination circuit A circuit consisting of both parallel and series connected components. Combination circuits are sometimes referred to as series-parallel circuits.

Component diagram (pictorial diagram) A component diagram is a drawing that shows the interconnection of system components by using photographs or drawings of the actual components. This is also known as a pictorial diagram.

Composition carbon resistor A resistor that derives its resistance from a combination of carbon graphite and a resin bonding material.

Compound A material made from the chemical combination of two or more elements.

Conductor A material that easily passes electrical current. Examples include silver, copper, and aluminum.

Conventional current flow The theory in which current flows from a positive charge to a negative charge; also known as hole flow or positive current flow.

Coulomb The unit of electrical charge equal to the total charge possessed by 6.25×10^{18} electrons. Abbreviated C.

Current The flow of electrons through a material. Measured in amperes.

Current divider A circuit of parallel circuit components that will split the total current according to the opposition of the branch, in inverse proportion to the total opposition.

Current source A source that keeps its output current constant regardless of the load applied. In the early days of electric power, street lighting circuits were often series circuits supplied with constant current transformers. In modern power systems, current sources are rarely if ever encountered. Current sources are more commonly used in electronics systems.

Diamagnetic Materials diametrically opposite to magnetic materials. When exposed to magnetizing force, they produce a weak field that opposes the magnetizing force.

Domain theory This theory relates to the effects of magnetic domains. If enough domains are aligned so that the small magnetic fields have the same magnetic polarity, the entire material will exhibit magnetic effects.

Electricity A class of phenomena that results from the interaction of objects that exhibit a charge (electrons and protons). In its static form, electricity exhibits many similarities with another naturally occurring force-magnetism.

Electrolyte Any material that will dissolve into ions when immersed in a liquid. The liquid thus becomes an electrical conductor.

Electromagnet An electromagnetic temporary magnet created when current flows through a coiled conductor. The magnet is usually an iron core that is easily magnetized and easily demagnetized when the electric current stops.

Electromagnetic induction The generation of electricity by passing a conductor through a magnetic field. The same thing can be accomplished by moving a magnetic field across a conductor.

Electromotive force (EMF) Electrical pressure created between a region of positive charge (fewer electrons) and a region of negative charge (more electrons), measured in volts and represented in formulas with "*E*".

Electron One of the three main components of an atom. The electron is a fundamental particle, and by definition has a negative electrical charge (from the Greek word *elektron*, meaning "to be like amber").

Electron current flow The theory in which current flows from a negative charge to a positive charge.

Element The simplest form of matter. There are more than 103 known elements, 92 of which occur naturally. All matter is made from chemical combinations of elements.

Ferromagnetic Properties displayed by certain substances (such as iron, nickel, cobalt, and various alloys) that exhibit extremely high magnetic capability.

Field The winding placed on a stator that creates the magnetic field.

Flux density Flux density is a measure of the pole strength of a magnet. It indicates how many lines of flux are either leaving or entering the pole face per unit area.

Generator A machine that converts mechanical energy into electrical energy.

Gluon The particle that mediates or transmits the strong nuclear force between quarks. The fundamental particle responsible for the strong nuclear force.

Hysteresis The lagging effect of moving the magnetic domains when magnetizing or demagnetizing a material. The difficulty encountered in aligning the domains is due to the hysteresis of the material, and likewise, the difficulty in allowing the domains to go back to random alignment is due to hysteresis.

Hysteresis loop The graph of the flux density versus magneto-motive force as applied to a particular material. The graph forms a loop showing magnetic saturation, residual magnetism, and coercive force as the MMF changes in strength and direction.

Hysteresis loss Losses in the core material of magnetic materials are partly caused by the wattage loss in overcoming hysteresis. The amount of energy lost is related to the movement of domains from one direction to another that causes heat.

Insulator A material that does not allow electrical current to flow easily. Examples include rubber, plastic, and mica.

Ions An atom or molecule that has gained or lost one or more electrons. A positive ion has lost electrons, and a negative ion has gained one or more electrons.

Isotope One of two or more atoms with the same number of protons but different number of neutrons.

Joule A joule refers to the amount of energy. One joule of energy used each second is equal to 1 W of work. One joule is the amount of energy used when 0.737 pounds is lifted a distance of 1 foot.

Law of proportionality Current and resistance are inversely proportional for a series circuit, and voltage drop and resistance are directly proportional in a series circuit. The voltage of a parallel circuit is constant, and current flows as an inverse proportion to the resistance values compared to the total equivalent resistance.

Left-hand rule for a conductor Using your left hand, grasp a conductor with your thumb in the direction of the electron flow. Your fingers will wrap around the conductor in the direction of the magnetic field.

Left-hand rule for generators Use your left hand, with thumb, first finger, and center finger at right angles. The thumb indicates the direction of thrust. The first finger indicates the direction of flux (north to south), and the middle finger indicates the direction of current flow.

Magnetic domains As the electrons in the atoms of a material spin in the same direction to create small magnetic fields, the fields of many atoms add together to create atomic level magnets called magnetic domains.

Magnetic flux Flux is the name given to the invisible magnetic lines that encircle the magnet.

Magnetic lines of force The force of a magnetic field expressed as lines, also called magnetic flux.

Magnetite The mineral form of black iron oxide. Magnetite is a naturally occurring magnet and was possibly the first type of magnet studied by early scientists.

Matter The material from which all known physical objects are made.

Metal film resistor A resistor that derives its resistance from a thin metal film applied to a ceramic rod.

Metal glaze resistor Similar to a metal film resistor except that the film is much thicker and is made of metal and glass.

Molecule The chemical combination of two or more atoms. The smallest particle of a compound that has the same chemical characteristics of the compound.

Motor A machine that converts electrical energy into mechanical energy.

Neutron One of the three main components of an atom. The neutron has no electrical charge and is therefore classed as electrically neutral. The neutron has been shown to be composed of even smaller particles, called quarks.

Ohm The unit of electrical resistance, often shown as the Greek letter omega (Ω).

Ohm's law The mathematical relationship among current, electrical potential, and electrical resistance, measured in amperes (amps), volts, and ohms, respectively.

Parallel circuit An electrical circuit that provides more than one possible path for current to flow from the source and back.

Parallel circuit voltage rule Each voltage across every branch of a parallel circuit is the same.

Parallel current rule In a parallel circuit, the total current is the sum of the individual branch currents.

Paramagnetic Materials that have weak magnetic abilities. They can become magnets with a large magnetizing force, but do not generally exhibit magnetic capability.

Permeability The degree to which a material focuses magnetic lines of flux, or the ease with which magnetic lines of force distribute themselves through a material. Permeability is a factor compared to air (permeability factor of 1). There is no unit of measure, just a factor compared to air. The symbol used for magnetic permeability in a formula is the lowercase Greek letter mu (μ).

Piezoelectric Electricity created by stress or pressure in a material—especially a crystalline material.

Primary cell A cell that cannot be recharged after it has depleted all of its stored chemical energy in the form of electricity.

Product over sum method (Product over sum equation) An equation that can be used to calculate the equivalent resistance of two resistors in parallel. It is a simplification of the reciprocal equation:

$$R_T = \frac{(R_T \times R_2)}{(R_T + R_2)}$$

Proton One of the three main components of an atom. By definition, the proton has a positive electrical charge. The proton has been shown to be composed of even smaller particles, called quarks.

Quark One of the fundamental particles of matter. There are six different types of quarks that are assembled in different combinations to create larger particles, such as protons and neutrons.

Relay An electromagnetic coil mechanism that mechanically opens or closes a contact to control a separate electrical circuit.

Reluctance The opposition that materials present to the flow of magnetic lines of flux (similar to electrical resistance). Permanent magnets have a high reluctance and temporary magnets have a low reluctance. The symbol is the Greek letter rel (R) and the force is measured in ampere-turns per weber.

Residual magnetism The magnetic effect left in a magnetic material after the MMF has been removed. The remaining magnetic effect is due to residual magnetism because the domains do not go back to random arrangement.

Resistance The physical opposition to electrical current. Resistance (measured in ohms) is caused by the energy loss that occurs when an electron displaces other electrons in a valence ring.

Resistivity Specific resistance is defined for both American and IEC units. Under American (ANSI) standards, the resistance of a conductor is based on a conductor 1 foot long and 1 millimeter in diameter. Under IEC standards, the resistance of a conductor is based on a conductor 1 meter long and 1 millimeter in diameter.

Retentivity The ability of a material to hold magnetism after the magnetizing force is removed.

Rule for parallel circuit resistance As resistance branches are added in parallel, the total resistance decreases and the total resistance for the circuit is smaller than the smallest branch resistor.

Saturation The point at which the magnetic domains are all lined up. Any further magneto-motive force will not produce a stronger magnet.

Schematic diagram A schematic is a structural or procedural diagram, especially of an electrical or mechanical system, using special symbols to represent the actual physical components. The schematic shows the "scheme" of the current flow in a systematic representation.

Secondary cell A cell whose chemical energy can be restored by forcing electrical energy into it.

Semiconductor A material that falls between conductors and insulators in terms of electrical conductivity.

Static electricity An electrical charge that is stationary or nonmoving. Sometimes called triboelectricity.

Strong nuclear force The force that holds quarks together to make up neutrons and protons. The residual strong nuclear force is also responsible for holding protons and neutrons together in the nucleus despite the electrical repulsion trying to force the protons apart.

Thermocouple A junction of two dissimilar metals that creates an electrical potential when heated.

Thermoelectricity Electricity created by heat.

Triboelectricity An electrical charge created by rubbing two materials together. Sometimes called static electricity.

Utilization equipment As defined by the *National Electrical Code® (NEC®)*, utilization equipment is equipment that utilizes (uses) electric energy for electronic, electromechanical, chemical, heating, lighting, or similar purpose.

Valence electrons The electrons that make up the valence shell. Valence electrons are free to participate in current flow.

Valence ring or shell The outermost shell of electrons in an atom.

Voltage divider An electrical circuit (usually made up of resistive elements) that can be used to break down, or divide, a supply voltage into two or more smaller voltages.

Voltage source A source that keeps its output voltage constant regardless of the load applied. Such a device does not exist in reality. However, batteries and other such power sources approximate them.

Watt (W) The unit of electrical power. One horsepower is equal to 745.7 W or approximately 746 W.

Weber (Wb) Field strength equal to 100,000,000 lines.

Wire-wound resistor A resistor made by winding resistive wire around an insulating form.

INDEX